Structural Analysis 2

In memory of my father, Tahar
In memory of my wife, Mrs Nadjet Khalfallah-Boudaa
To my family
20 Ramadhan 1437 AH

Series Editor
Gilles Pijaudier-Cabot

Structural Analysis 2

Statically Indeterminate Structures

Salah Khalfallah

ISTE

WILEY

Preface

The main objective of this volume is to provide students of civil, mechanical, aeronautical and marine engineering as well as those interested in structural analysis with the essentials for analyzing statically indeterminate structures. This book allows them to acquire sufficient knowledge to study and analyze statically indeterminate structures. The reader will find a series of exercises at the end of each chapter that can be used to deepen their knowledge and improve their ability to master statically indeterminate structural analysis methods.

This book is the second volume of structural analysis. The first volume is devoted to analyzing statically determinate structures, whereas this volume is exclusively analyzes statically indeterminate structures. Every chapter is designed in a specific way: an illustration of the objectives and the parts covered, a general introduction, a theory of the proposed approach, a numerical study of some examples and a summary at the end. Each chapter concludes with a series of problems. The primary objective is to meet the needs of mechanics students, allowing them to acquire the necessary knowledge and to apply statically indeterminate structural analysis methods. In addition, the numerical examples used are common across different numerical methods.

The book is divided into seven chapters dealing with statically indeterminate structures. It offers a comprehensive presentation of statically indeterminate structural analysis methods.

Chapter 1 is a general introduction to analyzing statically indeterminate structures. Chapter 2 presents the analysis of single span statically indeterminate beams (the method of three moments), the study of continuous beams (the Clapeyron method) and the focus method. In Chapter 3, the method of forces is described in the analysis of statically indeterminate beams, statically indeterminate by mechanical loads, or subjected to deflections, and/or undergoing a support settlement. The analysis methods for statically indeterminate structures are used here to make the number of unknowns equal to the number of equations, which allows the problem to be solved.

This book is particularly devoted to the analysis of statically indeterminate structures. To present the differences between the analysis methods for statically indeterminate structures, the problems we consider generally have a common object across the analysis methods. In this context, each chapter illustrates the theoretical foundation of the analytical method, presented in detail and accompanied by a series of numerical examples.

1.2. External static indeterminacy

The purpose of structural analysis is to determine the support reactions and the variation of internal actions in the elements of a statically indeterminate structure. The static indeterminacy of a structure can be internal, external or internal and external simultaneously. It is called externally statically indeterminate if the number of support reactions exceeds the number of independent equations. The plane structures are externally statically indeterminate if the number of support reactions is greater than 3 (Figure 1.1) and it is greater than 6 if the structure is spatial (Figure 1.2).

From this explanation, we define the degree of static indeterminacy of a system by the difference between the number of support reactions and the number of independent equations that can be constructed. The degree of external static indeterminacy f of a plane structure [1.1] or a space structure [1.2] is deduced by

$$f = r - (3 + k) \tag{1.1}$$

$$f = r - (6 + k) \tag{1.2}$$

We calculate the degree of static indeterminacy of the beam and frame (Figure 1.1).

Beam, f = 5 – 3 =2

Frame, f = 10 – (3 + 1) = 6

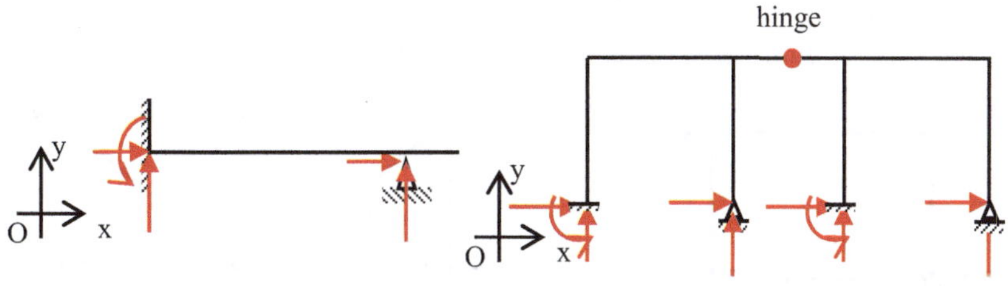

Figure 1.1. *Statically indeterminate externally of plane structures*[1]

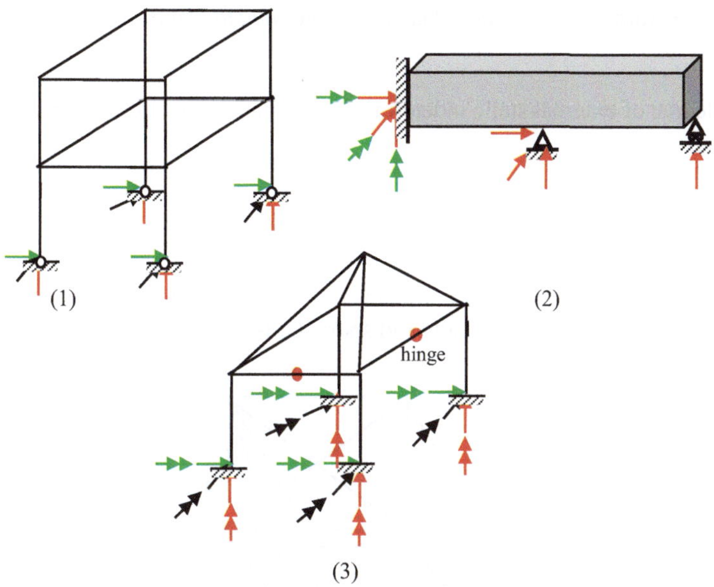

Figure 1.2. *Statically indeterminate externally of space structures*

For space structures (Figure 1.2), the degree of static indeterminacy is

Frame (1), f = 12 – 6 = 6

Beam (2), f = 10 – 6 = 4

Frame (3), f = 24 – (6 + 12) = 6

1 All of the figures in this chapter are available to view in full color at www.iste.co.uk/khalfallah/analysis2.zip.

The degree of static indeterminacy of trusses is calculated by using relationships [1.1] and [1.2]. Figure 1.3 presents plane and space trusses.

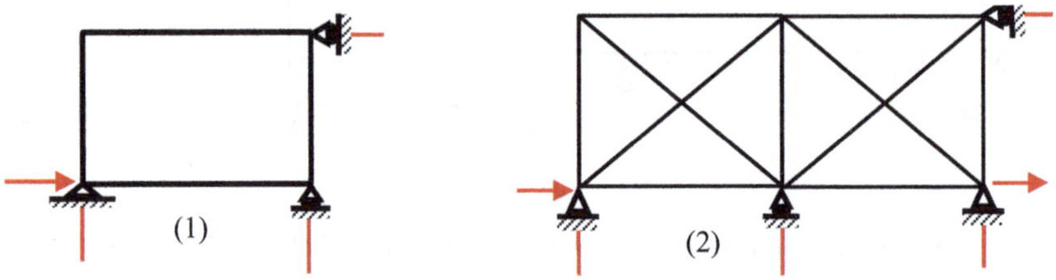

Figure 1.3. *Externally statically indeterminate plane structures*

The degree of external static indeterminacy of plane structures (Figure 1.3) is

Structure (1): f = 4 – 3 = 1

Structure (2): f = 6 – 3 = 3

In the same way, space truss structures (Figure 1.4) are the most used in the construction of large exhibition halls and sports halls, etc.

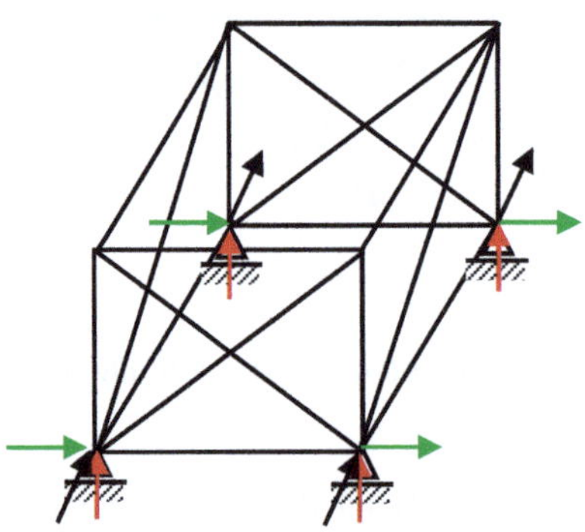

Figure 1.4. *Statically indeterminate space truss*

The degree of static indeterminacy of the structure is

f = 12 – 6 = 6

1.3. Internal static indeterminacy

In this section, we describe how to calculate the degrees of static indeterminacy of trusses, frames, beams and crossbeams.

1.3.1. *Truss structures*

Consider a truss structure with r support reactions, b bars and n joints including support joints. The number of unknowns $(b + r)$ of the problem is the support reactions and the forces in the bars of the truss.

At each joint of the truss, it is possible to write the following equations:

$$\sum F_x = 0 \qquad [1.3a]$$

$$\sum F_y = 0 \qquad [1.3b]$$

So, the total number of independent equations is $2n$.

We define the degree of internal static indeterminacy by

$$f = (b + r) - 2n \qquad [1.4]$$

EXAMPLE 1.1.–

Calculate the degree of internal static indeterminacy of the structure (Figure 1.5).

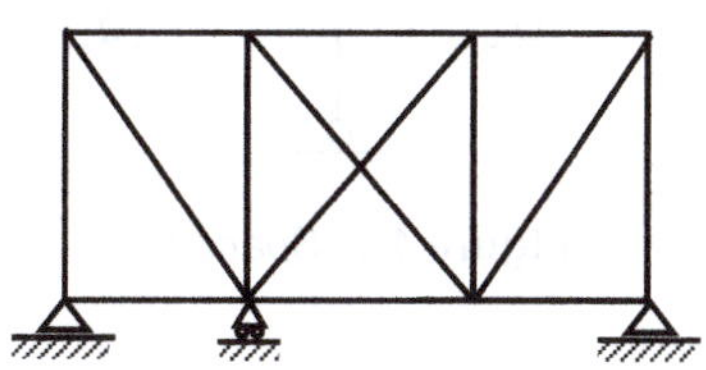

Figure 1.5. *Given truss*

Applying relationship [1.4] allows us to calculate the degree of static indeterminacy.

where n is the number of rigid joints including the support joints. If the frame or beam contains k hinges, the relationship [1.6] is written as

$$f = (3b + r) - (3n + k) \quad [1.9]$$

EXAMPLE 1.4.–

Determine the degree of internal static indeterminacy of the structures (Figure 1.8).

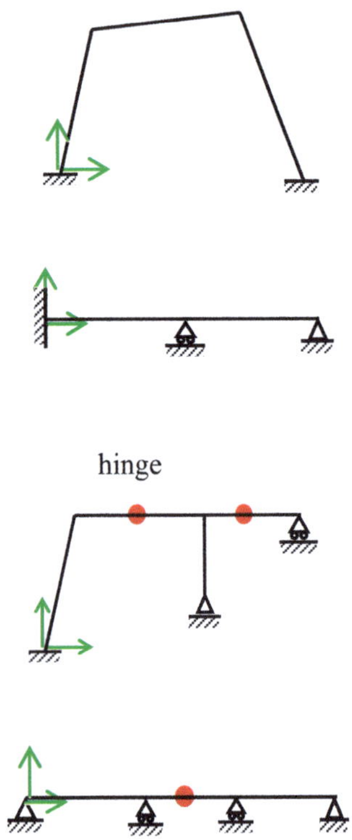

b = 3, r = 6, n = 4
f = (3.3 + 6) - 3.4 = 3

b = 2, r = 6, n = 3
f = (3.2 + 6) - 3.3 = 3

$b = 4,\ r = 6,\ n = 5,\ k = 2$
$f = (3.4 + 6) - (3.5 + 2) = 1$

$b = 3,\ r = 6,\ n = 4,\ k = 1$
$f = (3.3 + 6) - (3.4 + 1) = 2$

Figure 1.8. *Static indeterminacy of frames and beams*

For space structures, we can write six equilibrium equations per joint and each bar has six unknowns. The degree of internal static indeterminacy can be deduced by

$$f = (6b + r) - 6n \quad [1.10]$$

If the frame or beam contains k hinges, the degree of internal static indeterminacy is written as

$$f = (6b + r) - (6n + k) \qquad [1.11]$$

EXAMPLE 1.5.–

Determine the degree of internal static indeterminacy of the structures (Figure 1.9).

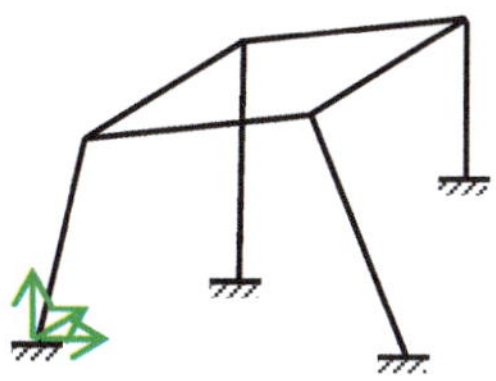

$b = 8, r = 24, n = 8$
$f = (6.8 + 24) - 6.8 = 24$

$b = 2, r = 9, n = 3$
$f = (6.2 + 9) - 6.3 = 3$

Figure 1.9. *Static indeterminacy of space beams and frames*

1.3.3. *Crossbeams*

There is a layer of orthogonal beams linked together at the levels of the rigid joints. At each joint, we can write the following three equations:

$$\sum F_z = 0 \qquad [1.12a]$$

$$\sum M_x = 0 \qquad [1.12b]$$

$$\sum M_z = 0 \qquad [1.12c]$$

At each end of a bar, we consider a vertical force along the axis (zz), bending and torsion moments (M_{xx}, M_{yy}, M_{xy}) (Figure 1.10).

The internal forces at any section can be determined when three out of six actions of a beam element are known. Therefore, each member presents three unknowns and the degree of internal static indeterminacy is obtained by

$$f = (3b + r) - 3n \qquad [1.13]$$

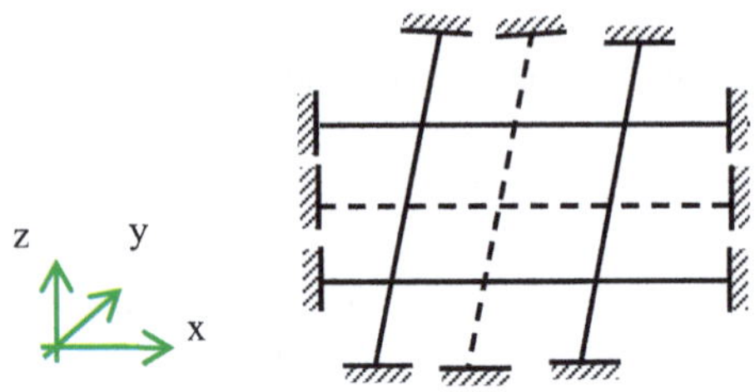

Figure 1.10. *Crossbeams*

Especially when the links between the bars are joints, the degree of internal static indeterminacy becomes

$$f = (2b + r) - (3n + 2n^*) \qquad [1.14]$$

where n^* is the number of articulated joints and n is the number of joint supports.

EXAMPLE 1.6.–

Determine the degree of static indeterminacy of the crossbeams (Figure 1.11).

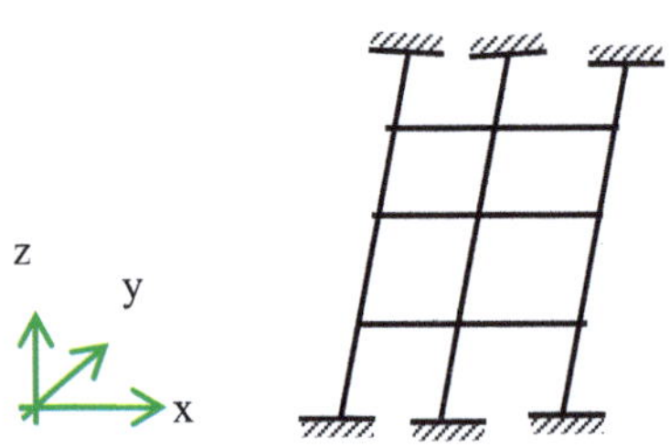

$b = 18$, $r = 18$, $n = 15$

$f = (3.18 + 18) - 3.15 = 27$

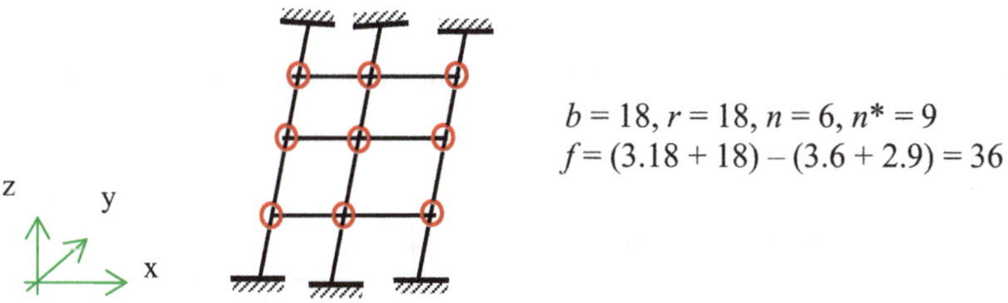

Figure 1.11. *Crossbeam layer*

1.4. Kinematic static indeterminacy

When a structure is stressed by one or more external actions, the joints of this structure undergo deflections. The method of displacements advocates rotations and displacements of the joints of the structure as unknowns of the problem.

The continuous beam (Figure 1.12), for example, is fixed at A and simply supported at B and C. It is solicited by a load q(x), which causes a deflection. Fixing A prevents rotation and displacement while supports B and C do not come from displacements following the vertical direction by allowing rotations ω_B and ω_C. In the case of vertical loads, there aren't longitudinal forces; for this reason, the horizontal displacements are neglected.

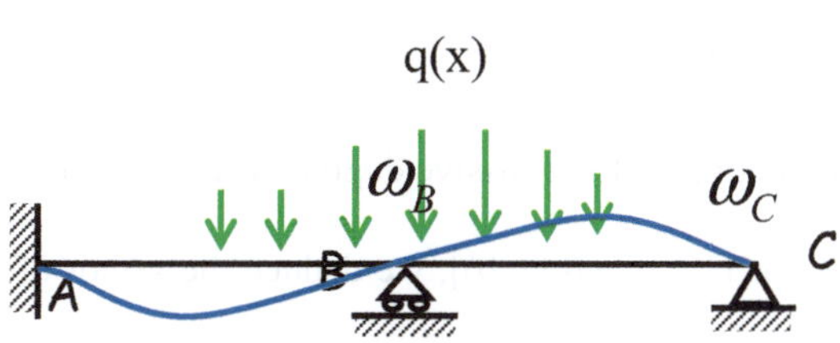

Figure 1.12. *Deflection of beam (AC)*

The unknowns of the problem are effectively independent rotations ω_B and ω_C. The number of independent deflections is called the degree of kinematic static indeterminacy or the number of active degrees of freedom. It encompasses all displacements and rotations of movable joints. The determination of the degree of kinematic static indeterminacy is briefly established in the following examples.

EXAMPLE 1.7.–

Determine the number of degrees of kinematic static indeterminacy of the structures (Figure 1.13).

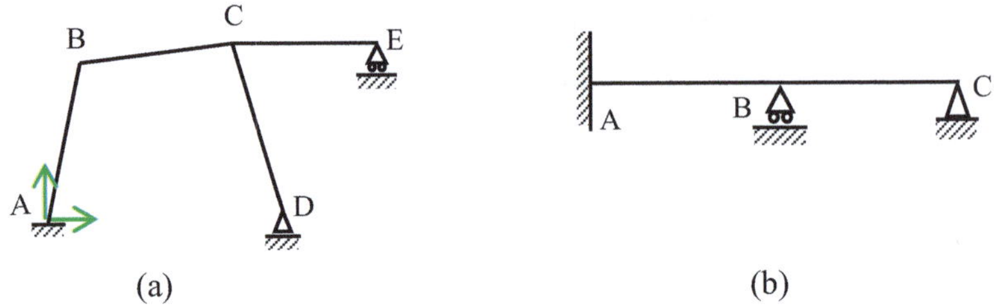

Figure 1.13. *Given structures*

The structure (Figure 1.13(a)) is isolated from the external environment and we deduce the vector components from the kinematic degrees of static indeterminacy.

$$\{q_0\} = \langle u_i \ v_i \ \omega_i \rangle^t \ : i = A, B, C, D \tag{1.15}$$

Linking the structure to the supports makes it possible to quantify a few degrees of freedom (Figure 1.14). The vector grouping these degrees of freedom is called the passive degree of freedom vector, which is

$$\{q_p\} = \langle u_A \ v_A \ \omega_A, u_D \ v_D, v_E \rangle^t \tag{1.16}$$

Note that any component of the passive degree of freedom vector is zero.

Using relationships [1.15] and [1.16], we deduce the vector of active degrees of freedom by

$$\{q_a\} = \langle u_B \ v_B \ \omega_B, u_C \ v_C \ \omega_C, \omega_D, u_E \ \omega_E \rangle^t \tag{1.17}$$

The degrees of freedom vector of the structure (Figure 1.13(b)) is given by

$$\{q_0\} = \langle u_i \ v_i \ \omega_i \rangle^t \ : i = A, B, C$$

The passive degrees of freedom vector is given as

$$\{q_p\} = \langle u_A \; v_A \; \omega_A, v_B, u_C v_C \rangle^t$$

The active degrees of freedom vector is deduced by

$$\{q_a\} = \langle u_B \; \omega_B, \omega_C \rangle^t$$

For strusses, the forces are applied to the joints and each bar is subjected to a normal force. The joints undergo translations according to the orthogonal directions and this allows us to work out the structural deformation.

EXAMPLE 1.8.–

Determine the active and passive degrees of freedom of the truss structure (Figure 1.14).

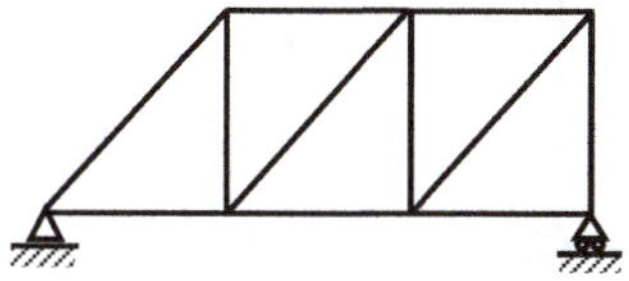

Figure 1.14. *Plane truss*

We isolate the structure from its supports (Figure 1.15).

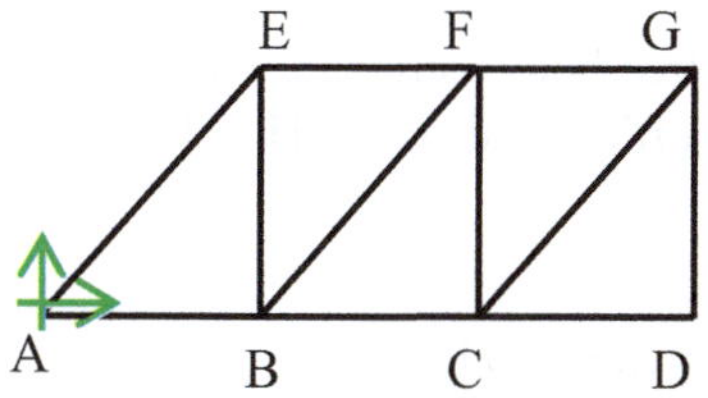

Figure 1.15. *Free-body structure*

Each joint of the structure (Figure 1.15) can experience displacements along the axes (xx) and (yy).

$$\{q_0\} = \langle u_i \; v_i \rangle^t \; : i = A, B, C, D, E, F, G$$

The passive degrees of freedom vector is given as

$$\{q_p\} = \langle u_A \ v_A, v_D \rangle^t$$

Hence, the active degrees of freedom vector is given as

$$\{q_a\} = \langle u_B \ v_B, u_c \ v_c, u_D, u_E \ v_E, u_F \ v_F, u_G \ v_G \rangle^t$$

EXAMPLE 1.9.–

The truss structure (Figure 1.16) is three-dimensional. Each joint moves in the three orthogonal directions. So each joint has 3 degrees of freedom.

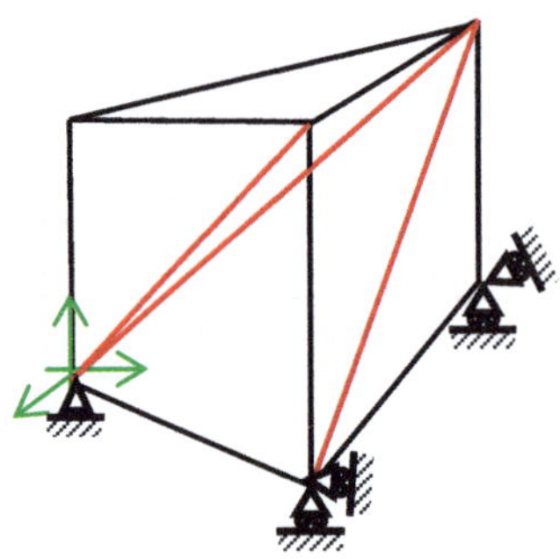

Figure 1.16. *Space truss*

Each joint of the free-body structure (Figure 1.17) can undergo three displacements along the orthogonal axes. The degrees of freedom vector of the structure is given as

$$\{q_0\} = \langle u_i \ v_i \ w_i \rangle^t : i = A, B, C, D, E, F, G \qquad [1.18]$$

The passive degrees of freedom vector is given as

$$\{q_p\} = \langle u_A \ v_A \ w_A, v_B \ w_B, u_C \ w_C \rangle^t$$

Hence, the active degrees of freedom vector can be deduced as

$$\{q_a\} = \langle u_B, v_c, u_D \ v_D \ w_D, u_E \ v_E \ w_E, u_F \ v_F \ w_F \rangle^t$$

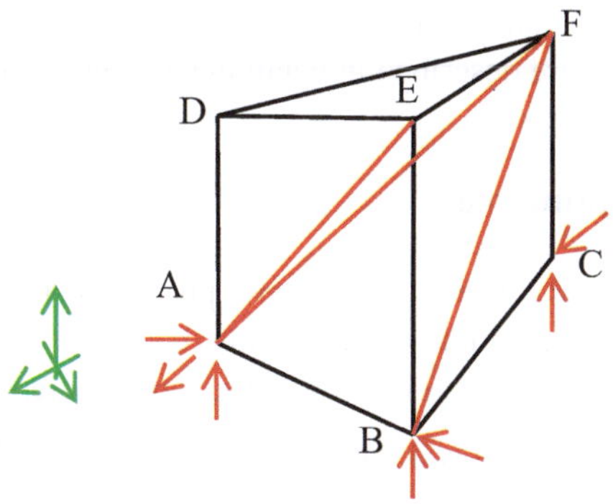

Figure 1.17. *Free-body structure*

1.5. Statically indeterminate structural analysis methods

Since the mid-1800s, several methods have been developed to analyze statically indeterminate structures. The static equations are insufficient to analyze this type of structure. In order to do so, it is necessary to use a statically indeterminate structural analysis method, regrouped in two categories: (1) the method of forces or flexibility and (2) the method of displacements or rigidity.

In the method of forces, the number of external actions must be eliminated, making the fundamental (statically determinate) structure equal to the degree of system static indeterminacy. The obtained structure is called a free-body structure, which undergoes inconsistent deflections at the site of the eliminated actions. To correct these deflections, we must apply unit actions instead of ignored forces. For example, in the method of forces, the solution is obtained by constructing a system of equations whose number is equal to the degree of static indeterminacy. In this case, the unknowns of the problem are the redundant actions and the final solution is obtained by a superposition of the effects of the eliminated actions.

In the displacement method, the fictitious supports must be introduced to prevent the joints from having a freedom to move. Then, we can calculate their corresponding support reactions. This allows us to calculate the displacements at the level of the fictitious supports, which create actions at the ends of each bar. Lastly, we use the superposition principle to calculate the final actions at the ends of each bar.

The unknowns of this method are displacements. In this case, the number of fictitious supports added must be equal to the number of possible displacements of the structure. The bases and principles of the methods for analyzing statically indeterminate structures are described in detail in Chapters 2–6.

1.6. Superposition principle

When the deflections of a structure are proportional to the actions which generate them, the superposition principle can be applied. Note that all actions applied to point i generate deflections at point j; force F_i creates a displacement δ_{ji} at point j (moment M_i generates a rotation ω_{ji} at point j). In particular, force F_i (moment M_i) creates a displacement δ_{ii} (a rotation ω_{ii}) at the point of application i.

We first apply force F_i to beam (ABCD). This force generates a deflection field of the beam including displacements δ_{ii} and δ_{ji}, at joints i and j, respectively. In the same way, we apply force F_j to point j that generates displacements δ_{ij} and δ_{jj}, respectively, at joints i and j (Figure 1.18). This approach is clarified in Figure 1.18. The relationship between the applied force and the resulting displacement is given as

$$\delta_{ii} = f_{ii} F_i \qquad [1.19]$$

where f_{ii} is the displacement at point i when a unit force is applied at the same point in the direction of force F_i. We apply force F_j at point j that generates a displacement δ_{ij} at point i.

$$\delta_{ij} = f_{ij} F_j \qquad [1.20]$$

If both forces are simultaneously applied, the displacement δ_i at point i is given as

$$\delta_i = f_{ii} F_i + f_{ij} F_j \qquad [1.21]$$

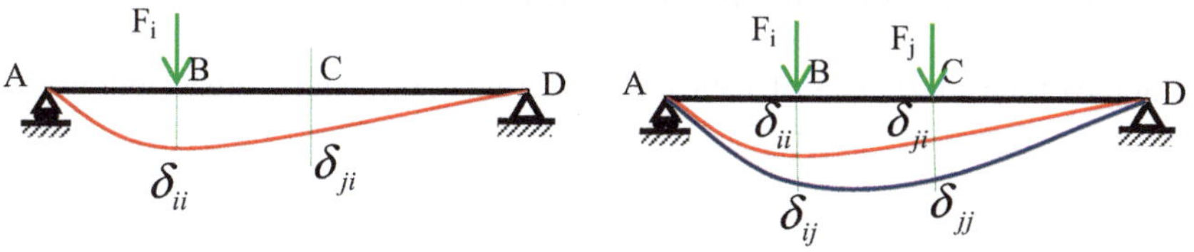

Figure 1.18. *Superposition of displacements and forces*

The relationship [1.21] can be generalized for several simultaneously applied forces. The total displacement at point i can be evaluated by

$$\delta_i = f_{i1}F_1 + f_{i2}F_2 + f_{i3}F_3 + \dots\dots\dots + f_{in}F_n \qquad [1.22]$$

In matrix form, relationship [1.22] is written as

$$\delta_i = \left[f_{ij} \right]\left\{ F_j \right\} \qquad [1.23]$$

In general, all actions A_i can be a support reaction, a bending moment, a shear stress or a normal force; due to the effect of several external actions, they can be obtained by a superposition of effects.

$$A_i = \left[A_{uij} \right]\left\{ F_j \right\} \qquad [1.24]$$

where A_{uij} is the intensity of the action at point i when a unit action is applied at joint j.

The relationship [1.24] evaluates the magnitude of the action of a statically indeterminate structure by using the principle of the superposition of effects of this action of statically determinate systems due to the effect of unit actions (see Chapter 3 for more details).

1.7. Advantages and disadvantages of statically indeterminate structures

Obviously, statically indeterminate structures have advantages and disadvantages, which are shown in the following section.

1.7.1. Advantages of statically indeterminate structures

Some of the advantages of statically indeterminate structures are as follows:

– *Rigidity*

Statically indeterminate structures are known for their rigidities (small deflections) compared to statically determinate structures (Figure 1.19).

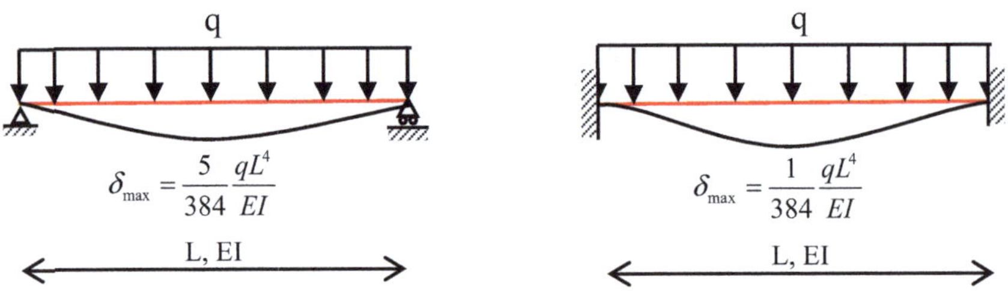

Figure 1.19. *Maximum displacements*

Moreover, we can see that the maximum displacement of a statically indeterminate structure is $\frac{1}{5}$ compared to a statically determinate structure.

– *Stress*

Maximum normal stresses in statically indeterminate structures are smaller than those generated in statically determinate structures by keeping the same geometry and applied loading (Figure 1.20). This figure shows that the maximum moment of the statically indeterminate beam is weaker than that of the statically determinate beam.

– *Distribution of actions*

Statically indeterminate structures have a great ability to distribute actions such as the slope-deflection method (Chapter 4) and the moment-distribution method (Chapter 5). This property has not been considered during the analysis of statically determinate structures.

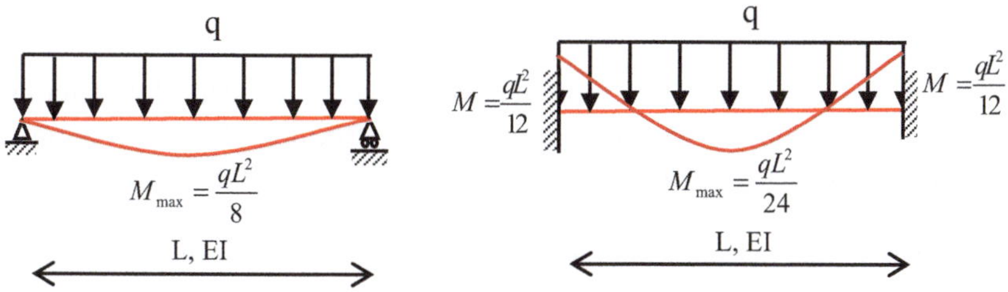

Figure 1.20. *Maximum bending moment*

The advantage of the distribution of actions is observed when a portion of a structure undergoes a localization of stresses or damage due to an earthquake, an impact force, a crack, an explosion or another force. In addition, statically indeterminate structures contain several structural elements and supports that make the structure more rigid, which statically determinate structures do not have.

Furthermore, if a bar or support deteriorates, the statically indeterminate structure will not necessarily be damaged and the load can be distributed to adjacent elements. The destruction or elimination, for example, of the interior support of statically indeterminate and statically determinate beams, damages the statically determinate beam, while the statically indeterminate beam becomes a statically determinate beam (Figure 1.21). In this case, the damage is an elimination or destruction of the interior support.

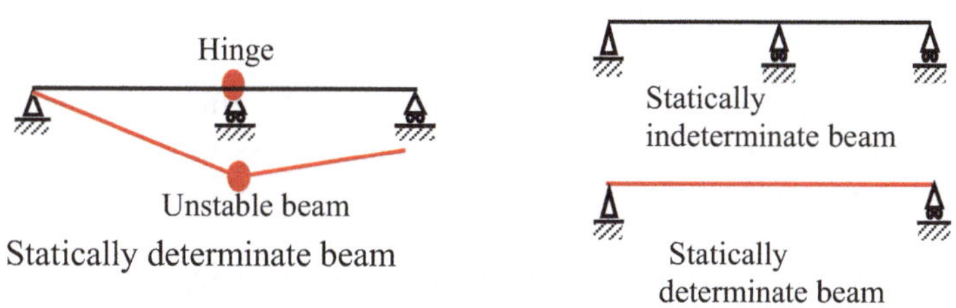

Figure 1.21. *Destruction of the interior support*

1.7.2. *Disadvantages of statically indeterminate structures*

The disadvantages of statically indeterminate structures compared to statically determinate structures are as follows:

– *Support settlements*

A support settlement does not generate stresses in statically determinate structures, but it does in statically indeterminate structures. For this reason, it is necessary to take this effect into account during the design phase.

To clarify this phenomenon, consider a support settlement of a statically determinate beam with a hinge and another statically indeterminate beam (see Figure 1.22). The members of the statically determinate beam undergo the movement of the rigid body because of the presence of the hinge, whereas the support settlement of a statically indeterminate beam generates additional moments to those of the mechanical loads at the ends of each bar (Chapters 4 and 5).

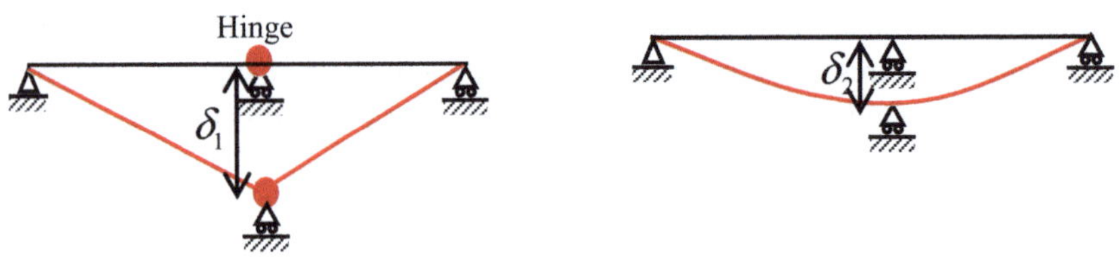

Statically determinate beam   Statically indeterminate beam

Figure 1.22. *Support settlement*

– *Temperature change and manufacturing errors*

Identical to the effect of the support settlement, the effect of temperature or a manufacturing error does not cause any additional stress in statically determinate structures but can generate significant stresses in statically indeterminate structures.

When a statically determinate beam is subjected to a temperature change ΔT, it lengthens by $\delta = \alpha.\Delta T.L$ (α coefficient of thermal expansion) and no stress can develop inside the beam because of the freedom of movement given by the hinge (Figure 1.23).

However, the statically indeterminate beam is impeded at its ends vis-à-vis the axial deflection. The temperature effect creates an axial force of intensity equal to $\alpha.\Delta T.E.\Omega$. The effect of a manufacturing error is very similar to that of the temperature change.

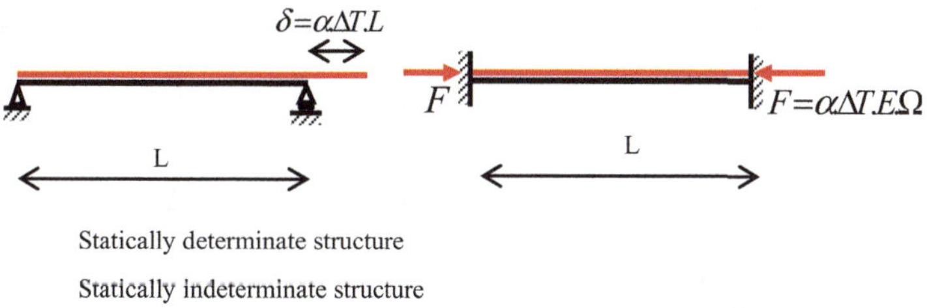

Figure 1.23. *Temperature effect*

1.8. Conclusion

In this chapter, we first presented in detail the external, internal and kinematic static indeterminacy of statically indeterminate structures. The static indeterminacy of a structure plays a very important role in choosing the appropriate method of analysis.

In the second part, we illustrated statically indeterminate structural analysis methods that are classified in two categories: (1) the method of forces or flexibility and (2) the method of displacements or rigidity.

As it is important and commonly used in the analysis of statically indeterminate structures, the principle of superposition of effects is described.

Finally, at the end of this chapter, we have illustrated the advantages and disadvantages of statically indeterminate structures compared to statically determinate structures. In this, several parameters are cited showing their impacts on statically indeterminate and statically determinate structures.

1.9. Problems

Exercise 1

Determine the degree of static indeterminacy of the following plane structures:

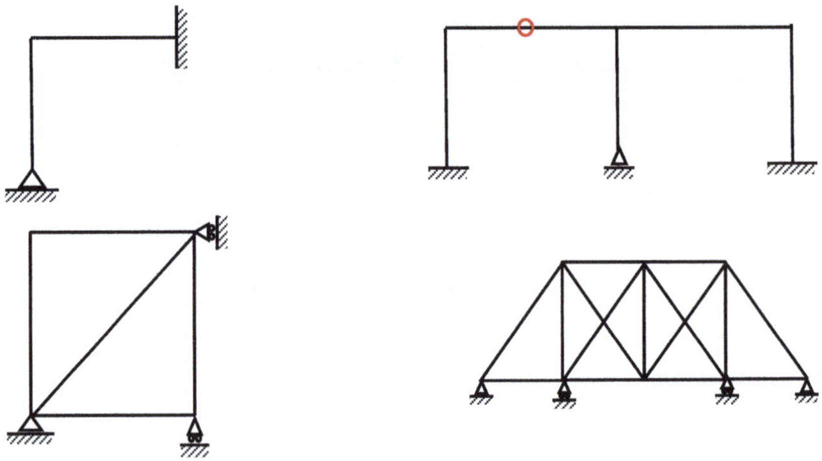

Exercise 2

Determine the degree of static indeterminacy of the following plane structures:

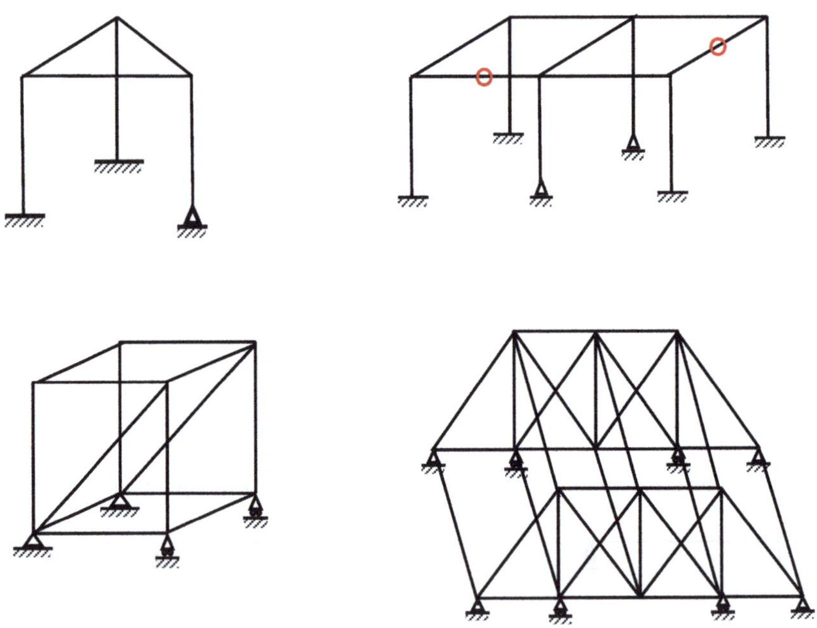

Exercise 3

Determine the degrees of internal and external static indeterminacy of the following plane structures:

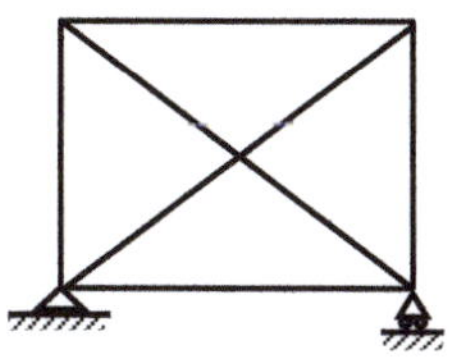

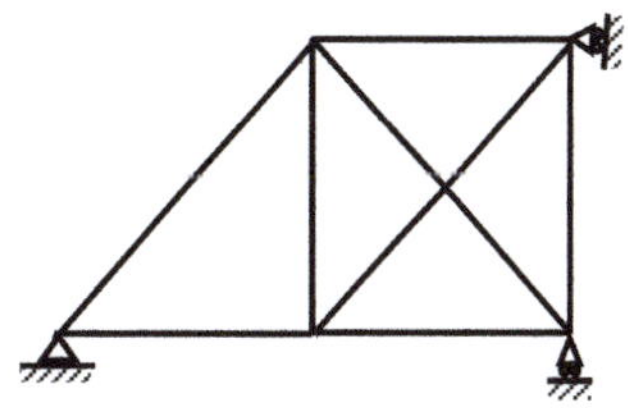

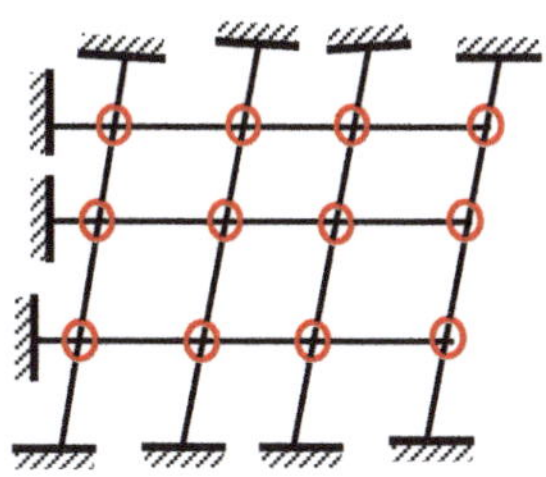

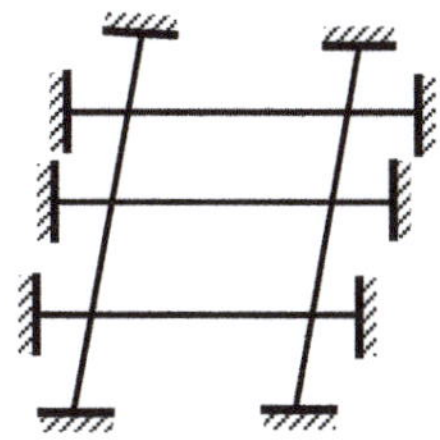

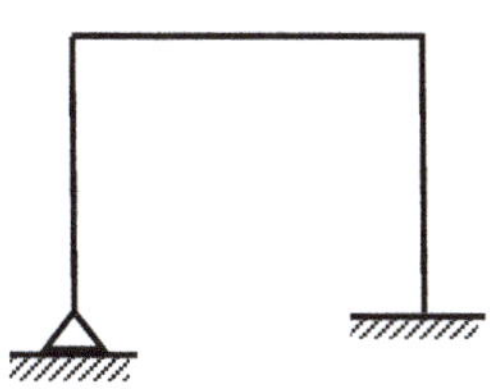

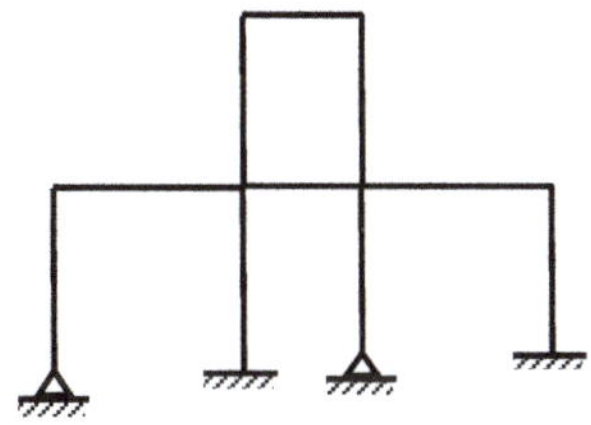

Exercise 4

Determine the degree of internal and external static indeterminacy of the space structures given in Exercise 2.

Exercise 5

Give the vector of the degrees of freedom of the free-body structure, the vector of the passive degrees of freedom and the vector of the active degrees of freedom of the structures in Exercise 1.

Exercise 6

Give the vector of the degrees of freedom of the free-body structure, the vector of the passive degrees of freedom and the vector of the active degrees of freedom of the structures in Exercise 2.

2

Method of Three Moments

The teaching objectives of this chapter are as follows:

– formulating the three moments equation at which the support settlement can be taken into consideration;

– applying the compatibility of the deformations to the supports of continuous beams;

– calculating the actions at the ends of each bar and deducing the support reactions;

– analyzing continuous beams with variable moments of inertia.

This chapter presents the formulation of the three moments equation for the analysis of single span statically indeterminate beams. The successive application of the three moments equation leads to the formulation of the Clapeyron method. This derivation is formulated in the general case taking into account the support settlements. Directly applying the three moments equation to the continuous beams leads to the formulation of the focus method.

2.1. Simple beams

To describe the method of three moments, particular attention is given to the condition of compatibility of the deformations at the supports. We therefore consider a statically indeterminate beam with a single span of length L, flexural rigidity (EI) and subjected to a load q (x) (Figure 2.1). For a general description of the analysis method, the beam considered is fixed at its ends.

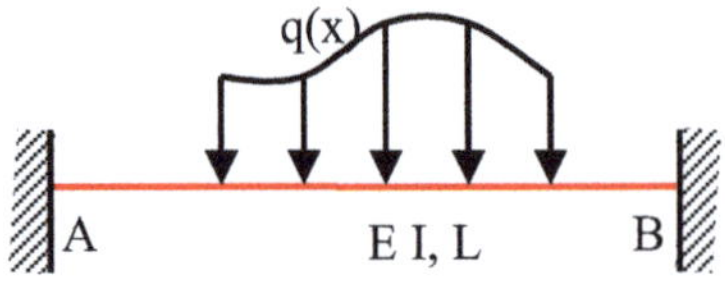

Figure 2.1. *Single span beam*[1]

The statically indeterminate beam can be divided into several statically determinate systems.

1) *A fundamental or basic structure*

We ignore the effect of fixing at supports A and B to make the beam statically determinate (Figure 2.2). The resulting system facilitates the calculation of slopes or rotations at supports A and B, denoted by $\omega_A^{'}$ and $\omega_B^{'}$. Using a method for calculating elastic deformations, such as the virtual work method, makes it possible to calculate the slope expression at support A.

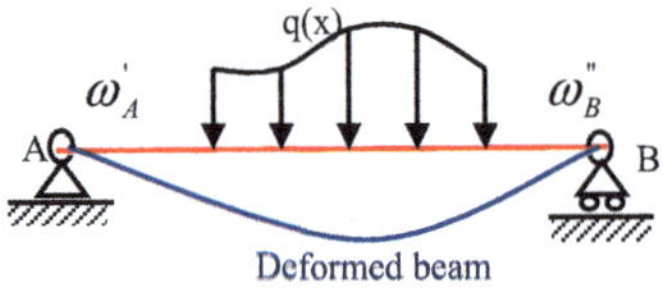

Figure 2.2. *Statically determinate structure*

$$\omega_A^{'} = \int_0^L \frac{\mu(x).m(x)}{EI} dx \qquad [2.1]$$

where $\mu(x)$ and m(x) are, respectively, the bending moment of the fundamental system and the unit moment applied to joint A, respecting the sign convention of the positive moments (Figure 2.3).

Figure 2.3. *Unit moment at joint A*

1 All of the figures in this chapter are available to view in full color at www.iste.co.uk/khalfallah/analysis2.zip.

$$m(x) = -(1 - \frac{x}{L}) \quad [2.2]$$

The slope at joint A can be calculated by

$$\omega_A^{'} = -\int_0^L \frac{\mu(x)\,(1 - \frac{x}{L})}{EI}\, dx \quad [2.3]$$

Figure 2.4. *Unit moment at joint B*

In the same way, we can calculate the slope $\omega_B^{''}$ by applying the unit moment in the positive direction at joint B (Figure 2.4). The bending moment is written as

$$m(x) = \frac{x}{L} \quad [2.4]$$

The slope at joint B is then given as

$$\omega_B^{''} = \int_0^L \frac{\mu(x)\,(\frac{x}{L})}{EI}\, dx \quad [2.5]$$

Relationships [2.3] and [2.5] give the formulas for slopes at the end joints of bar (AB) of the fundamental system.

2) *Fixing effect*

The fixing of bar (AB) generates moments M_A and M_B at supports A and B. Potentially, each moment generates slopes at the ends of the bar. Either $\theta_A(M_A)$ or $\theta_B(M_A)$ are the slopes at points A and B due to the effect of the fixing moment M_A (Figure 2.5).

$$\text{M(x)} = -M_A(1 - \frac{x}{L}) \quad [2.6]$$

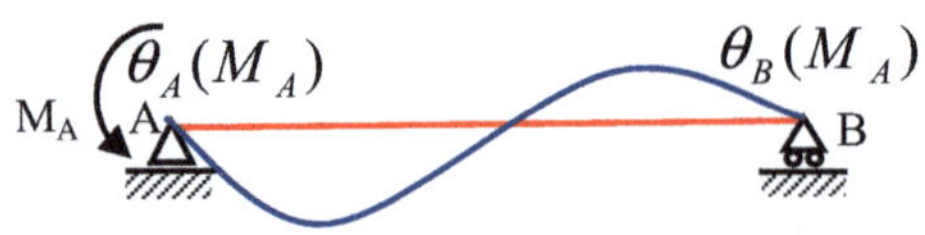

Figure 2.5. *Effect of fixing moment M_A*

The variation of the bending moment along the beam due to the effect of the unit moment applied to joint A is written as

$$m(x) = -(1-\frac{x}{L}) \quad [2.7]$$

Using the theory of virtual work, the slope expression at support A is given as

$$\theta_A(M_A) = \int_0^L \frac{M(x).m(x)}{EI}dx = M_A\int_0^L \frac{(1-\frac{x}{L})^2}{EI}dx \quad [2.8]$$

Similarly, the slope at joint B due to the effect of M_A can be deduced by applying a unit moment at support B. In this case, the expression of the bending moment along the beam is written as

$$m(x) = \frac{x}{L} \quad [2.9]$$

$$\theta_B(M_A) = \int_0^L \frac{M(x)m(x)}{EI}dx = -M_A\int_0^L \frac{(1-\frac{x}{L})\frac{x}{L}}{EI}dx \quad [2.10]$$

The same procedure is used to express the slopes at the ends of bar (AB) due to the fixing effect at joint B (Figure 2.6).

The expression of the bending moment is given by

$$M(x) = M_B\frac{x}{L} \quad [2.11]$$

The slopes at the ends of bar (AB) are expressed by

$$\theta_A(M_B)=\int_0^L \frac{M(x).m(x)}{EI}dx=-M_B\int_0^L \frac{(1-\frac{x}{L})\frac{x}{L}}{EI}dx \quad [2.12]$$

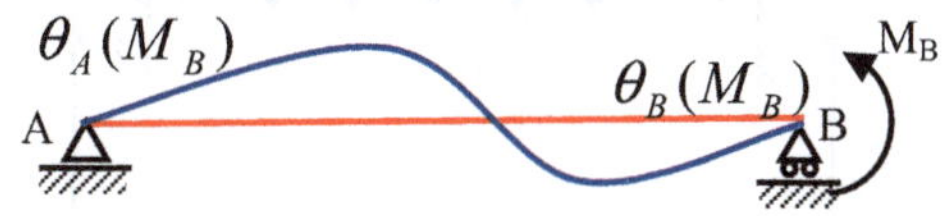

Figure 2.6. *Effect of fixing moment M_B*

$$\theta_B(M_B)=\int_0^L \frac{M(x).m(x)}{EI}dx=M_B\int_0^L \frac{(\frac{x}{L})^2}{EI}dx \quad [2.13]$$

3) *Support settlement effect*

The support settlements occur when one or more structural supports displaces with different values. If this phenomenon is taken into account in the analysis, it is necessary to introduce the slopes of the bar into the general formulation, which can be obtained from the beam's deformation (Figure 2.7). The settlements are generally considered very small and the corresponding slopes at the ends of the bar denoted by β_A and β_B have the following expressions:

$$\beta_A=-\frac{\lambda_B-\lambda_A}{L} \quad [2.14a]$$

$$\beta_B=-\frac{\lambda_B-\lambda_A}{L} \quad [2.14b]$$

The negative sign is introduced in expressions [2.14a] and [2.14b] to respect the sign convention used. It is remarkable that these two slopes have the same value and the same direction of rotation.

$$\beta=\beta_A=\beta_B \quad [2.15]$$

Using the superposition of effects principle and the kinematic conditions at supports A and B, the slopes at supports A and B are written as:

$$\omega_A = \omega_A^{'} + \theta_A(M_A) + \theta_A(M_B) + \beta = 0 \quad [2.16a]$$

$$\omega_B = \omega_B^{''} + \theta_B(M_A) + \theta_B(M_B) + \beta = 0 \quad [2.16b]$$

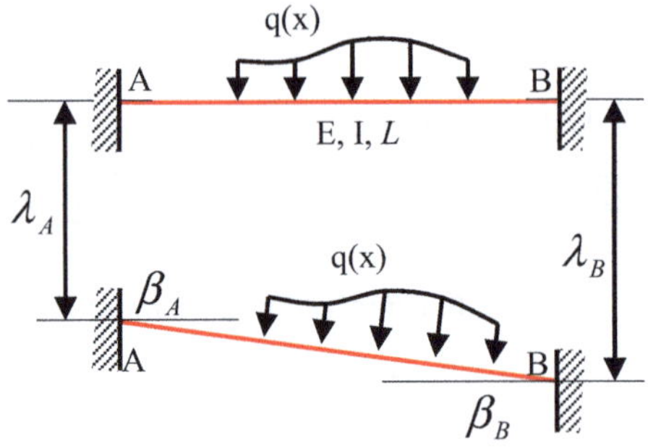

Figure 2.7. *Support settlement effect*

Substituting the slope expressions [2.3], [2.5], [2.8], [2.10] and [2.12–2.14] with the equation system [2.16], we obtain

$$-\int_0^L \frac{\mu(x)(1-\frac{x}{L})}{EI} dx + M_A \int_0^L \frac{(1-\frac{x}{L})^2}{EI} dx - M_B \int_0^L \frac{(1-\frac{x}{L})\frac{x}{L}}{EI} dx + \beta = 0 \quad [2.17a]$$

$$\int_0^L \frac{\mu(x)\frac{x}{L}}{EI} dx - M_A \int_0^L \frac{(1-\frac{x}{L})\frac{x}{L}}{EI} dx + M_B \int_0^L \frac{(\frac{x}{L})^2}{EI} dx + \beta = 0 \quad [2.17b]$$

we put

$$a = \int_0^L \frac{(1-\frac{x}{L})^2}{EI} dx$$

$$b = \int_0^L \frac{(1-\frac{x}{L})\frac{x}{L}}{EI} dx$$

$$c = \int_0^L \frac{(\frac{x}{L})^2}{EI} dx$$

The integrated quantities a, b and c are called the mechanical constants or the flexibility coefficients, which characterize the mechanical properties of the beam. They depend on the geometry and the flexural rigidity of the beam.

The relationship [2.17] is written as

$$-\int_0^L \frac{\mu(x).(1-\frac{x}{L})}{EI} dx + aM_A - bM_B + \beta = 0 \qquad [2.18a]$$

$$\int_0^L \frac{\mu(x)\frac{x}{L}}{EI} dx - bM_A + cM_B + \beta = 0 \qquad [2.18b]$$

In case of a constant inertia of the beam, the mechanical constants a, b and c have the following expressions:

$$a = \frac{1}{EI}\int_0^L (1-\frac{x}{L})^2 dx = \frac{L}{3EI} \qquad [2.19a]$$

$$b = \frac{1}{EI}\int_0^L (1-\frac{x}{L})\frac{x}{L} dx = \frac{L}{6EI} \qquad [2.19b]$$

$$c = \frac{1}{EI}\int_0^L (\frac{x}{L})^2 dx = \frac{L}{3EI} \qquad [2.19c]$$

Expressions [2.19] conclude that a = 2b = c = $\frac{L}{3EI}$.

The system of equations [2.18] becomes

$$-\frac{1}{EI}\int_0^L \mu(x).(1-\frac{x}{L}).dx + aM_A - bM_B + \beta = 0 \qquad [2.20a]$$

$$\frac{1}{EI}\int_0^L \mu(x)\frac{x}{L} dx - bM_A + cM_B + \beta = 0 \qquad [2.20b]$$

The relationship [2.20] can also be written as

$$aM_A - bM_B = \frac{1}{EI}\int_0^L \mu(x)(1-\frac{x}{L})dx - \beta \qquad [2.21a]$$

$$-bM_A + cM_B = -\frac{1}{EI}\int_0^L \mu(x)\frac{x}{L}dx - \beta \qquad [2.21b]$$

Resolving the system of equations [2.21] allows us to calculate fixed-end moments M_A and M_B.

If the support settlement is neglected in the calculation, the system of equations [2.21] is reduced to

$$aM_A - bM_B = \frac{1}{EI}\int_0^L \mu(x)(1-\frac{x}{L})dx \qquad [2.22a]$$

$$-bM_A + cM_B = -\frac{1}{EI}\int_0^L \mu(x)\frac{x}{L}dx \qquad [2.22b]$$

The fixed-end moments of bar (AB) potentially have the formulas

$$M_A = \frac{c}{EI(ac-b^2)}\int_0^L \mu(x)(1-\frac{x}{L})dx - \frac{b}{EI(ac-b^2)}\int_0^L \mu(x)(\frac{x}{L})dx \qquad [2.23a]$$

$$M_B = \frac{b}{EI(ac-b^2)}\int_0^L \mu(x)(1-\frac{x}{L})dx - \frac{a}{EI(ac-b^2)}\int_0^L \mu(x)(\frac{x}{L})dx \qquad [2.23b]$$

The resulting bending moments are a combination of the bending moment of the fundamental system, the effect of the fixing moment M_A [2.2] and the fixing moment M_B [2.4]. At any section x, the bending moment is written as

$$M(x) = \mu(x) - M_A(1-\frac{x}{L}) + M_B(\frac{x}{L}) \qquad [2.24]$$

The expression of the shear force is obtained by deriving the relationship [2.24].

$$T(x) = \frac{d\mu(x)}{dx} + \frac{M_A + M_B}{L} \qquad [2.25]$$

EXAMPLE 2.1.–

Analyze the beam (Figure 2.8) of constant flexural rigidity EI and length L. The beam is considered fixed at one end and simply supported at the other. The beam is subjected to a concentrated load of intensity P located at $\frac{L}{3}$ of support A.

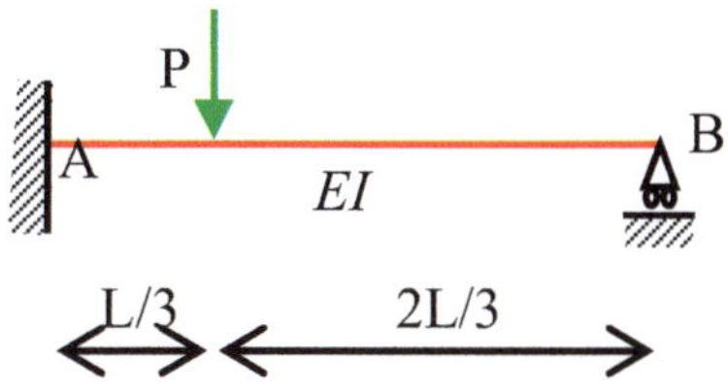

Figure 2.8. *Studied beam*

– *Statically determinate beam (Figure 2.9)*

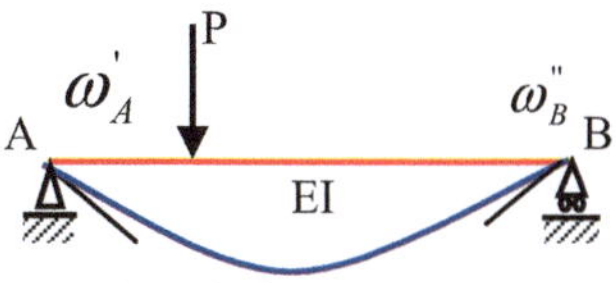

Figure 2.9. *Basic statically determinate beam*

The support reactions are given as

$$\text{VA} = \frac{2}{3}P$$

$$\text{VB} = \frac{1}{3}P$$

To calculate the rotations at A and B of the basic system (Figure 2.9), the differential equation of the elastic line deflection in each region is used.

$$y_1''(x) = \frac{\mu_1(x)}{EI} = \frac{2P}{3EI}x \qquad x \le \frac{L}{3}$$

$$y_2''(x) = \frac{\mu_2(x)}{EI} = \frac{P}{3EI}(L - x) \qquad \frac{L}{3} \le x \le L$$

The first and second integration of the differential equations lead to

$$y_1'(x) = \frac{2P}{3EI}(\frac{x^2}{2} + c_1)$$

$$y_1(x) = \frac{2P}{3EI}(\frac{x^3}{6} + c_1 x + c_2)$$

$$y_2'(x) = \frac{P}{3EI}(Lx - \frac{x^2}{2} + c_3)$$

$$y_2(x) = \frac{P}{3EI}(\frac{Lx^2}{2} - \frac{x^3}{6} + c_3 x + c_4)$$

where c_i are the integration constants that can be determined using the limit conditions.

$$y_1(x = 0) = 0$$

$$y_2(x = L) = 0$$

Continuity of rotation and displacement at point $x = \frac{L}{3}$ allows us to write

$$y_1'(x = \frac{L}{3}) = y_2'(x = \frac{L}{3})$$

and

$$y_1(x = \frac{L}{3}) = y_2(x = \frac{L}{3})$$

Therefore, the integration constants c_i (i = 1–4) are given as

$$c_1 = -\frac{5}{54}L^2,\ c_2 = 0,\ c_3 = -\frac{19}{54}L^2 \text{ and } c_4 = \frac{1}{54}L^3$$

The expressions of the slopes of the elastic beam become

$$0 \le x \le \frac{L}{3}$$

$$y_1'(x) = \frac{2P}{3EI}(\frac{x^2}{2} - \frac{5}{54}L^2)$$

$$\frac{L}{3} \leq x \leq L$$

$$y_2'(x) = \frac{P}{3EI}(Lx - \frac{x^2}{2} - \frac{19}{54}L^2)$$

In particular, the slopes at the ends of the beam are given as

$$\omega_A' = y_1'(x=0) = -\frac{5PL^2}{81EI}$$

$$\omega_B'' = y_2'(x=L) = \frac{8PL^2}{162EI}$$

Applying equation [2.22a] to beam (AB), we obtain

$$\omega_A' + aM_{AB} = 0$$

The expression of the fixing moment is

$$\text{MA} = -\frac{\omega_A'}{a} = \frac{5}{27}qL^2$$

The variations of the bending moment along the beam are given as

$$0 \leq x \leq \frac{L}{3}$$

$$\text{M(x)} = \frac{2P}{3}x - \frac{5}{27}PL(1-\frac{x}{L}) = \frac{23P}{27}x - \frac{5}{27}PL$$

$$\frac{L}{3} \leq x \leq L$$

$$\text{M(x)} = \frac{PL}{3} - \frac{P}{3}x - \frac{5}{27}PL(1-\frac{x}{L}) = \frac{4PL}{27} - \frac{4P}{27}x$$

The moments at the ends of the beam are given by

$$M(x=0) = \frac{5}{27}PL$$

$$M(x=\frac{L}{3}) = \frac{8}{81}PL$$

The diagrams of the bending moment and the shear force are given in Figure 2.10. The expressions giving the variation of the shear force are given by

$$T(x) = \frac{dM(x)}{dx}$$

$\frac{2}{3}L \leq x \leq L$

$$T(x) = \frac{dM(x)}{dx} = \frac{4}{27}P$$

$0 \leq x \leq \frac{2}{3}L$

$$T(x) = \frac{dM(x)}{dx} = -\frac{23}{27}P$$

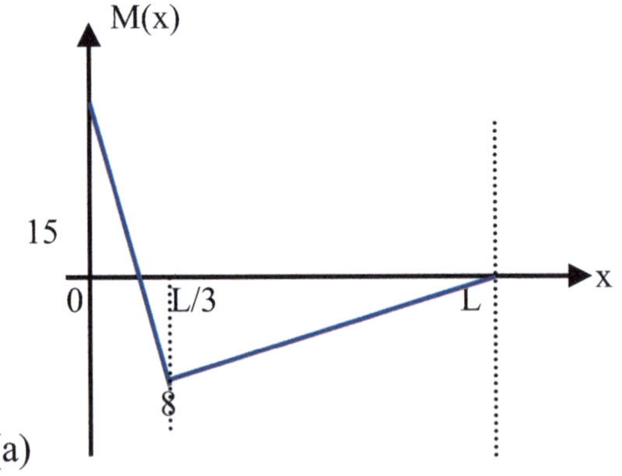

(a)

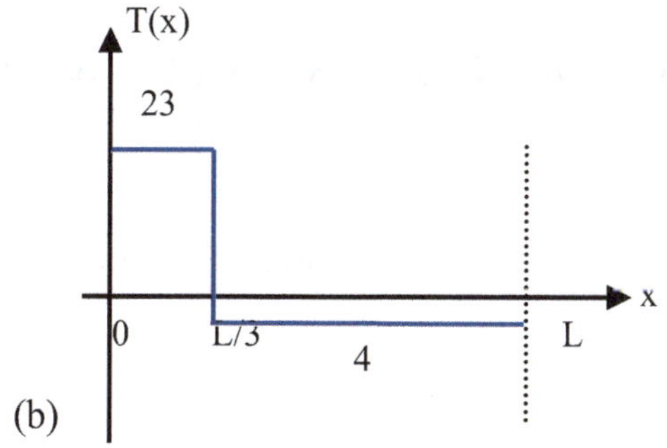

Figure 2.10. *Diagrams a) of bending moment* ($\frac{1}{81}PL$) and b) of shear force * ($\frac{1}{27}P$)*

2.2. Continuous beam

The successive application of the three moments formula [2.22], used to analyze a single span statically indeterminate beam, leads to the formulation of a system of algebraic equations whose unknowns are the moments at the supports. The number of unknowns, which is the moments at the supports, must be equal to the number of equations established.

Consider a continuous beam $(n - 1)$ with spans of length L_i and flexural rigidity (EI), which we assume to be constant for simplifying reasons. It is subject to a set of external loads (Figure 2.11).

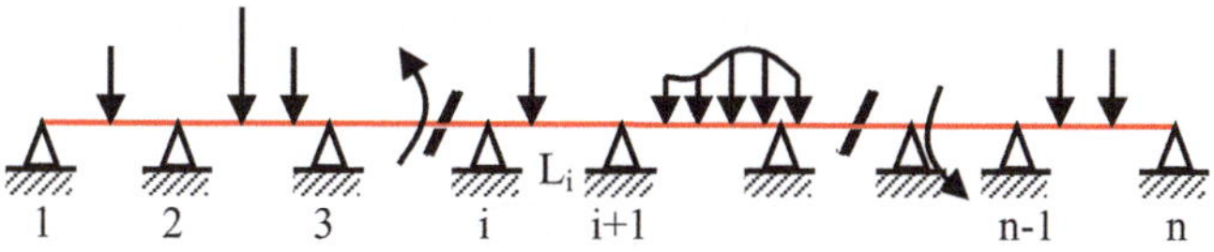

Figure 2.11. *Beam with several spans*

To build the basic system, hinges are added to each of the continuous beam's supports (Figure 2.11) to make each span statically determinate (Figure 2.12).

Consider $(i - 1)$ span delimited by successive supports $[i - 1, i]$ (Figure 2.13). The left support settlement i due to the load applied on span $(i - 1)$ can be deduced from equation [2.5].

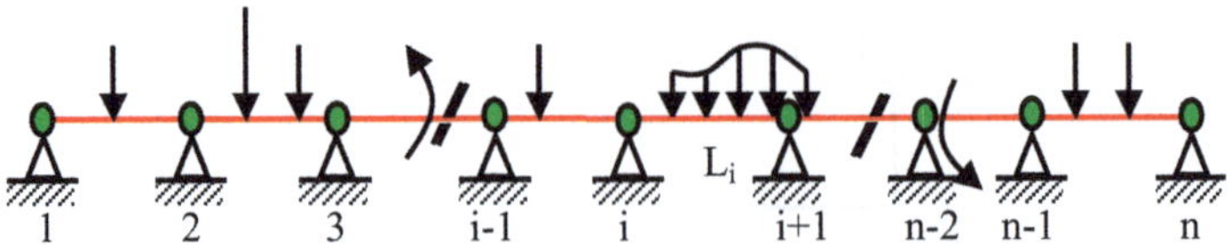

Figure 2.12. *Continuous statically determinate beam*

$$\omega_i^{''} = \int_0^{L_{i-1}} \frac{\mu_{i-1}(x)\frac{x}{L_{i-1}}}{EI} dx \qquad [2.26a]$$

Similarly, consider span (i) delimited by [i, i + 1] (Figure 2.14), the right support settlement i of span (i) can be formulated using relationship [2.3].

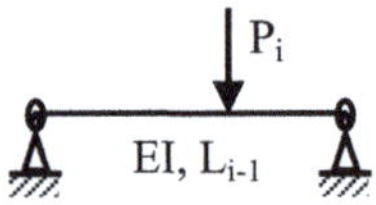

Figure 2.13. *(i – 1)th span*

$$\omega_i^{'} = -\int_0^{L_i} \frac{\mu_i(x)(1-\frac{x}{L_i})}{EI} dx \qquad [2.26b]$$

With $\mu_{i-1}(x)$ and $\mu_i(x)$ are the statically determinate moments, respectively, of spans (i – 1) and (i).

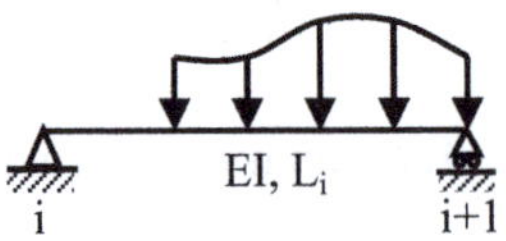

Figure 2.14. *ith span*

The resulting slope at joint i is written as

$$\omega_i = \omega_i^{''} - \omega_i^{'} \qquad [2.27]$$

Moments due to geometric continuity cause slopes at the ends of each span (Figures 2.15 and 2.16).

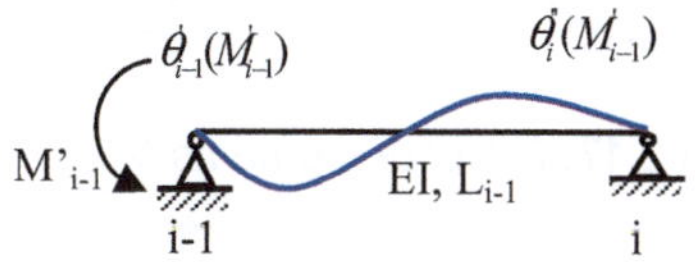

Figure 2.15. *Effect of left moment M'i-1*

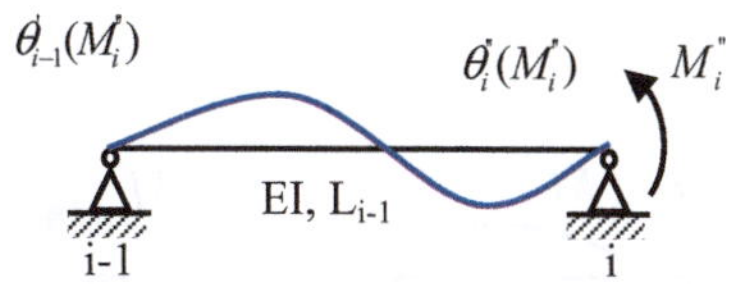

Figure 2.16. *Effect of left moment M"i.*

$$\theta''_i(M'_{i-1}) = -M'_{i-1}\int_0^{L_{i-1}} \frac{(1-\frac{x}{L_{i-1}})\frac{x}{L_{i-1}}}{EI}dx \quad [2.28a]$$

$$\theta''_i(M''_i) = M''_i\int_0^{L_{i-1}} \frac{(\frac{x}{L_{i-1}})^2}{EI}dx \quad [2.28b]$$

$$\theta'_i(M'_i) = M'_i\int_0^{L_i} \frac{(1-\frac{x}{L_i})^2}{EI}dx \quad [2.28c]$$

In the same way, moments M'_i and M''_{i+1} generate slopes at joints i and $(i + 1)$ (Figures 2.17 and 2.18).

$$\theta'_i(M''_{i+1}) = -M''_{i+1}\int_0^{L_i} \frac{(1-\frac{x}{L_i})\frac{x}{L_i}}{EI}dx \quad [2.28d]$$

The resulting slope θ_i due to the geometric continuity of the beam or due to the effect of the moments M'_{i-1}, M''_i, M'_i and M''_{i+1} is

$$\theta_i = \theta''_i(M'_{i-1}) + \theta''_i(M''_i) - \theta'_i(M'_i) - \theta'_i(M''_{i+1}) \quad [2.29]$$

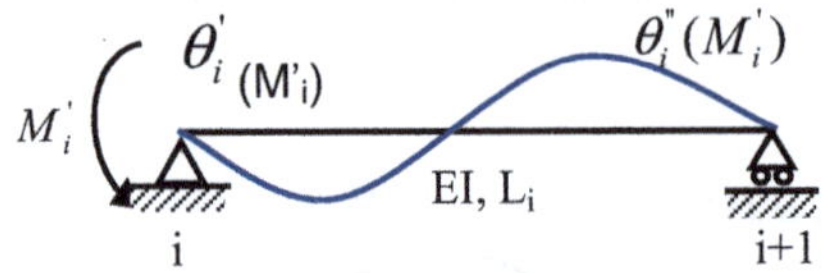

Figure 2.17. *Effect of moments M'i*

If the support settlement is considered, the slope at joint *i* becomes

$$\omega_i + \theta_i + \beta_i = 0 \qquad [2.30]$$

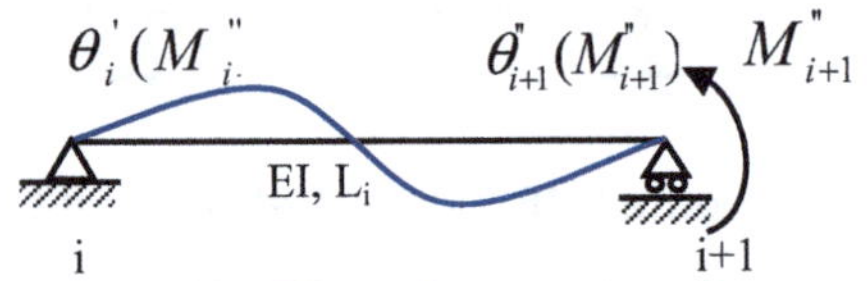

Figure 2.18. *Effect of moments* $M_{i+1}^{"}$

Substituting equations [2.27] and [2.29] in relationship [2.30], we obtain

$$\omega_i^{"} - \omega_i^{'} + \theta_i^{"}(M_{i-1}^{'}) + \theta_i^{"}(M_i^{"}) - \theta_i^{'}(M_i^{'}) - \theta_i^{'}(M_{i+1}^{"}) + \beta_i^{"} - \beta_i^{'} = 0 \qquad [2.31a]$$

knowing that $\beta_i = \beta_i^{"} - \beta_i^{'}$

Similarly, substituting relationships [2.26] and [2.28] in equation [2.31a], we have

$$M_{i-1}^{'} \int_0^{L_{i-1}} \frac{(1-\frac{x}{L_{i-1}})\frac{x}{l_{i-1}}}{EI}dx - M_i^{"} \int_0^{L_{i-1}} \frac{(\frac{x}{L_{i-1}})^2}{EI}dx + M_i^{'} \int_0^{L_i} \frac{(1-\frac{x}{L_i})^2}{EI}dx - \int_0^{L_i} \frac{(1-\frac{x}{L_i})\frac{x}{L_i}}{EI}dx = \qquad [2.31b]$$
$$M_{i+1}^{"}$$
$$\int_0^{L_{i-1}} \frac{\mu_{i-1}(x)(\frac{x}{L_{i-1}})}{EI}dx - \int_0^{L_i} \frac{\mu_i(x)(1-\frac{x}{L_i})}{EI}dx + \beta_i^{'} - \beta_i^{"}$$

(this equation was established by B.P. Clapeyron)

with

$$\beta_i^{'} = -\frac{\lambda_{i+1} - \lambda_i}{L_i} \quad [2.32a]$$

$$\beta_i^{"} = -\frac{\lambda_{i-1} - \lambda_i}{L_{i-1}} \quad [2.32b]$$

λ_{i-1}, λ_i and λ_{i+1} are, respectively, the support settlements $i - 1$, i and $i + 1$.

When all the spans of the beam have a constant inertia (EI), equation [2.31b] is written as

$$b_{i-1}M_{i-1}^{'} - c_{i-1}M_i^{"} + a_i M_i^{'} - b_i M_{i+1}^{"} = \omega_i^{"} - \omega_i^{'} + \beta_i^{"} - \beta_i^{'} \quad [2.33]$$

Knowing that $M_i^{"} + M_i^{'} = 0$ and $M_{i+1}^{"} + M_{i+1}^{'} = 0$, we can therefore write

$$b_{i-1}M_{i-1}^{'} + (a_i + c_{i-1})M_i^{'} + b_i M_{i+1}^{'} = \omega_i^{"} - \omega_i^{'} + \beta_i^{"} - \beta_i^{'} \quad [2.34]$$

Equation [2.34] is called Clapyeron's formula with constant inertia of the spans.

Particular cases

– *Beam with two spans*

Beam composed of two spans of identical lengths and constant inertia (Figure 2.19); calculate the moment of intermediate support.

We apply the Clapeyron equation for $i = 2$ and without taking into account the support settlement, the result is

$$b_1 M_1^{'} + (a_2 + c_1)M_2^{'} + b_2 M_3^{'} = \omega_2^{"} - \omega_2^{'}$$

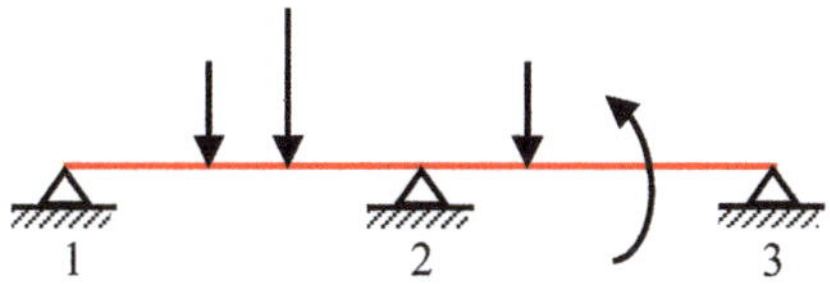

Figure 2.19. *Beam with two spans*

Knowing that $M_1^{'} = M_3^{''} = 0$, the moment of intermediate support, M_2, is written as

$$M_2^{'} = -M_2^{''} = \frac{\omega_2^{''} - \omega_2^{'}}{(a_2 + c_1)}$$

– *Fixed beam*

The Clapeyron formula is applied to spans 1 and 2 knowing that the support settlement is neglected (Figure 2.20).

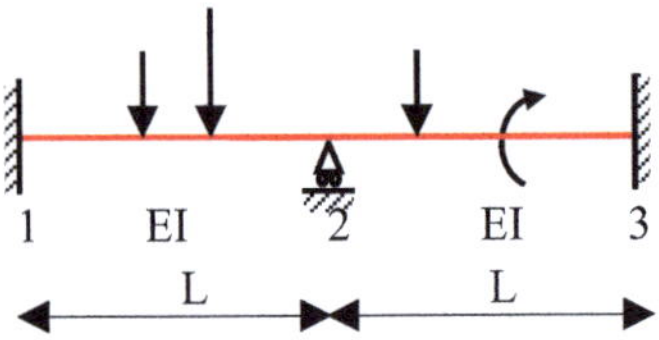

Figure 2.20. *Beam with two spans*

$$-\omega_1^{'} = a_1 M_1^{'} - b_1 M_2^{''}$$

$$\omega_2^{''} - \omega_2^{'} = b_1 M_1^{'} + (a_2 + c_1) M_2^{'} - b_2 M_3^{''}$$

$$\omega_3^{''} - \omega_3^{'} = b_2 M_2^{'} - c_2 M_3^{''}$$

We have

$$M_2^{'} + M_2^{''} = 0$$

Hence

$$-\omega_1^{'} = a_1 M_1^{'} + b_1 M_2^{'}$$

$$\omega_2^{''} - \omega_2^{'} = b_1 M_1^{'} + (a_2 + c_1) M_2^{'} - b_2 M_3^{''}$$

$$\omega_3^{''} = b_2 M_2^{'} - c_2 M_3^{''}$$

Resolving the system of equations allows us to calculate moments M'_1, M'_2 and M_3''.

$$M_1^{'} = -\frac{1}{a_1}\omega_1^{'} + \frac{b_1 c_2}{b_1^2 c_2 - b_2^2 a_1 - a_1 c_1 c_2 - a_1 a_2 c_2}(\omega_2^{''} - \omega_2^{'} + \frac{b_1}{a_1}\omega_1^{'} - \frac{b_2}{c_2}\omega_3^{''})$$

$$M_2^{'} = -M_2^{''} = -\frac{a_1 c_2}{b_1^2 c_2 - b_2^2 a_1 - a_1 c_1 c_2 - a_1 a_2 c_2}(\omega_2^{''} - \omega_2^{'} + \frac{b_1}{a_1}\omega_1^{'} - \frac{b_2}{c_2}\omega_3^{''})$$

$$M_3^{''} = -\frac{1}{c_2}\omega_3^{''} + \frac{a_1 b_2}{b_1^2 c_2 - b_2^2 a_1 - a_1 c_1 c_2 - a_1 a_2 c_2}(\omega_2^{''} - \omega_2^{'} + \frac{b_1}{a_1}\omega_1^{'} - \frac{b_2}{c_2}\omega_3^{''})$$

In the case of a beam with constant inertia: a = 2b = c = $\frac{L}{3EI}$

$$M_1^{'} = \frac{1}{6b}(\omega_2^{''} - \omega_2^{'} + \frac{7}{2}\omega_1^{'} - \frac{1}{2}\omega_3^{''})$$

$$M_2^{'} = -M_2^{''} = \frac{1}{3b}(\omega_2^{''} - \omega_2^{'} + \frac{1}{2}\omega_1^{'} - \frac{1}{2}\omega_3^{''})$$

$$M_3^{''} = \frac{1}{6b}(\omega_2^{''} - \omega_2^{'} + \frac{1}{2}\omega_1^{'} - \frac{7}{2}\omega_3^{''})$$

Calculating the slopes of statically determinate beams, $\omega_1^{'}, \omega_2^{''}, \omega_2^{'}$ and $\omega_3^{''}$ leads to determining the moments at joints 1–3 according to the applied loads.

2.3. Applying Clapeyron's theorem

The steps to apply Clapeyron's method are as follows:

– selecting moments at the interior supports as unknowns of the problem;

– calculating the slopes at each joint of the basic system, due to the effect of the fixed-end moments and the effect of the support settlement, if it is taken into consideration;

– we solve the constructed system of equations whose unknowns are the moments at the joints;

– expressions of bending moment and shear force are obtained using the superposition of effects principle;

– eventually, it is possible to deduce the support reactions by using the equilibrium equations.

2.3.1. *Beam with two spans*

Analyze the beam (Figure 2.21) subjected to a uniformly distributed load of intensity q and a concentrated load P. The flexural rigidity (EI) is considered constant on the entire length of the beam.

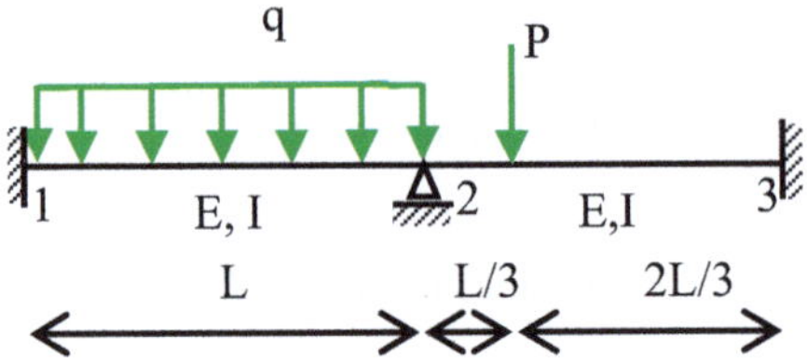

Figure 2.21. *Beam with two spans*

The expressions of the moments at the ends of each bar according to the slopes are given as

$$M_1^{'} = \frac{1}{6b}(\omega_2^{''} - \omega_2^{'} + \frac{7}{2}\omega_1^{'} - \frac{1}{2}\omega_3^{''})$$

$$M_2^{'} = -M_2^{''} = \frac{1}{3b}(\omega_2^{''} - \omega_2^{'} + \frac{1}{2}\omega_1^{'} - \frac{1}{2}\omega_3^{''})$$

$$M_3^{''} = \frac{1}{6b}(\omega_2^{''} - \omega_2^{'} + \frac{1}{2}\omega_1^{'} - \frac{7}{2}\omega_3^{''})$$

The slopes at the joints of the statically determinate elements are written as

$$\omega_1^{'} = -\omega_2^{''} = -\frac{qL^3}{24.EI}$$

$$\omega_2^{'} = -\frac{5.PL^2}{81.EI}$$

$$\omega_3^{''} = \frac{4.PL^2}{81.EI}$$

$$M_1^{'} = \frac{5}{48}qL^2 - \frac{3}{81}PL$$

$$M_2^{'} = -M_2^{''} = \frac{2}{48}qL^2 + \frac{6}{81}PL$$

$$M_3^{''} = \frac{1}{48}qL^2 - \frac{9}{81}PL$$

For the particular case: P = q.L

$$M_1^{'} = 0.067PL$$

$$M_2^{'} = -M_2^{''} = 0.116PL$$

$$M_3^{''} = -0.09PL$$

The variation of the bending moment is given by Figure 2.22.

$$M(x) = \mu(x) - M_{AB}(1 - \frac{x}{L}) + M_{BA}(\frac{x}{L})$$

Span 1

$$M_1(x) = \frac{1}{2}P.x - \frac{P}{2L}x^2 - 0.097PL(1 - \frac{x}{L}) - 0.116PL(\frac{x}{L})$$

Span 2

$$x \leq \frac{L}{3}$$

$$M_2(x) = \frac{2}{3}P.x - 0.116PL(1 - \frac{x}{L}) + 0.09PL(\frac{x}{L})$$

$$\frac{L}{3} \leq x \leq L$$

$$M_2(x) = \frac{1}{3}P.L - \frac{1}{3}P.x - 0.116PL(1 - \frac{x}{L}) + 0.09PL(\frac{x}{L})$$

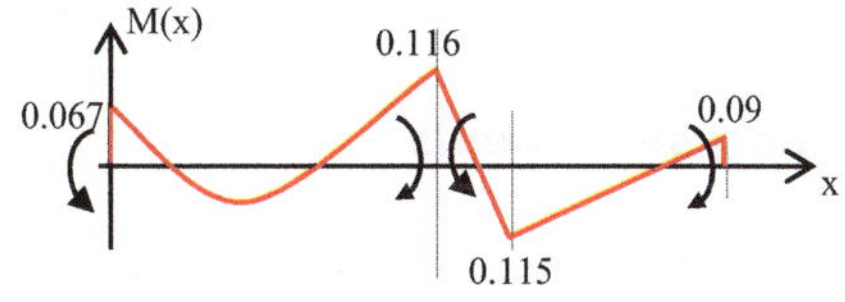

Figure 2.22. *Bending moment diagram *(P.L)*

Thus, the variation in shear force (Figure 2.23) is given as

$$T(x) = \frac{d\mu(x)}{dx} + \frac{M_2^{"} + M_1^{'}}{L}$$

Span 1

$$T_1(x) = \frac{1}{2}P - \frac{P}{L}x - 0.049P = 0.451P - \frac{P}{L}x$$

Span 2

$$x \leq \frac{L}{3}$$

$$T_2(x) = \frac{2}{3}P + 0.026P = 0.693P$$

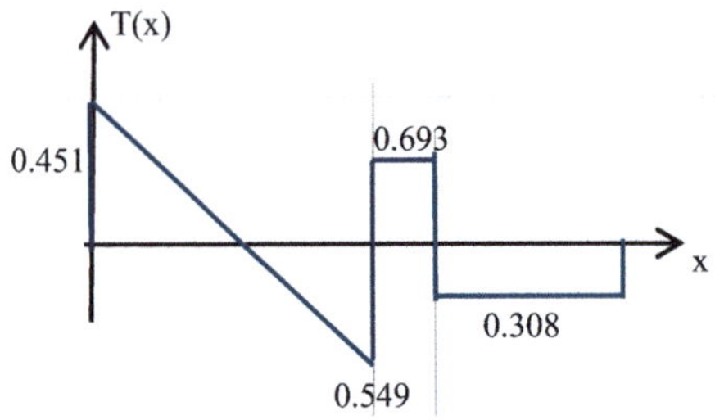

Figure 2.23. *Shear force diagram *(P)*

$$\frac{L}{3} \leq x \leq L$$

$$T_2(x) = -\frac{1}{3}P + 0.026P = -0.308P$$

2.3.2. *Beam with support settlements*

Calculate the moments at the ends of each span due to the supports settlements B and C (Δ_1 and Δ_2). The beam has a constant flexural rigidity (EI) (Figure 2.24).

We apply equation [2.35].

– *Joint 1*

$$\omega_1 = -\beta_1^{'} + a_1 M_1^{'} - b_1 M_2^{"} = 0$$

The support settlement is

$$\beta_1^{'} = a_1 .M_1^{'} - b_1 .M_2^{"}$$

– *Joint 2*

$$\omega_2 = \beta_2^{'} + a_2 M_2^{'} - b_2 M_3^{"}$$

$$\omega_2 = -\beta_2^{"} - b_1 M_1^{'} + c_1 M_2^{"}$$

We can deduce

$$-\beta_2^{"} - \beta_2^{'} = b_1 M_1^{'} + (a_2 + c_1) M_2^{'} + b_2 M_3^{'}$$

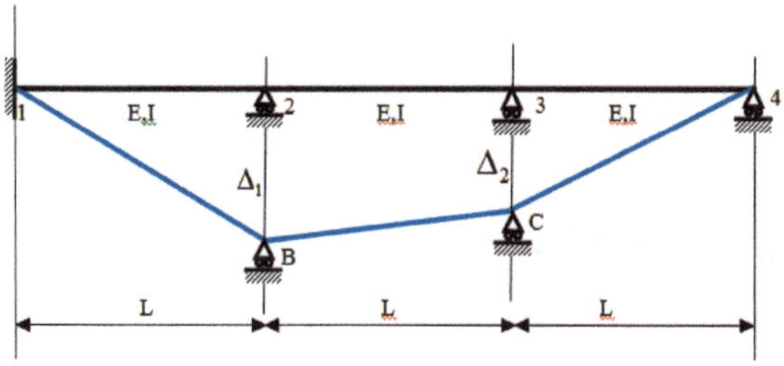

Figure 2.24. *Continuous beam with support settlement*

– *Joint 3*

$$\omega_3 = \beta_3^{'} + a_3 M_3^{'}$$

$$\omega_3 = \beta_3^{"} - b_2 M_2^{'} + c_2 M_3^{"}$$

$$\beta_3^{"} - \beta_3^{'} = b_2 M_2^{'} + (a_3 + c_2) M_3^{'}$$

$$\beta_1^{'} = \beta_2^{"} = \frac{\Delta_1}{L}$$

$$\beta_2^{'} = \beta_3^{"} = \frac{\Delta_1 - \Delta_2}{L}$$

$$\beta_3^{'} = \frac{\Delta_2}{L}$$

We replace the slope expressions in the preceding equations.

$$\frac{\Delta_1}{L} = \frac{L}{3EI}M_1^{'} + \frac{L}{6EI}M_2^{'}$$

$$-\frac{\Delta_1}{L} - \frac{\Delta_1 - \Delta_2}{L} = \frac{L}{6EI}M_1^{'} + \frac{2L}{3EI}M_2^{'} + \frac{L}{6EI}M_3^{'}$$

$$\frac{\Delta_1 - \Delta_2}{L} - \frac{\Delta_2}{L} = \frac{L}{6EI}M_2^{'} + \frac{3L}{3EI}M_3^{'}$$

Resolving the aforementioned equations makes it possible to calculate the moments at the joints of the continuous beam according to support settlements. For $\Delta_1 = 2\Delta_2$, we have

$$M_1^{'} = 9.70\frac{EI}{L^2}.\Delta_2$$

$$M_2^{'} = -M_2^{"} = -7.38\frac{EI}{L^2}.\Delta_2$$

$$M_3^{'} = 1.85\frac{EI}{L^2}.\Delta_2$$

The diagram of the bending moment of the given beam is shown in Figure 2.25.

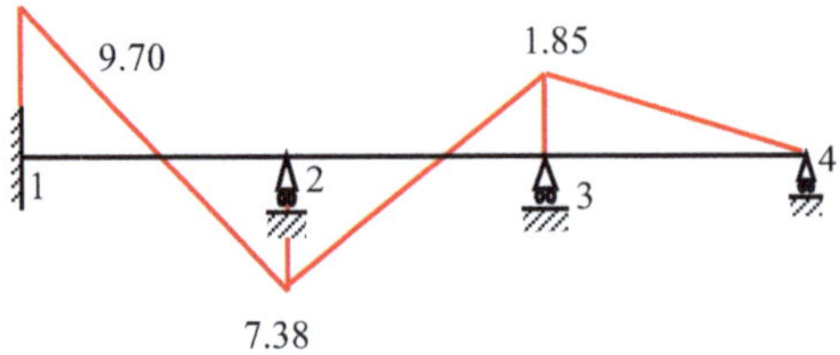

Figure 2.25. *Bending moment diagram* ($\frac{EI}{L^2}\Delta_2$)*

Similarly, the diagram of shear force is shown in Figure 2.26.

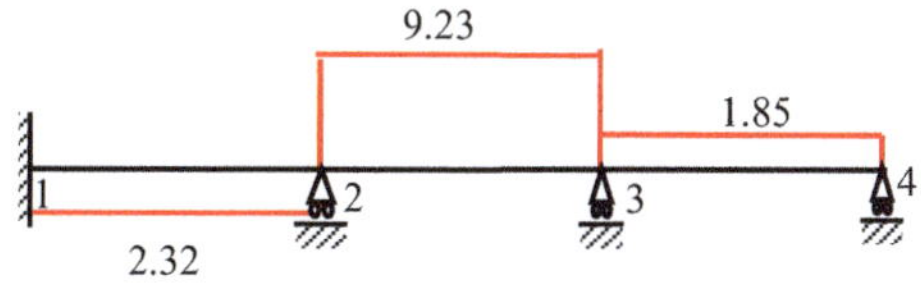

Figure 2.26. *Shear force diagram* $*(\frac{EI}{L^3}\Delta_2)$

2.3.3. *Beam with cantilever*

We analyze the beam (Figure 2.27) subjected to a uniformly distributed load of intensity q and a concentrated load P applied to the end of a cantilever. The flexural rigidity (EI) is assumed to be constant along the entire length of the beam.

The Clapeyron method is applied to each span of the beam.

$$\omega_A = \omega_A^{'} + aM_{AB} + bM_{BC} = 0$$

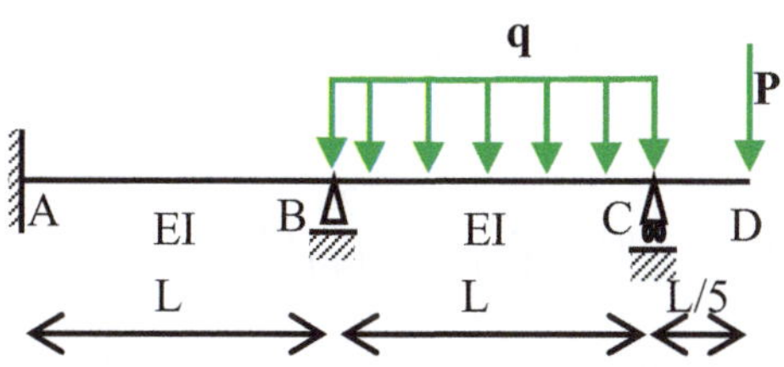

Figure 2.27. *Beam with cantilever*

$$\omega_B = \omega_B^{''} - bM_{AB} - cM_{BC}$$

$$\omega_B = \omega_B^{'} + aM_{BC} + bM_{CD}$$

With $M_{CD} = -M_{CB} = \frac{PL}{5}$, the slopes at joints A and B are given as

$$\omega_A^{'} = 0$$

$$\omega_B^{'} = -\omega_B^{''} = -\frac{qL^3}{24EI}$$

When P = qL, resolving the equations leads to

$$M_{AB} = -\frac{PL}{140}$$

$$M_{BC} = -M_{BA} = \frac{PL}{140}$$

The bending moment expressions at each span are given as:

Span (AB)

$$M_{AB}(x) = \mu_{AB}(x) - M_{AB}(1-\frac{x}{L}) + M_{BA}(\frac{x}{L})$$

$$M_{AB}(x) = \frac{1}{140}PL(1-\frac{x}{L}) - \frac{2}{140}PL(\frac{x}{L}) = \frac{1}{140}PL - \frac{1}{140}Px$$

Span (BC)

$$M_{BC}(x) = \frac{1}{2}qL.x - \frac{1}{2}qx^2 - \frac{2}{140}PL(1-\frac{x}{L}) - \frac{28}{140}PL(\frac{x}{L})$$

$$M_{BC}(x) = \frac{1}{2}qL.x - \frac{1}{2}qx^2 - \frac{2}{140}PL - \frac{26}{140}Px$$

The bending moment diagram is shown by Figure 2.28.

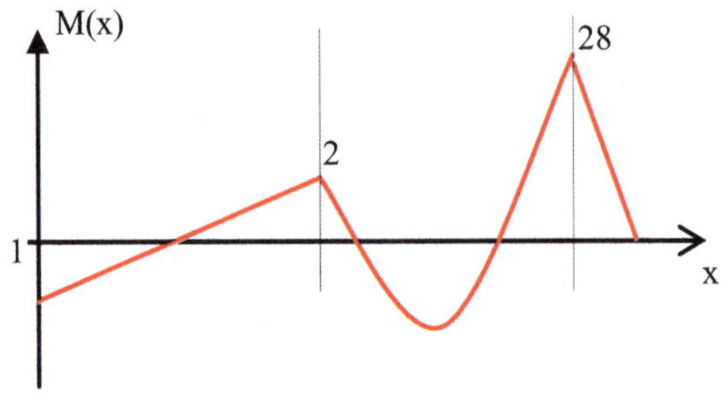

Figure 2.28. *Bending moment diagram* *$(\frac{PL}{140})$

Thus, the variation in shear force is given as

$$T(x) = \frac{d\mu(x)}{dx} + \frac{M_2^{"} + M_1^{'}}{L}$$

Span (AB)

$$T_{AB}(x) = -\frac{1}{140}P$$

Span (BC)

$$T_{BC}(x) = \frac{1}{2}qL - qx - \frac{26}{140}P$$

Cantilever (CD)

$$T_{CD}(x) = q.L$$

The shear force diagram is shown in Figure 2.29.

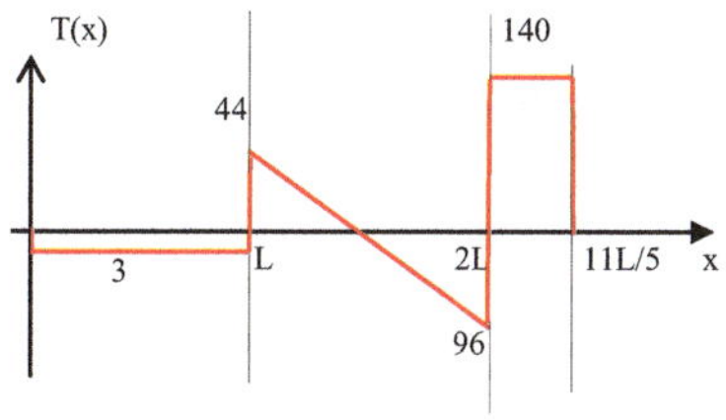

Figure 2.29. *Shear force diagram* $*(\frac{P}{140})$

2.4. Focus method

The focus method can be used when the following hypotheses are verified:

– no load is applied to the $(n - 1)$ spans of the continuous beam;

– the beam is subjected only to a moment applied at one end.

– the support settlement is neglected.

The focus method is distinguished by the position of the point of zero moment at each span. So, we distinguish two different procedures.

2.4.1. *Left focus method*

Consider a continuous beam of n spans (1, 2, …, n) with a constant inertia (EI) and subjected to the effect of a moment applied to a joint (n + 1) (Figure 2.30).

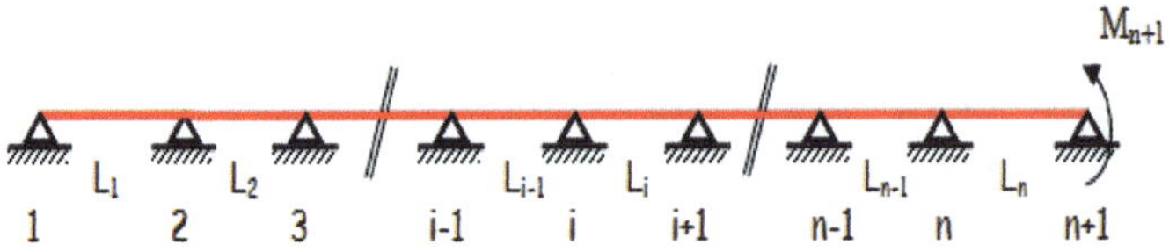

Figure 2.30. *Continuous beam under M_{n+1}*

The Clapeyron method [2.34] is used to calculate the support moments.

We have $M_1^{'} = 0$ (roller of shore)

i = 2

$$b_1 M_1^{'} + (a_2 + c_1) M_2^{'} + b_2 M_3^{'} = 0 \quad [2.35]$$

The focus of the second span is defined by

$$\varphi_2 = -\frac{M_2^{'}}{M_3^{'}} = \frac{b_2}{a_2 + c_1} \quad [2.36]$$

Hence

$$M_2^{'} = -\varphi_2 M_3^{'} \quad [2.37]$$

For i = 3

$$b_2 M_2^{'} + (a_3 + c_2) M_3^{'} + b_3 M_4^{'} = 0 \quad [2.38]$$

Substituting equation [2.37] with [2.38], we obtain

$$(a_3 + c_2 - \varphi_2 b_2) M_3^{'} + b_3 M_4^{'} = 0 \quad [2.39]$$

In the same way, we establish the focus of the third span by

$$\phi_3 = -\frac{M_3^{'}}{M_4^{'}} = \frac{b_3}{a_3 + c_2 - \phi_2 b_2} \qquad [2.40]$$

For the first span, we can write the expression giving the focus of the first span by

$$\phi_i = -\frac{M_i^{'}}{M_{i+1}^{'}} = \frac{b_i}{a_i + c_{i-1} - \phi_{i-1} b_{i-1}} \qquad [2.41]$$

In particular, the focus of the second span is

$$\phi_n = -\frac{M_n}{M_{n+1}} = \frac{b_n}{a_n + c_{n-1} - \phi_{n-1} b_{n-1}} \qquad [2.42]$$

From the results [2.34]–[2.41], calculating the moments at the supports is carried out as follows

$$M_n^{'} = -\phi_n M_{n+1}^{'} \qquad [2.43]$$

If moment M_{n+1} is assumed to be positive, moment M_n is the opposite sign.

Applying equation [2.41] to $(n-1)$th span could lead to

$$M_{n-1}^{'} = -\phi_{n-1} M_n^{'} \qquad [2.44]$$

Substituting equation [2.43] with [2.44], we obtain

$$M_{n-1}^{'} = \phi_n \phi_{n-1} M_{n+1}^{'} \qquad [2.45]$$

In general, we can write the expression of the moment at joint i by using relationship [2.45].

$$M_i^{'} = -\phi_i M_{i+1} = (\pm 1)\phi_n \phi_{n-1} \phi_{n-2} \ldots\ldots \phi_i M_{n+1}^{'} \qquad [2.46]$$

or

$$M_i^{'} = (\pm 1) \prod_{j=i}^{n} \phi_j M_{n+1}^{'} \qquad [2.47]$$

The moment at support 2 is calculated by

$$M_2^{'} = (\pm 1)\prod_{j=2}^{n}\phi_j M_{n+1}^{'} \qquad [2.48]$$

To establish the variation of the bending moment at each span, we use the equation of three moments [2.24].

$$M_i(x) = -M_i^{'}(1 - \frac{x}{L_i}) + M_{i+1}^{'}(\frac{x}{L_i}) \qquad [2.49]$$

As well as that of the shear force [2.24] which is:

$$T_i(x) = \frac{M_i^{'} + M_{i+1}^{'}}{L_i} \qquad [2.50]$$

After calculating the moments at the supports and using relationships [2.49] and [2.50], it is possible to draw the bending moment diagram and the shear force diagram for the continuous beam (Figure 2.31 and 2.32).

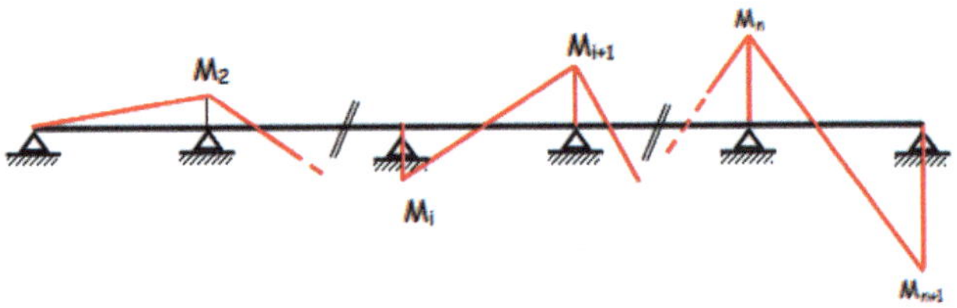

Figure 2.31. *Bending moment diagram*

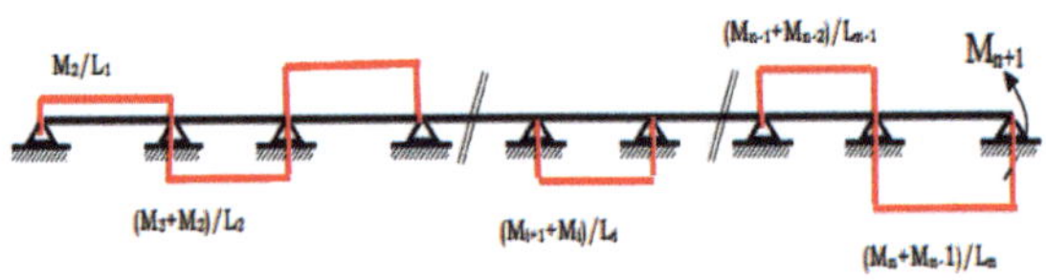

Figure 2.32. *Shear force diagram*

2.4.2. *Right focus method*

In the opposite case, we assume that the bending moment is now applied to joint 1 (Figure 2.33).

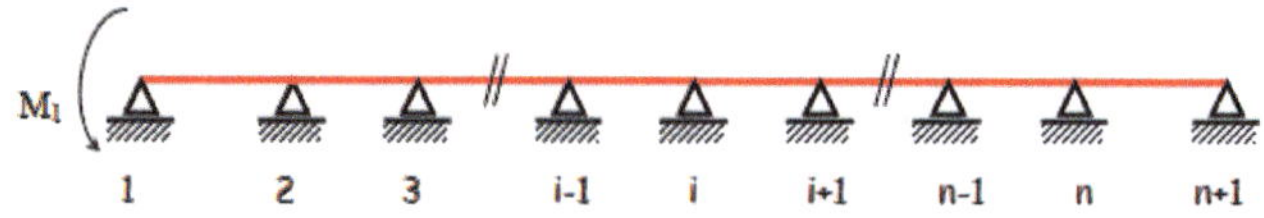

Figure 2.33. *Continuous beam under M_s*

The Clapeyron method [2.34] is used to calculate the moments at the support of the beam (Figure 2.33).

Support (n + 1) is a roller of shore, so

$Mn + 1 = 0$

We follow the same procedure described in the section on the left focus.

$i = n - 1,$

$$b_{n-1}M'_{n-1} + (a_n + c_{n-1})M'_n = 0 \qquad [2.51]$$

We define the right focus relative to (n – 1)th span by

$$\varphi'_{n-1} = -\frac{M'_n}{M'_{n-1}} = \frac{b_{n-1}}{a_n + c_{n-1}} \qquad [2.52]$$

Hence, we can deduce the moment at joint n

$$M'_n = -\varphi'_{n-1}M'_{n-1} \qquad [2.53]$$

Similarly for i = n – 2

$$b_{n-2}M'_{n-2} + (a_{n-1} + c_{n-2})M'_{n-1} + b_{n-1}M'_n = 0 \qquad [2.54]$$

Substituting the expression [2.53] with [2.54], we obtain

$$b_{n-2}M_{n-2} + (a_{n-1} + c_{n-2} - b_{n-1}\phi'_{n-1})M_{n-1} = 0 \qquad [2.55]$$

In the same way, we establish the focus of the studied span

$$\varphi'_{n-2} = -\frac{M'_{n-1}}{M'_{n-2}} = \frac{b_{n-2}}{a_{n-1} + c_{n-2} - \varphi'_{n-1}b_{n-1}} \qquad [2.56]$$

Generally, for the *i*th span, we can write

$$\phi_i^{'} = -\frac{M_{i+1}^{'}}{M_i^{'}} = \frac{b_i}{a_{i+1} + c_i - \phi_{i+1}^{'} b_{i+1}} \tag{2.57}$$

In particular for the first span, the focus is

$$\phi_1^{'} = -\frac{M_2^{'}}{M_1^{'}} = \frac{b_1}{a_2 + c_1 - \phi_2^{'} b_2} \tag{2.58}$$

Using the results obtained [2.51]–[2.58], the calculation of the support moments is

$$M_2^{'} = -\varphi_1^{'} M_1^{'} \tag{2.59}$$

If moment M_1 is assumed to be negative, moment M_n is the opposite sign.

$$M_3^{'} = -\phi_2^{'} M_2^{'} = +\phi_1^{'} \phi_2^{'} M_1^{'} \tag{2.60}$$

From this result, the support moment *i* becomes

$$M_{i+1}^{'} = -\varphi_i^{'} M_i^{'} = (\pm 1)\, \varphi_1^{'}.\varphi_2^{'}.\varphi_3^{'}.......\varphi_i^{'}.M_1^{'} \tag{2.61}$$

or

$$M_{i+1}^{'} = (\pm 1)\prod_{j=1}^{i} \phi_j^{'} M_1^{'} \tag{2.62}$$

In particular, the moment at support *n* is written according to the applied moment M_1 by

$$M_n^{'} = (\pm 1)\prod_{j=1}^{n-1} \phi_j^{'} M_1^{'} \tag{2.63}$$

To establish the variation of the bending moment and the shear force at each span, we use the equation of three moments ([2.24] and [2.25]).

$$M_i(x) = -M_i^{'}(1 - \frac{x}{l_i}) + M_{i+1}^{"}(\frac{x}{l_i})$$

The shear force is written as

$$T_i(x) = \frac{M_i^{'} + M_{i+1}^{''}}{l_i}$$

The bending moment diagram is shown in Figure 2.34.

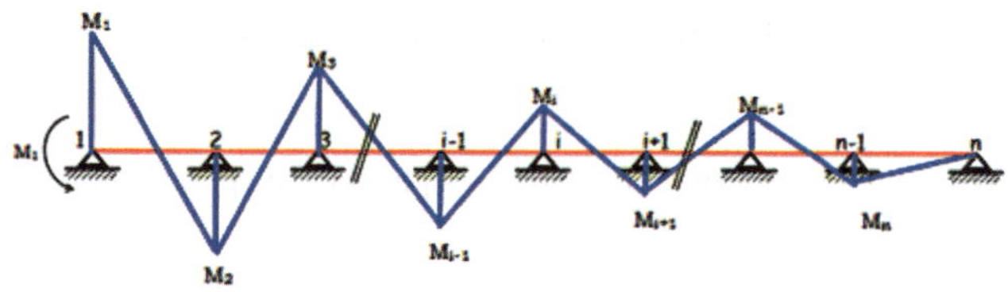

Figure 2.34. *Bending moment diagram*

Similarly, the bending moment diagram is shown in Figure 2.35.

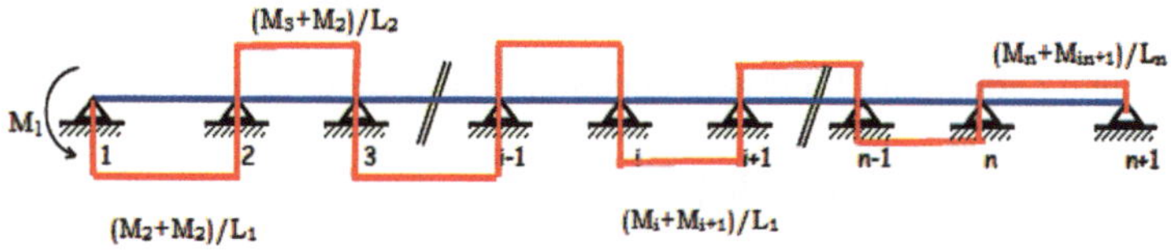

Figure 2.35. *Shear force diagram*

2.4.3. *Focus method with loaded bays*

We have described the left and right focus method with no loading applied on each span. Only a couple is applied to one end of the continuous beam. We now want to extend this method to the analysis of continuous beams with loaded bays. Consider a continuous beam with constant inertia (EI), which is subjected to a set of external loads, as shown in Figure 2.36.

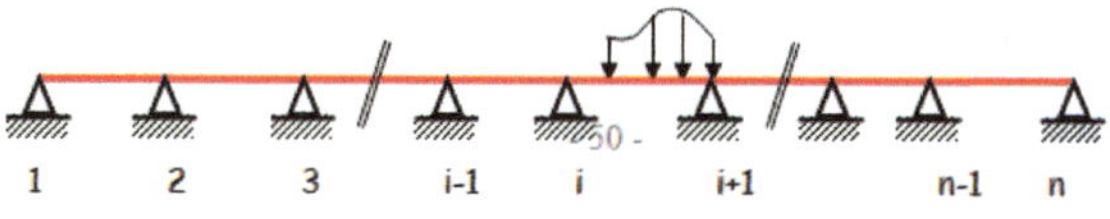

Figure 2.36. *Continuous beam with loaded bays*

Consider the *i*th span with the corresponding load (Figure 2.37). Applying the Clapeyron method allows us to write

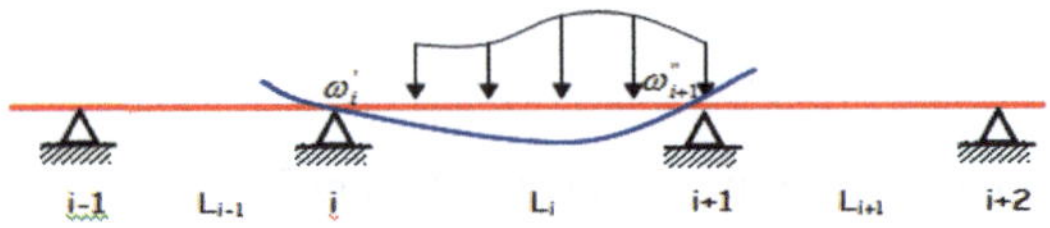

Figure 2.37. *Study of the ith span*

We can write the Clapeyron equation [2.34] to the *i*th span (Figure 2.37).

$$b_{i-1}M'_{i-1} + (a_i + c_{i-1})M'_i + b_i M'_{i+1} = -\omega'_i \quad [2.64]$$

Equation [2.40] gives a relationship between the moments of two consecutive supports by replacing the index i by $(i-1)$.

$$M'_{i-1} = -\phi_{i-1}M'_i \quad [2.65]$$

Substituting equation [2.65] with equation [2.64], we obtain

$$(a_i + c_{i-1} - \phi_{i-1}b_{i-1})M'_i + b_i M'_{i+1} = -\omega'_i \quad [2.66]$$

Using relationship [2.40], we can deduce that

$$\frac{1}{\varphi_i}M'_i + M'_{i+1} = -\frac{1}{b_i}\omega'_i \quad [2.67]$$

In the same way, we apply the Clapeyron method to the support $(i+1)$.

$$b_i M'_i + (a_{i+1} + c_i)M'_{i+1} + b_{i+1}M'_{i+2} = \omega''_{i+1} \quad [2.68]$$

Equation [2.65] allows us to write

$$M'_{i+2} = -\phi'_{i+1}M'_{i+1} \quad [2.69]$$

Substituting equation [2.69] with [2.68], we write

$$b_i M'_i + (a_{i+1} + c_i - \varphi'_{i+1}b_{i+1})M'_{i+1} = \omega''_{i+1} \quad [2.70]$$

Using equation [2.56] leads to

$$M_i^{'} + \frac{1}{\varphi_i^{'}} M_{i+1}^{'} = \frac{1}{b_i} \omega_{i+1}^{''} \quad [2.71]$$

Resolving the system of equations [2.67] and [2.71] gives the support moments of the loaded bay.

$$M_i^{'} = -\frac{1}{b_i} \frac{\varphi_i}{1 - \varphi_i \varphi_i^{'}} (\varphi_i^{'} \omega_{i+1}^{''} + \omega_i^{'}) \quad [2.72]$$

$$M_{i+1}^{'} = \frac{1}{b_i} \frac{\varphi_i^{'}}{1 - \varphi_i \varphi_i^{'}} (\varphi_i \omega_i^{'} + \omega_{i+1}^{''}) \quad [2.73]$$

After determining moments M_i and M_{i+1}, the moments of the remaining supports can now be deduced by simultaneously using the left and right focus method.

From the relationship [2.65], we can write

$$M_{i+2}^{'} = -\varphi_{i+1}^{'} M_{i+1}^{'} \quad [2.74]$$

$$M_{i+3}^{'} = -\phi_{i+2}^{'} M_{i+2}^{'} = \phi_{i+1}^{'} \phi_{i+2}^{'} M_{i+1}^{'} \quad [2.75]$$

Hence

$$M_n^{'} = (\pm 1) \phi_{i+1}^{'} \phi_{i+2}^{'} \varphi_{i+3}^{'} \ldots\ldots \varphi_{n-1}^{'}) M_{i+1}^{'} \quad [2.76]$$

or

$$M_n^{'} = (\pm 1) \prod_{j=i+1}^{n-1} \phi_j^{'} M_{i+1}^{'} \quad [2.77]$$

Now, the left focus method is used to calculate the support moments at the supports 1, 2, 3, ……$(i-1)$. From equation [2.42], we can write

$$M_{i-1}^{'} = -\phi_{i-1} M_i^{'} \quad [2.78]$$

In the same way,

$$M'_{i-2} = -\phi_{i-2} M'_{i-1} = +\phi_{i-1}.\varphi_{i-2} M'_i \quad [2.79]$$

$$M'_{i-3} = -\phi_{i-3} M'_{i-2} = -\phi_{i-1}.\varphi_{i-2}.\phi_{i-3} M'_i \quad [2.80]$$

Hence

$$M'_2 = (\pm 1)\, \phi_{i-1}.\varphi_{i-2}\phi_{i-3}.........\phi_2 .M'_i \quad [2.81]$$

or

$$M'_2 = (\pm 1)\prod_{k=2}^{i-1} \varphi_k M'_i \quad [2.82]$$

The layout of the bending moment diagram due to the effect of the loading applied on the ith span is given by Figure 2.38. Moments M'_i and M'_{i+1} have been evaluated using relationships [2.72] and [2.73] and the rest of the moments can be established by the focus method with loaded bays.

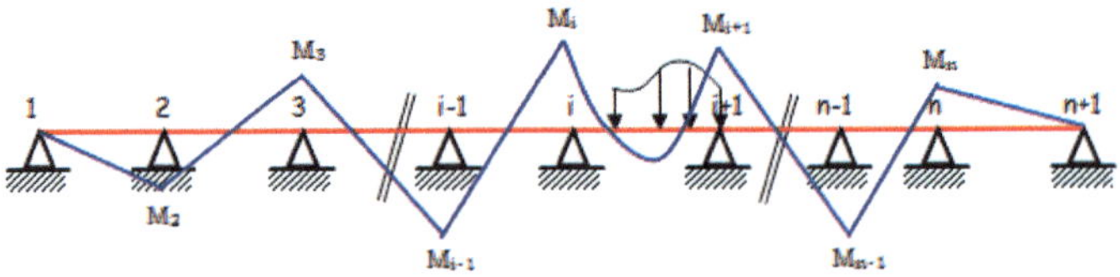

Figure 2.38. *Bending moment diagram*

The shear force diagram is shown in Figure 2.39.

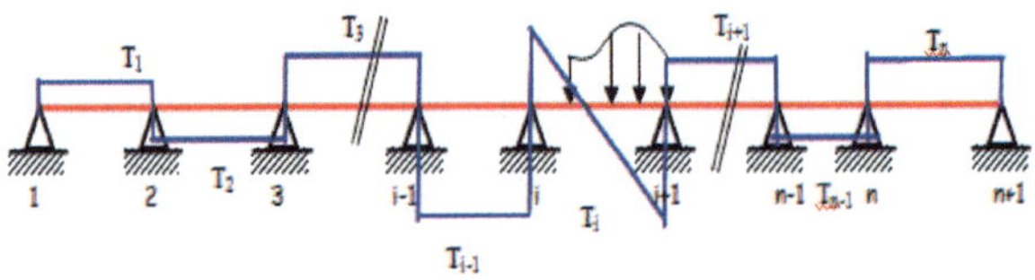

Figure 2.39. *Shear force diagram*

Finally, the procedure described can be used for all loads applied to each span. The final moment at each joint is deduced by the superposition of all the moments resulting from each applied loading.

EXAMPLE 2.2.– Continuous beam under moment M_0

Analyze the beam (Figure 2.40). We assume that the flexural rigidity EI and the length of the spans are constant. The goal is to present the moment distribution M_4 along the continuous beam.

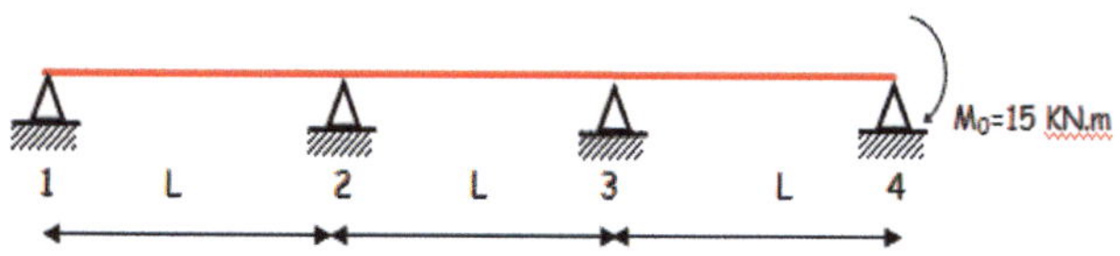

Figure 2.40. *Given beam*

Equation [2.34] is applied to the beam (Figure 2.40).

$$b_{i-1}M'_{i-1} + (a_i + c_{i-1})M'_i + b_i M'_{i+1} = \omega''_i - \omega'_i$$

Therefore, for $i = 2$, we have:

$$b_1 M'_1 + (a_2 + c_1)M'_2 + b_2 M'_3 = 0$$

or:

$$(a_2 + c_1)M'_2 + b_2 M'_3 = 0$$

The focus of the second span is defined by

$$\phi_2 = -\frac{M'_2}{M'_3} = \frac{b_2}{a_2 + c_1} = \frac{1}{4}$$

Similarly for $i = n - 3$, we obtain

$$b_2 M'_2 + (a_3 + c_2)M'_3 + b_3 M'_4 = 0$$

The equation that gives φ_2 is introduced in the last relationship, resulting in

$$\varphi_3 = -\frac{M_3^{'}}{M_4^{'}} = \frac{b_3}{a_3 + c_2 - \varphi_2 b_2} = \frac{4}{15}$$

Using the previous equations and knowing that M_4= -15 kN·m, we can deduce

$$M_3^{'} = -\frac{4}{15} M_4^{'} = 4\,\text{kN}\cdot\text{m}$$

$$M_2^{'} = -\frac{1}{4} M_3^{'} = -1\,\text{kN}\cdot\text{m}$$

$$M_1^{'} = 0$$

As the spans are not loaded, the expression of the bending moment is written only according to the moments at the end of each span. The bending moment diagram is shown in Figure [2.41].

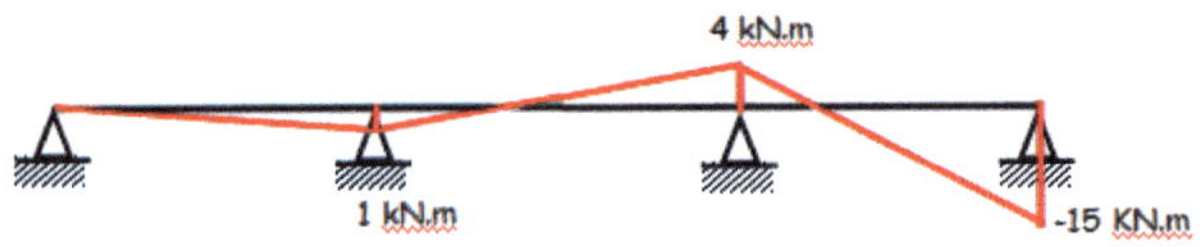

Figure 2.41. *Bending moment diagram*

The shear force diagram is shown in Figure 2.42.

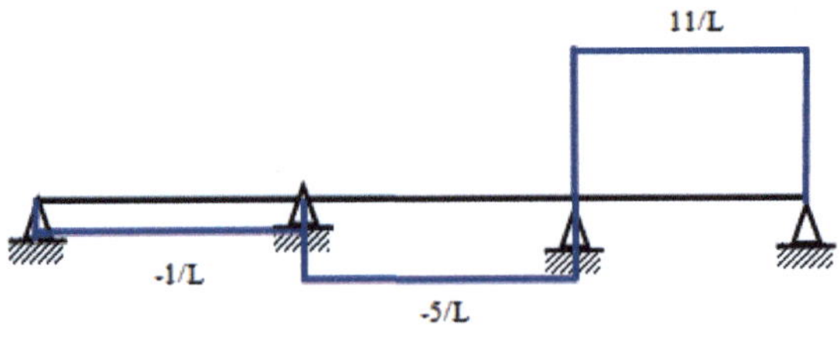

Figure 2.42. *Shear force diagram (KN)*

2.5. Conclusion

In this chapter, we have presented two ways of analyzing statically indeterminate beams: (1) the method of three moments and the Clapeyron method for the analysis of single span and continuous statically indeterminate beams and (2) the focus method.

The method of three moments is formulated based on calculating the rotations to the right of the beam's supports. This method is used for the analysis of single span statically indeterminate beams stressed by an external load or a support settlement. The successive application of the method of three moments leads to the formulation of the Clapeyron method, which can be used to analyze continuous beams. Resolving the equations makes it possible to determine the fixed-end and internal moments.

The bending moment as well as the shear force can be expressed at each span of the continuous beam by introducing the effects of the statically determinate moment and moments at the ends of the span.

2.6. Problems

Exercise 1

Calculate the mechanical constants *a*, *b* and *c* of the following beams, knowing that the Young's modulus, *E*, is constant:

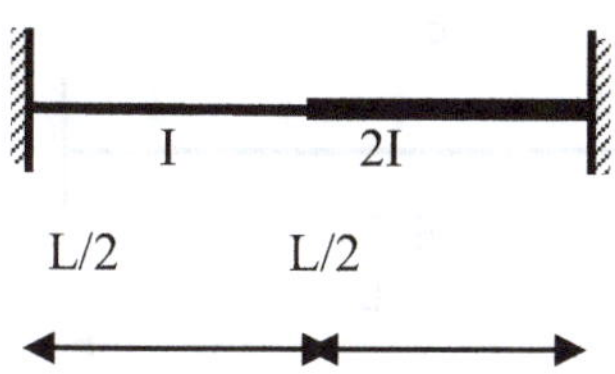

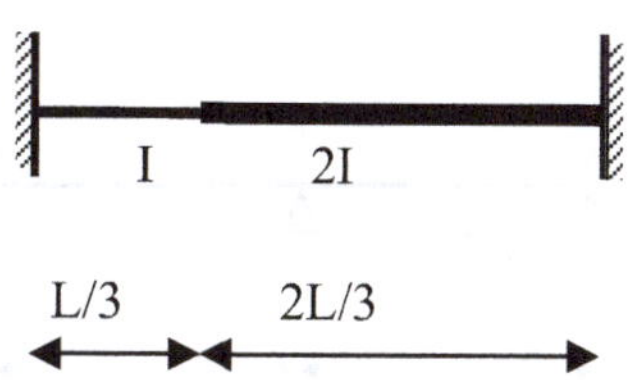

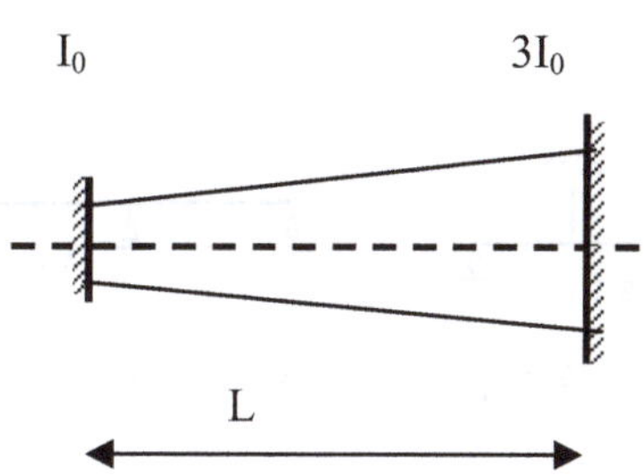

Exercise 2

Calculate support moments of the following beams:

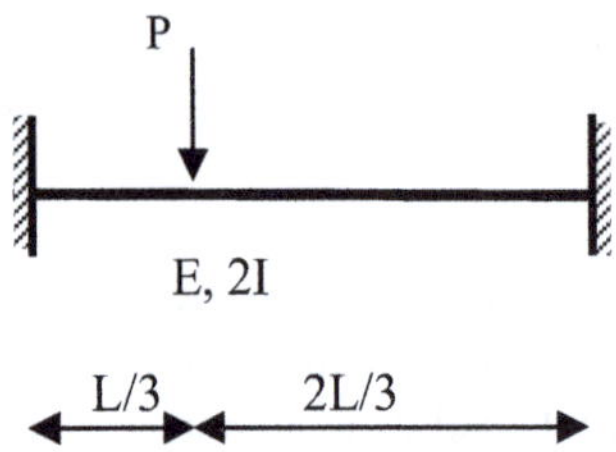

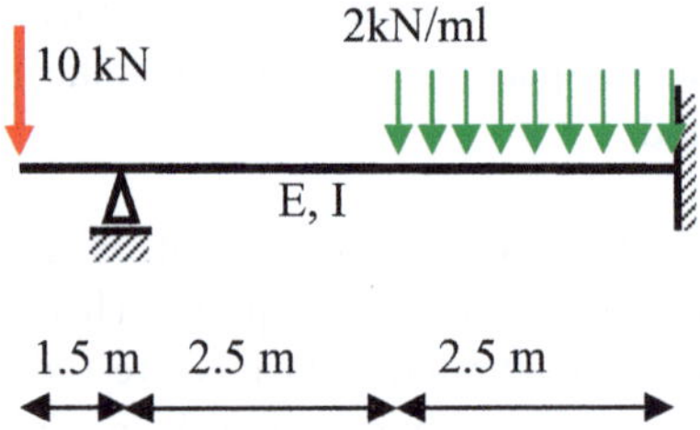

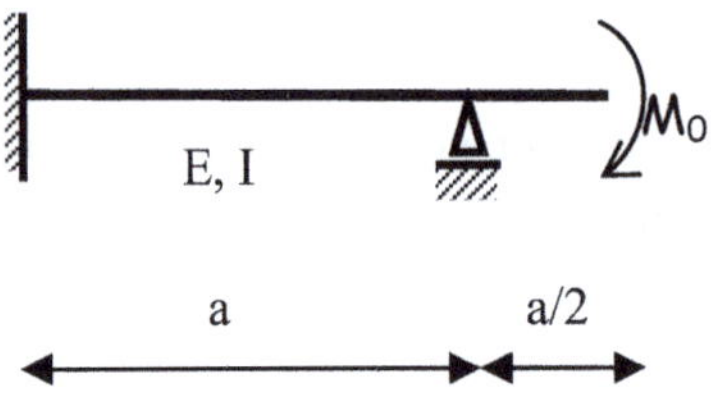

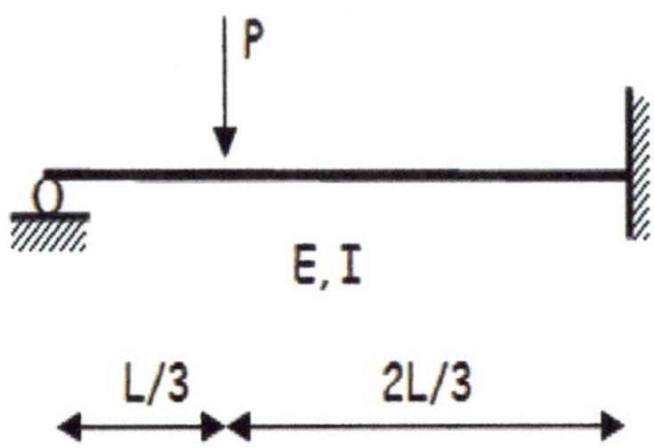

– Draw diagrams of the bending moment and the shear force.

Exercise 3

Give the moments at the ends of each bar of the following beams:

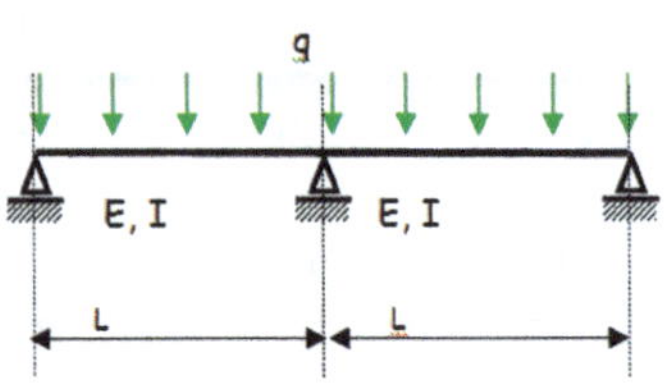

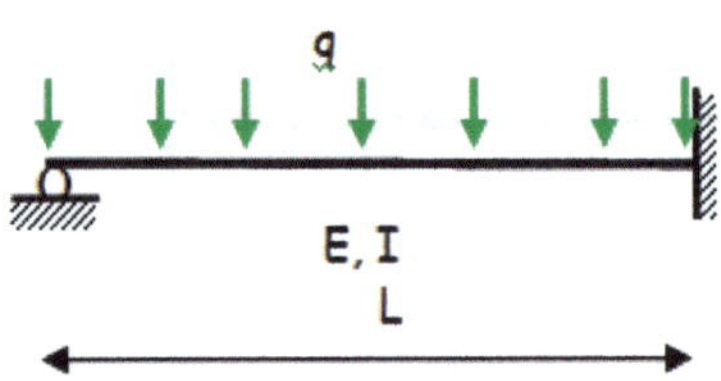

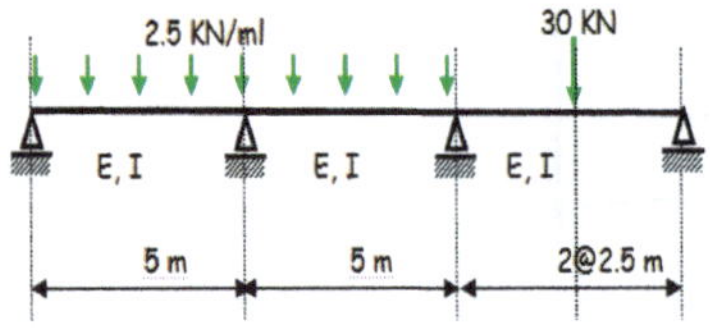

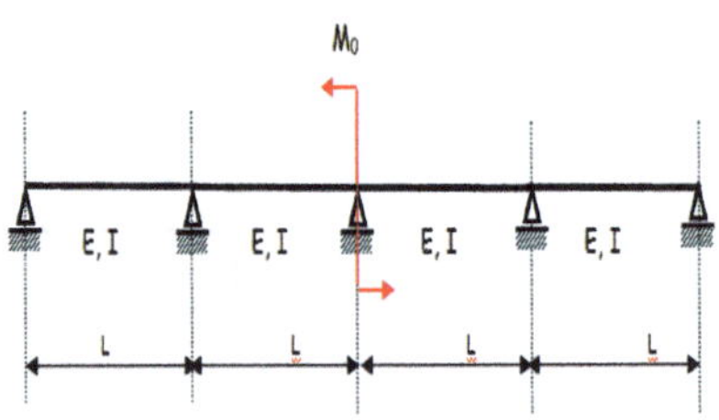

– Draw the bending moment diagram for each beam.

– Draw the shear force diagram for each beam.

Exercise 4

Determine moments in B and C of the beam subjected to support settlement B of a quantity Δ and a concentrated force of 10 kN applied at point A.

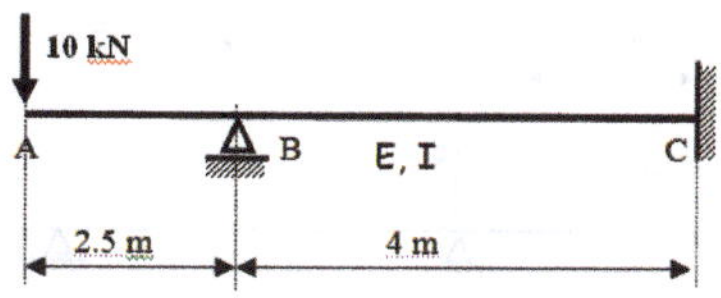

–Draw diagrams of the bending moment and the shear force.

Numerical application:

$E = 200$ GPa, $I = 5 \times 10^6$ mm^4 and $\Delta = 80$ mm.

Exercise 5

Using the focus method, calculate the moments at the joints of the following beams:

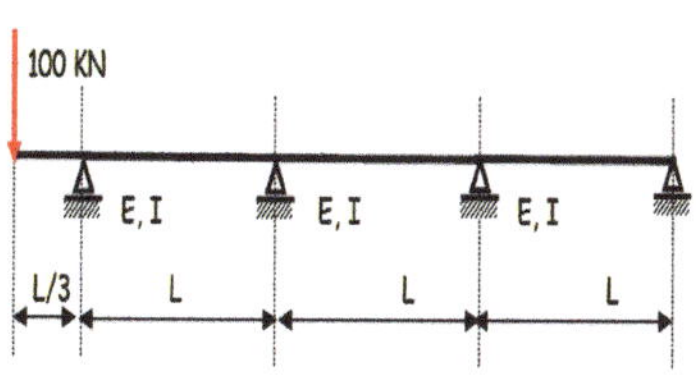

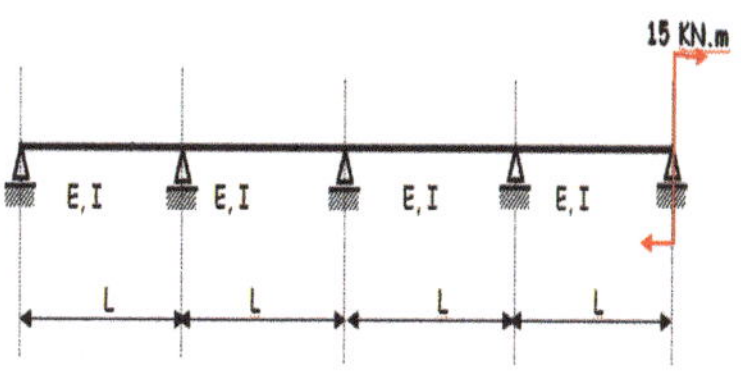

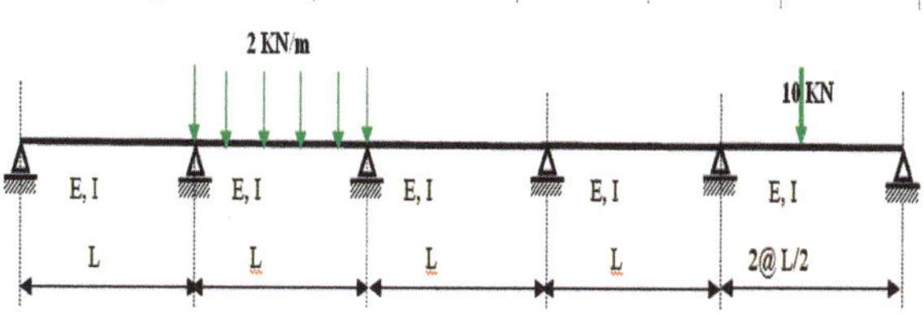

– Draw diagrams of the bending moment and the shear force for each beam.

Exercise 6

Construct diagrams of the bending moment and the shear force for the following beam. Suppose EI is constant.

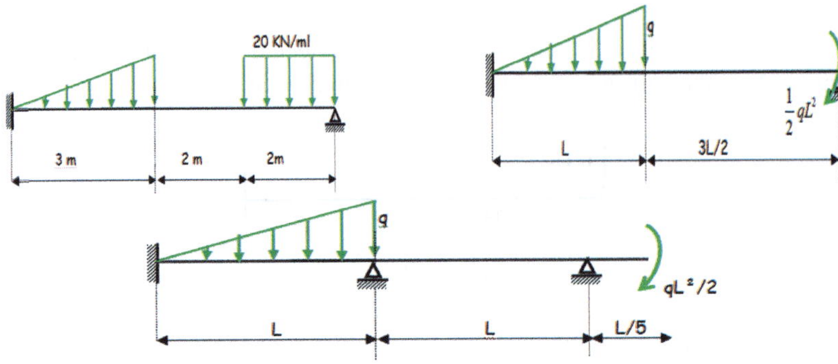

3

Method of Forces

The teaching objectives for the method of forces are as follows:

– analyzing structures with one or more degrees of static indeterminacy;

– applying deflection calculation methods;

– applying the method of forces to the analysis of trusses, beams and frames with or without support settlements;

– listing the benefits and limitations of using the method of forces.

In this chapter, we present the basics of the method of forces and its implementation in the analysis of common structures, such as single and continuous beams, frames and trusses. In this method, determining the degree of static indeterminacy plays a key role, and so we will give it special attention.

3.1. Beam with one degree of static indeterminacy

The degree of static indeterminacy plays a very important role in the analysis of statically indeterminate structures using the method of forces. In general, the formula for calculating the degree of static indeterminacy is given by:

$$f = (3.b + r) - (3.n + k) \quad [3.1]$$

Figure 3.1 shows a single span beam that is fixed at support A and is simply supported at B. This beam has already been studied using the method of three moments (Figure 2.8).

The degree of static indeterminacy of the beam (Figure 2.8) is calculated by formula [3.1].

$$f = (3.1 + 4) - (3.2 + 0) = 1 \qquad [3.2a]$$

The given structure is once statically indeterminate. To analyze a statically indeterminate structure using the method of forces, we use the following systems.

– *Equivalent system*

The equivalent structure is constructed, in which support reactions are replaced by actions whose number must be equal to the number of degrees of static indeterminacy of the system. In this case, the simple support B is replaced by an equivalent action, which is considered as the vertical reaction V_B (Figure 3.1).

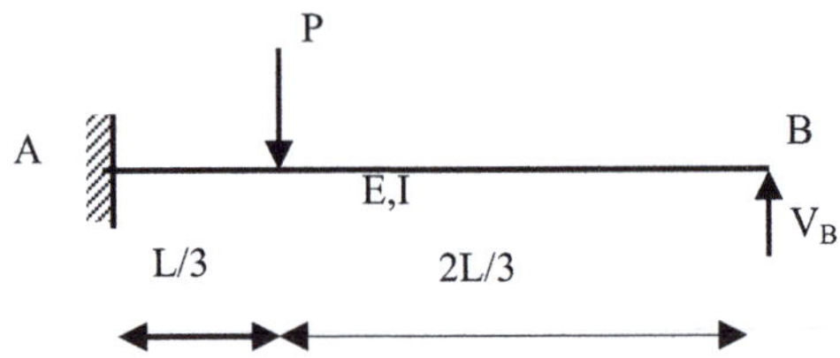

Figure 3.1. *Equivalent beam*

– *Fundamental system*

The fundamental system is obtained by ignoring the support reactions replaced in the equivalent structure (Figure 3.2). In this system, the effect of vertical reaction V_B is neglected. The resulting system is a statically determinate system.

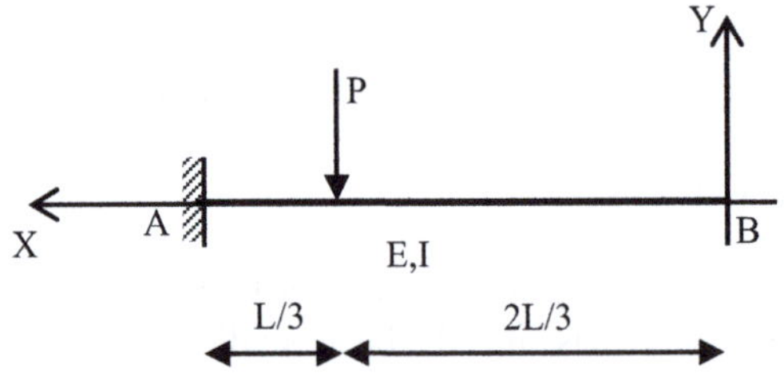

Figure 3.2. *Fundamental system*

The bending moment expressions (Figure 3.2) are given by:

$$\mu(x) = 0 \qquad 0 \le x \le \frac{2}{3}L \qquad [3.2b]$$

$$\mu(x) = -P(x - \frac{2}{3}L) \qquad \frac{3}{2}L \le x \le L \qquad [3.2c]$$

The vertical displacement at point B of the fundamental system using the virtual work method is:

$$\delta_B^0 = \int_0^L \frac{\mu(x)\, m(x)}{EI} dx \qquad [3.3]$$

$\mu(x)$ and m(x) are, respectively, the bending moment of the fundamental system (Figure 3.2) and the bending moment due to a unit action applied in the direction of the eliminated force (Figure 3.3). In this case, we write:

$$m(x) = 1.x = x \qquad 0 \le x \le L \qquad [3.4]$$

Substituting the expressions [3.2a] and [3.4] in equation [3.3], we obtain:

$$\delta_B^0 = \int_{2L/3}^{L} \frac{-P(x - \frac{2}{3}L).x}{EI} dx = -\frac{4}{81}\frac{PL^3}{EI} \qquad [3.5]$$

– *Unit system*

Now, a unit force is applied to point B of the free body beam (Figure 3.3).

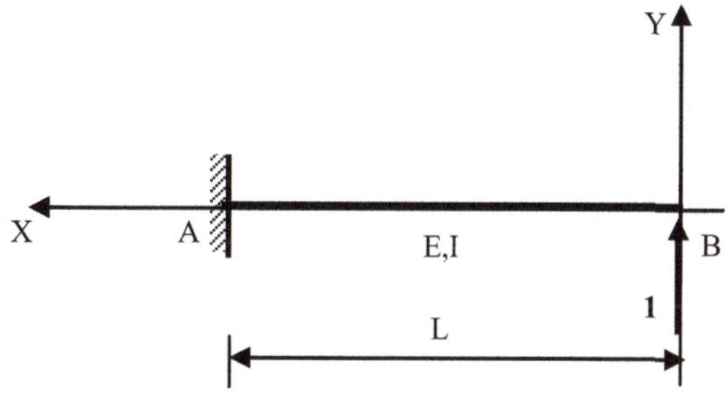

Figure 3.3. *Unit system*

To calculate the displacement, this time we use the equation of elastic line deflection.

$$y''(x) = \frac{m(x)}{EI} = \frac{x}{EI} \quad [3.6]$$

The first and second integration of the differential equation [3.6] allows us to write the slope and deflection expressions.

The slope expression is:

$$y'(x) = \frac{1}{EI}(\frac{x^2}{2} + c_1)$$

The deflection expression can be deduced by:

$$y(x) = \frac{1}{EI}(\frac{x3}{6} + c_1 x + c_2)$$

The limit conditions (Figure 3.1) are used to calculate the integration constants.

$$y'(x = L) = 0\,, \quad c_1 = -\frac{L^2}{2}$$

$$y(x = L) = 0\,, \quad c_2 = \frac{L^3}{2}$$

Finally, the equation of beam deflection is written as:

$$y(x) = \frac{1}{EI}(\frac{x3}{6} - \frac{L^2}{2}x + \frac{L^3}{3}) \quad [3.7]$$

In particular, the vertical displacement at point B is given as:

$$\delta_B^1 = y(x = 0) = \frac{1}{3}\frac{L^3}{EI} \quad [3.8]$$

Hence, the displacement due to the free body beam V_B becomes (V_B. δ_B^1).

Initially, the displacement at point B of the given system is zero.

$$\delta_B = 0 \qquad [3.9]$$

Applying the compatibility of displacements principle at point B makes it possible to write:

$$\delta_B = \delta_B^0 + \delta_B^1 = 0 \qquad [3.10]$$

Substituting the expressions of displacements [3.5] and [3.8] with [3.10], we obtain:

$$-\frac{4}{81}\frac{PL^3}{EI} + \frac{L^3}{3EI}V_B = 0$$

The reaction at support B is:

$$V_B = \frac{4}{27}P$$

The equilibrium of moments relating to point A gives the fixed moment.

$$\mathrm{M_A} = \frac{5}{27}PL$$

The variation of the bending moment along the beam is written as:

$$\mathrm{M(z)} = \mu(\mathrm{z}) - M_A(1-\frac{z}{L}) + M_B(\frac{z}{L}) \qquad [3.11]$$

with $z = L - x$,

or,

$$\mathrm{M(z)} = \frac{2P}{3}z - \frac{5}{27}PL(1-\frac{z}{L}) = \frac{23}{27}Pz - \frac{5}{27}PL \qquad 0 \le z \le \frac{L}{3}$$

$$\mathrm{M(x)} = \frac{2}{3}Pz - P(\mathrm{z}-\frac{L}{3}) - \frac{5}{27}PL(1-\frac{z}{L}) = \frac{4}{27}PL - \frac{4}{27}\mathrm{Pz} \qquad \frac{L}{3} \le z \le L$$

The variation in shear force is given as:

$$T(\mathrm{z}) = \frac{dM(\mathrm{z})}{dz} \qquad [3.12]$$

$$T(z) = \frac{dM(z)}{dz} = -\frac{4}{27}P \qquad \frac{2}{3}L \le x \le L$$

$$T(z) = \frac{dM(z)}{dz} = \frac{23}{27}P \qquad 0 \le x \le \frac{1}{3}L$$

The results obtained by the method of forces are identical to those obtained by the method of three moments (Example 2.1). The diagrams of the bending moment and the shear force are shown in Figure 2.10.

3.2. Beam with many degrees of static indeterminacy

We analyze the beam in Figure 2.21 knowing that the flexural rigidity (EI) is constant along the beam. The beam is stressed by a uniformly distributed load q and by a concentrated force P. The beam is analyzed using the Clapeyron method (Figure 2.21).

To analyze the beam, we follow the same procedure, which is established for the analysis of beams with a single degree of static indeterminacy.

The degree of static indeterminacy of the system is:

$$f = (3.b + r) - (3.n + k) = (3.1 + 6) - (3.2 + 0) = 3$$

– *Equivalent system*

We choose an equivalent structure; support 3 is replaced by an equivalent reaction, and the fixings 1 and 3 are replaced by fixed moments M_1 and M_3 (Figure 3.4).

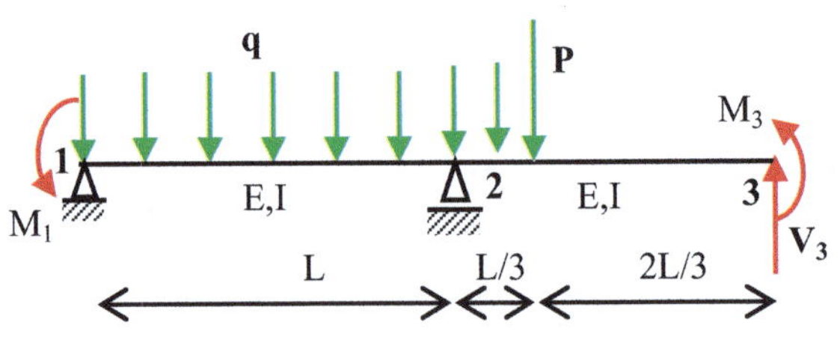

Figure 3.4. *Equivalent beam*[1]

1 All of the figures in this chapter are available to view in full color at www.iste.co.uk/khalfallah/analysis2.zip.

– *Fundamental system*

We ignore the effect of reaction V_3 and moments M_1 and M_2. The system thus becomes statically determinate. Static analysis of the fundamental system makes it possible to express the variation of the bending moment.

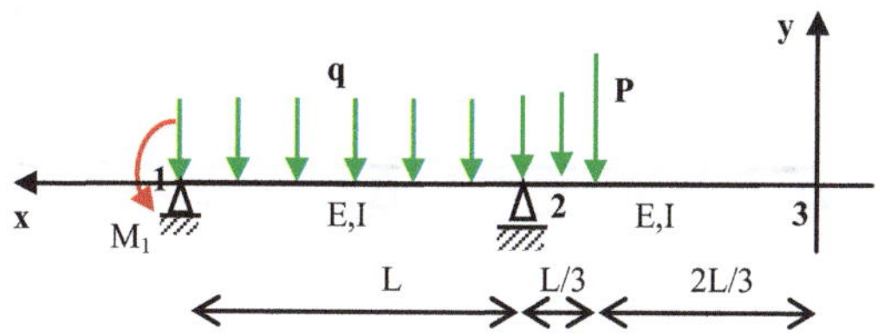

Figure 3.5. *Fundamental beam*

The expression of the bending moment of the basic system is given by:

$$V_2 = \frac{11}{6} P$$

$$\mu(x) = 0 \qquad 0 \le x \le \frac{2L}{3}$$

$$\mu(x) = -qL(x - \frac{2L}{3}) \qquad \frac{2L}{3} \le x \le L$$

$$\mu(x) = \frac{11}{6} qLx - \frac{10}{6} qL^2 - \frac{1}{2} qx^2 \qquad L \le x \le 2L$$

The diagram of the bending moment of the statically determinate beam is shown in Figure 3.6.

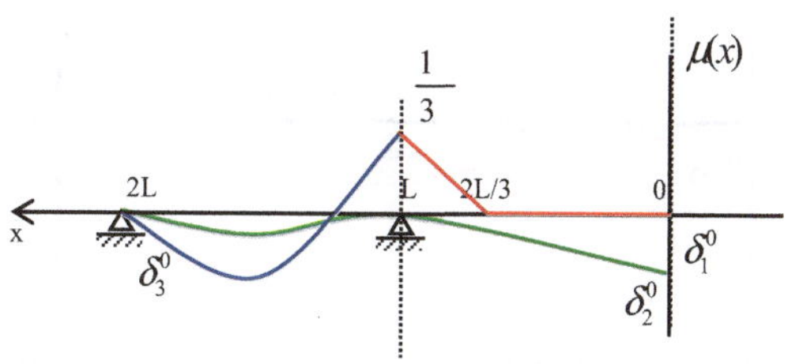

Figure 3.6. *Deflection and diagram of the fundamental moment *(* qL^2 *)*

To calculate the deformations of the fundamental system at joints 1 and 3, we use the method of virtual work. The bending moment expressions of the unit systems and the fundamental system are necessary. In this concept, a unit force is applied alternately to joint 3 (Figure 3.7) and two unit moments are applied, respectively, to joints 1 and 3 (Figures 3.8 and 3.9).

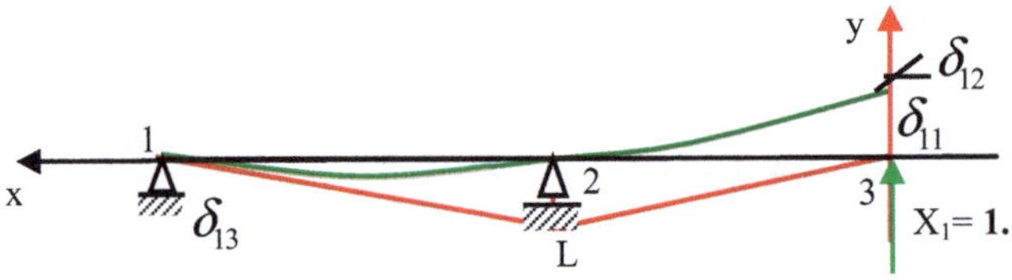

Figure 3.7. *Deflection and bending moment of the first unit system*

The bending moment expressions under the effect of a unit force (Figure 3.8) are given by:

$$m_1(x) = x \qquad 0 \le x \le L$$

$$m_1(x) = 2L - x \qquad L \le x \le 2L$$

In the same way, the bending moment of the beam under the effect of a unit moment applied to joint 3 is shown in Figure 3.8.

$$m_2(x) = 1 \qquad 0 \le x \le L$$

$$m_2(x) = 2 - \frac{x}{L} \qquad L \le x \le 2L$$

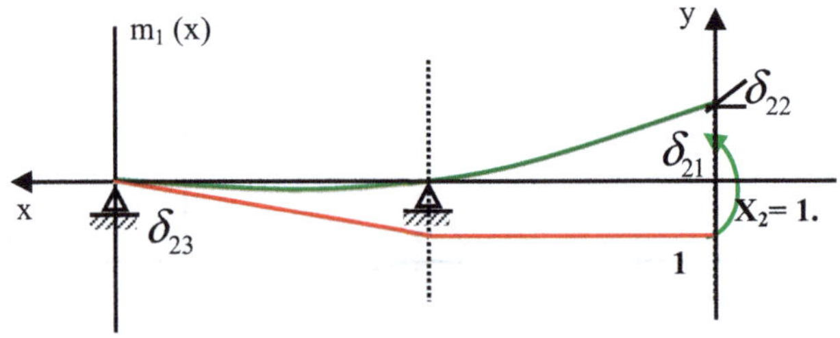

Figure 3.8. *Deflection and bending moment of the second unit system*

We apply a unit bending moment $X_3 = 1$ to point 1.

$$m_3(x) = 0. \qquad 0 \le x \le L$$

$$m_3(x) = -\frac{x}{L} \qquad L \le x \le 2L$$

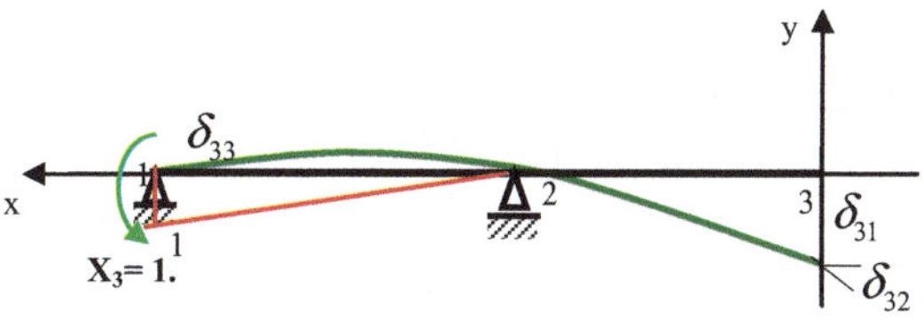

Figure 3.9. *Deflection and bending moment of the third unit system*

The displacements of the statically determinate system are given as:

$$\delta_1^0 = \frac{1}{EI}\int_0^{2L} \mu(x).m_1(x)dx = -0.085\frac{q.L^4}{EI}$$

$$\delta_2^0 = \frac{1}{EI}\int_0^{2L} \mu(x).m_2(x)dx = -0.0913\frac{q.L^3}{EI}$$

$$\delta_3^0 = \frac{1}{EI}\int_0^{2L} \mu(x).m_3(x)dx = -0.0116\frac{q.L^3}{EI}$$

The displacements at points 1 and 3 of unit systems (Figures 3.7–3.9) are calculated as follows:

$$\delta_{11} = \frac{1}{EI}\left[\int_0^L xdx + \int_L^{2L}(2L-x)^2 dx\right] = \frac{2}{3}\frac{L^3}{EI}$$

The slope at joints 2 and 1 is:

$$\delta_{22} = \frac{1}{EI}\left[\int_0^L dx + \int_L^{2L}(2-\frac{x}{L})^2 dx\right] = \frac{4}{3}\frac{L}{EI}$$

$$\delta_{33} = \frac{1}{EI}\left[\int_{L}^{2L} \frac{x^2}{L^2} dx\right] = \frac{7}{3}\frac{L}{EI}$$

$$\delta_{12} = \frac{1}{EI}\left[\int_{0}^{L} x dx + \int_{L}^{2L} (2L - x)(2 - \frac{x}{L}) dx\right] = \frac{5}{6}\frac{L^2}{EI}$$

$$\delta_{13} = \frac{1}{EI}\left[\int_{L}^{2L} (2L - x)(-\frac{x}{L}) dx\right] = -\frac{2}{3}\frac{L^2}{EI}$$

$$\delta_{23} = \frac{1}{EI}\left[\int_{L}^{2L} (2 - \frac{x}{L})(-\frac{x}{L}) dx\right] = -\frac{2}{3}\frac{L}{EI}$$

Applying kinematic conditions to joints 1 and 3, respectively, leads to the construction of the following system of equations:

$$\delta_1^0 + V_1\,\delta_{11} + M_3\,\delta_{12} + M_1\,\delta_{13} = 0 \qquad [3.13a]$$

$$\delta_2^0 + V_1\,\delta_{21} + M_3\,\delta_{22} + M_1\,\delta_{23} = 0 \qquad [3.13b]$$

$$\delta_3^0 + V_1\,\delta_{31} + M_3\,\delta_{32} + M_1\,\delta_{33} = 0 \qquad [3.13c]$$

By substituting the values of the above displacements in the system of equations [3.13], solving the system of equations leads to the determination of support reaction V_1 and moments M_1 and M_3.

$$V_3 = 0.307qL$$
$$M_3 = -0.09qL^2$$
$$M_1 = 0.0116qL^2$$

The moments at the joints can thus be calculated using the superposition principle of the bending moments of the fundamental system and the unit systems.

$$M_{ij} = M_{ij}^0 + \sum V_k .m_{ij} \qquad [3.14]$$

or

$$M_{12} = M_{12}^{0} + V_3\, m_1\,(0) \;\; + M_3 m_2(0) + M_1 m_3\,(0) = 0.067\, qL^2$$
$$M_{21} = M_{21}^{0} + V_3\, m_1\,(L) \;\; + M_3 m_2(L) + M_1 m_3\,(L) = -0.116\, qL^2$$

$$M_{23} = M_{23}^{0} + V_3\, m_1\,(L) \;\; + M_3 m_2(L) + M_1 m_3\,(L) = +0.0116\, qL^2$$
$$M_{32} = M_{32}^{0} + V_3\, m_1\,(2L) \;\; + M_3 m_2(2L) + M_1 m_3\,(2L) = -0.09\, qL^2$$

The diagrams of the bending moment and the shear force are shown, respectively, in Figures 2.21 and 2.22.

3.3. Continuous beam with support settlements

We analyze the beam of constant flexural rigidity (EI), subjected to the effect of support settlement B of quantity Δ_1 and that of support C of quantity Δ_2 (Figure 2.24).

The beam is not stressed by any external load. The moments at the ends of each element are due to the settlement of supports B and C.

$$\delta_B = -\Delta_1$$

$$\delta_C = -\Delta_2$$

– *Unit systems*

To calculate the displacements at points B, C and D, we use the graphical method here. We apply a unit force to point B, structure (AD) deforms (Figure 3.10) and therefore the displacements at points B, C and D are respectively δ_{11}, δ_{12} and δ_{13}.

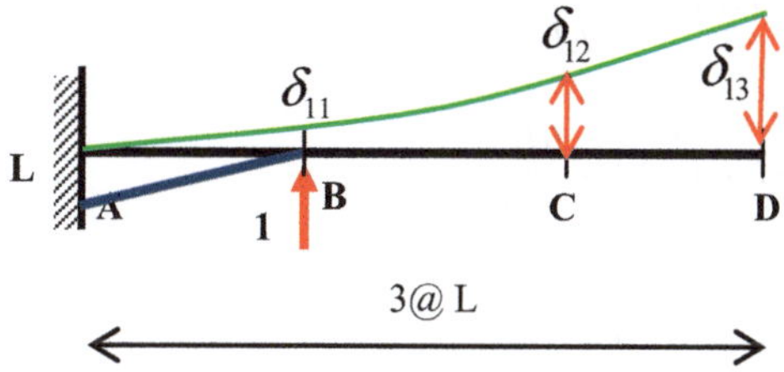

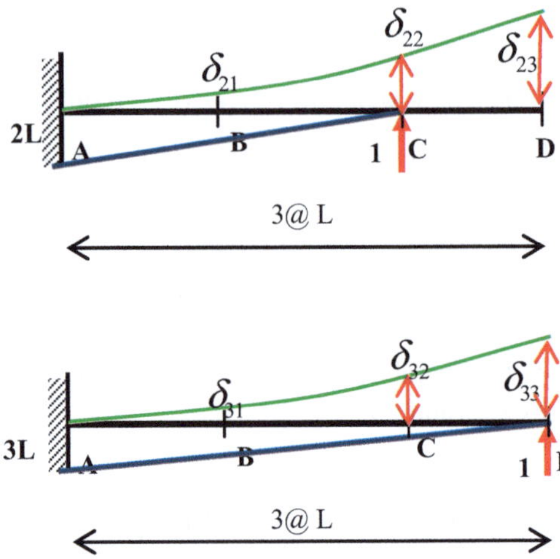

Figure 3.10. *Deflections of unit systems*

Applying a unit force at point C generates displacements δ_{21}, δ_{22} and δ_{23}, respectively, at joints B, C and D. Similarly, applying a unit force at joint D generates displacements δ_{31}, δ_{32} and δ_{33} at joints B, C and D.

In this case, the displacements are evaluated by the graphical method. They are given as:

$$\delta_{11} = \frac{1}{2} L.L.(\frac{2}{3} L) = \frac{L^3}{3EI}$$

$$\delta_{22} = \frac{1}{2} 2L.2L.(\frac{2}{3} 2L) = \frac{8L^3}{3EI}$$

$$\delta_{33} = \frac{1}{2} 3L.3L.(\frac{2}{3} 3L) = \frac{9L^3}{EI}$$

$$\delta_{12} = \delta_{21} = \frac{1}{2} L.L.(\frac{5}{3} L) = \frac{5L^3}{6EI}$$

$$\delta_{13} = \delta_{31} = \frac{1}{2} L.L.(\frac{8}{3} L) = \frac{8L^3}{6EI}$$

$$\delta_{23} = \delta_{32} = \frac{1}{2} 2L.2L.(\frac{7}{3} L) = \frac{14L^3}{3EI}$$

The kinematic conditions at joints B and C (Figure 2.24) can be applied.

$$\delta_B = \delta_{11}V_B + \delta_{21}V_C + \delta_{31}V_D = -\Delta_1$$

$$\delta_C = \delta_{12}V_B + \delta_{22}V_C + \delta_{32}V_D = -\Delta_2$$

$$\delta_D = \delta_{13}V_B + \delta_{23}V_C + \delta_{33}V_D = 0.$$

By substituting the expressions of the displacements calculated above, we obtain the following system of equations:

$$\frac{L^3}{3EI}V_B + \frac{5L^3}{6EI}V_C + \frac{8L^3}{6EI}V_D = -\Delta_1$$

$$\frac{5L^3}{6EI}V_B + \frac{8L^3}{3EI}V_C + \frac{14L^3}{3EI}V_D = -\Delta_2$$

$$\frac{8L^3}{6EI}V_B + \frac{14L^3}{3EI}V_C + \frac{9L^3}{EI}V_D = 0.$$

Resolving the system of equations $\Delta = \Delta_1 = 2\Delta_2$ leads to determining the reactions of supports B, C and D according to the support settlement.

$$V_B = -26.31\frac{EI}{L^3}\Delta$$

$$V_C = 11.08\frac{EI}{L^3}\Delta$$

$$V_C = -1.85\frac{EI}{L^3}\Delta$$

The expressions of the moment at the ends of each bar are evaluated using the principle of superposition.

$$M_{ij} = M_{ij}^0 + \sum V_k . m_{ij}$$

They are given as:

$$M_{AB} = M_{AB}^0 + V_B m_1(0) + V_C m_2(0) + V_D m_3(0) = -9.7\frac{EI}{L^2}$$

$$M_{BA} = M_{BA}^0 + V_B m_1(L) + V_C m_2(L) + V_D m_3(L) = 7.38\frac{EI}{L^2}$$

$$M_{BC} = M_{BC}^0 + V_B m_1(L) + V_C m_2(L) + V_D m_3(L) = -7.38\frac{EI}{L^2}$$

$$M_{CB} = M_{CB}^0 + V_B m_1(2L) + V_C m_2(2L) + V_D m_3(2L) = -1.85\frac{EI}{L^2}$$

$$M_{CD} = M_{CD}^0 + V_B m_1(2L) + V_C m_2(2L) + V_D m_3(2L) = 1.85\frac{EI}{L^2}$$

$$M_{DC} = 0.$$

The diagrams of the bending moment and the shear force are shown in Figures 2.24 and 2.25, respectively, and the results obtained by the method of forces are the same as those obtained by the Clapeyron method.

3.4. Analysis of a beam with two degrees of static indeterminacy

We analyze the continuous beam with a cantilever in Figure 2.27. Flexural rigidity (EI) is constant for both spans and P = qL.

The beam presents two degrees of static indeterminacy neglecting the effect of the normal force. In the first place, we build the equivalent system and then replace support C and the fixed moment M_A with their equivalent forces (Figure 3.11).

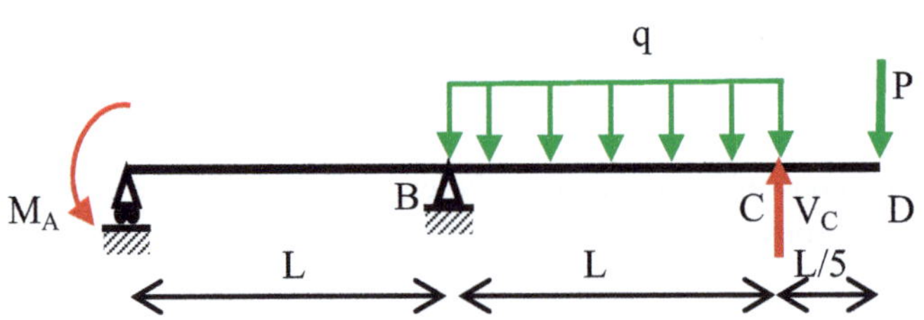

Figure 3.11. *Equivalent beam*

In the same way, we build a statically determinate beam (Figure 3.12) and unit systems (Figure 3.13).

– *Fundamental beam*

The expressions of bending moment variation of the fundamental system are given as:

$$\mu(x) = -Px = -qLx \qquad 0 \le x \le \frac{L}{5}$$

$$\mu(x) = -qLx - \frac{1}{2}q(x - \frac{L}{5})^2 \qquad \frac{L}{5} \le x \le \frac{6L}{5}$$

$$\mu(x) = \frac{17}{10}qLx - \frac{187}{50}qL^2 \qquad \frac{6L}{5} \le x \le \frac{11L}{5}$$

The deflection of the beam is shown in Figure 3.12.

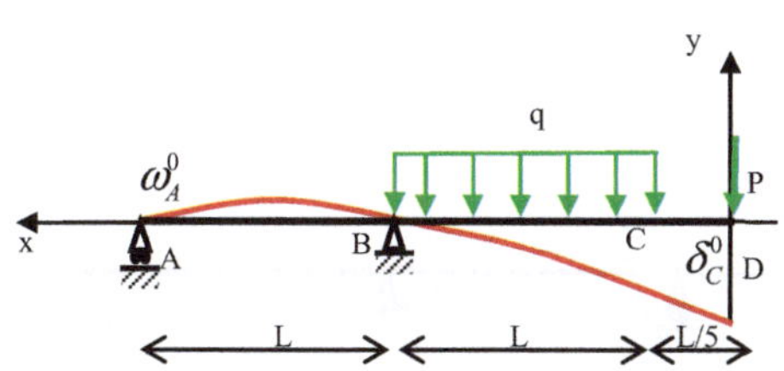

Figure 3.12. *Determinate beam*

– *Unit systems*

A unit force is applied sequentially at point C (Figure 3.13) and a unit moment at point A (Figure 3.14).

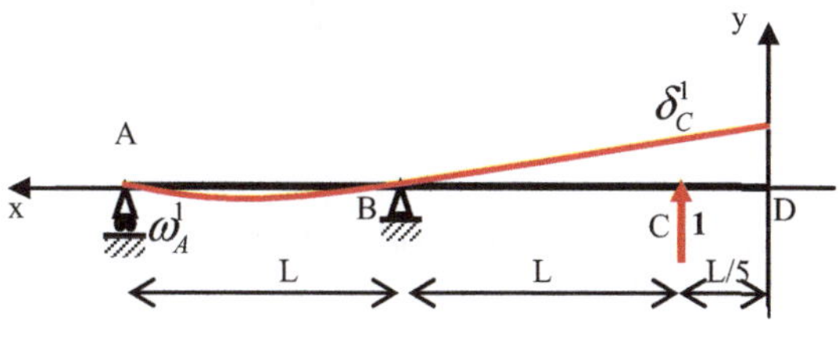

Figure 3.13. *First unit system*

$$m_1(x) = 0 \qquad 0 \le x \le \frac{L}{5}$$

$$m_1(x) = (x - \frac{L}{5}) \qquad \frac{L}{5} \le x \le \frac{6L}{5}$$

$$m_1(x) = x - 2(x - \frac{6L}{5}) = \frac{11L}{5} - x \qquad \frac{6L}{5} \le x \le \frac{11L}{5}$$

We apply a unit moment at joint 1 (Figure 3.14).

$$m_2(x) = 0 \qquad 0 \le x \le \frac{L}{5}$$

$$m_2(x) = 0 \qquad \frac{L}{5} \le x \le \frac{6L}{5}$$

$$m_2(x) = -\frac{1}{L}(x - \frac{6L}{5}) = \frac{6}{5} - \frac{x}{L} \qquad \frac{6L}{5} \le x \le \frac{11L}{5}$$

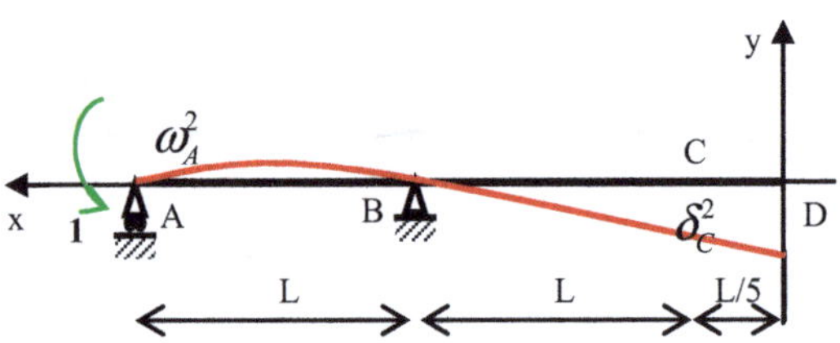

Figure 3.14. *Second unit system*

– Displacement calculation

$$\delta_C^0 = \frac{1}{EI}\int_0^{11L/5} \mu(x)m_1(x)dx = \frac{1}{EI}\left[\int_0^{L/5} \mu(x)m_1(x)dx + \int_{L/5}^{6L/5} \mu(x)m_1(x)dx + \int_{6L/5}^{11L/5} \mu(x)m_1(x)dx\right] = -\frac{4017}{2625}\frac{qL^4}{EI}$$

$$\omega_A^0 = \frac{1}{EI}\int_0^{11L/5} \mu(x)m_2(x)dx = \frac{1}{EI}\left[\int_0^{L/5} \mu(x)m_2(x)dx + \int_{L/5}^{6L/5} \mu(x)m_2(x)dx + \int_{6L/5}^{11L/5} \mu(x)m_2(x)dx\right] = \frac{3}{700}\frac{qL^3}{EI}$$

$$\delta_C^1 = \frac{1}{EI}\int_0^{11L/5} m_1^2(x)dx = \frac{1}{EI}\left[\int_0^{L/5} m_1^2(x)dx + \int_{L/5}^{6L/5} m_1^2(x)dx + \int_{6L/5}^{11L/5} m_1^2(x)dx\right] = \frac{68}{75}\frac{qL^3}{EI}$$

$$\omega_A^2 = \frac{1}{EI}\int_0^{11L/5} m_2^2(x)dx = \frac{1}{EI}\left[\int_0^{L/5} m_2^2(x)dx + \int_{L/5}^{6L/5} m_2^2(x)dx + \int_{6L/5}^{11L/5} m_2^2(x)dx\right] = \frac{1}{3}\frac{L}{EI}$$

$$\omega_A^1 = \frac{1}{EI}\int_0^{11L/5} m_1(x).m_2(x)dx = \frac{1}{EI}\left[\int_0^{L/5} m_1(x).m_2(x)dx + \int_{L/5}^{6L/5} m_1(x).m_2(x)dx + \int_{6L/5}^{11L/5} m_1(x).m_2(x)dx\right] = -\frac{20}{75}\frac{L^2}{EI}$$

$$\delta_C^2 = \frac{1}{EI}\int_0^{11L/5} m_2(x).m_1(x).dx = \frac{1}{EI}\left[\int_0^{L/5} m_2(x).m_1(x)dx + \int_{L/5}^{6L/5} m_2(x).m_1(x)dx + \int_{6L/5}^{11L/5} m_2(x).m_1(x)dx\right] = -\frac{20}{75}\frac{L^2}{EI}$$

The initial conditions of the beam (Figure 3.15) allow us to build the following system of equations:

$$\delta_c^0 + V_c\,\delta_c^1 + M_A\delta_c^2 = 0 \quad [3.15a]$$

$$\omega_A^0 + V_c\,\omega_A^1 + M_A\omega_A^2 = 0 \quad [3.15b]$$

Resolving the system of equations allows us to deduce vertical reaction V_C and fixed moment M_A.

$$V_C = \frac{59}{35}qL$$

$$M_A = -\frac{1}{140}qL^2$$

The moments at the ends of each bar are given by:

$$M_{AB} = M_{AB}^0 + V_C.m_1\left(\frac{11L}{6}\right) + M_A.m_2\left(\frac{11L}{6}\right) = 0 + 0 + (-1)\left(-\frac{1}{140}qL^2\right) = \frac{1}{140}qL^2$$

$$M_{BA} = M_{BA}^0 + V_C.m_1\left(\frac{6L}{5}\right) + M_A.m_2\left(\frac{6L}{5}\right) = -\frac{17}{10}qL^2 + L^2\left(\frac{59}{35}qL\right) + 0 = -\frac{1}{70}qL^2$$

$$M_{BC}=M_{BC}^{0}+V_{C}.m_{1}(\frac{6L}{5})\ +M_{A}.m_{2}(\frac{6L}{5})=\frac{17}{10}qL^{2}-L^{2}(\frac{59}{35}qL)+0=\frac{1}{70}qL^{2}$$

$$M_{CB}=M_{CB}^{0}(\frac{L}{5})+V_{C}.m_{1}(\frac{L}{5})\ +M_{A}.m_{2}(\frac{L}{5})=-\frac{1}{5}qL^{2}+0+0=-\frac{1}{5}qL^{2}$$

$$M_{CD}=M_{CD}^{0}(\frac{L}{5})+V_{C}.m_{1}(\frac{L}{5})\ +M_{A}.m_{2}(\frac{L}{5})=\frac{1}{5}qL^{2}+0+0=\frac{1}{5}qL^{2}$$

$$M_{DC}=0.$$

The diagrams of the bending moment and the shear force are shown, respectively, in Figures 2.27 and 2.28.

3.5. Analysis of a beam subjected to a moment

We analyze the beam in Figure 2.40. We assume that the flexural rigidity EI and the length of the spans are constant. The purpose of this example is to present the distribution of moment $M_D = M_0$ along the continuous beam using the method of forces.

– *Equivalent system (Figure 3.15)*

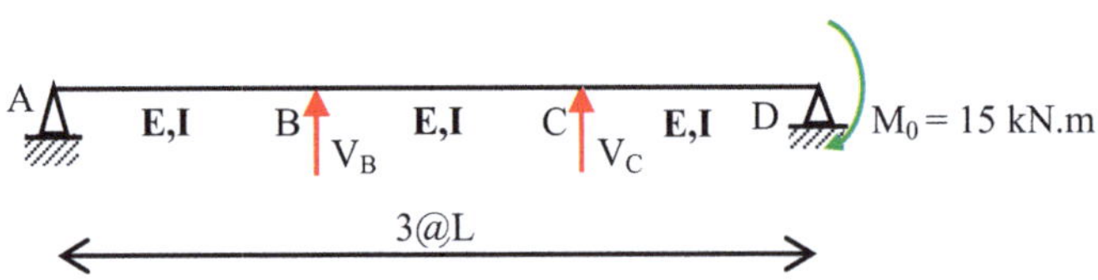

Figure 3.15. *Equivalent beam*

– *Fundamental system (Figure 3.16)*

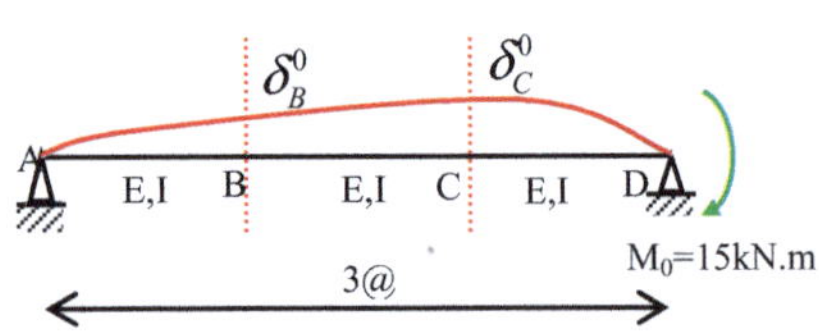

Figure 3.16. *Basic beam*

$$\mu(x)=-\frac{M_0}{3L}x \qquad\qquad 0\le x\le 3L$$

– *Unit systems (Figures 3.17 and 3.18)*

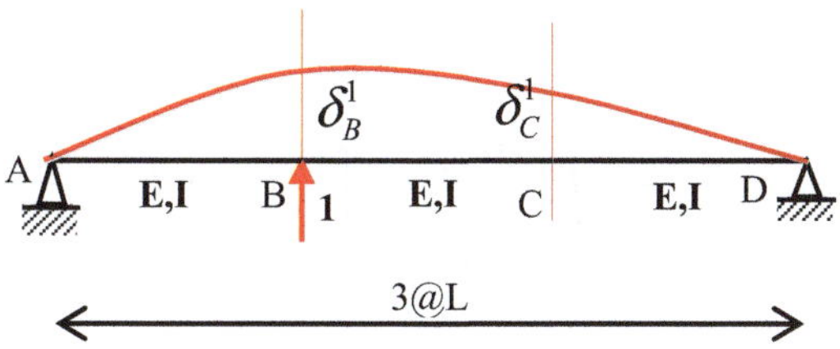

Figure 3.17. *First unit system*

$$m_1(x) = -\frac{2}{3}x \qquad 0 \le x \le L$$

$$m_1(x) = \frac{x}{3} - L \qquad L \le x \le 3L$$

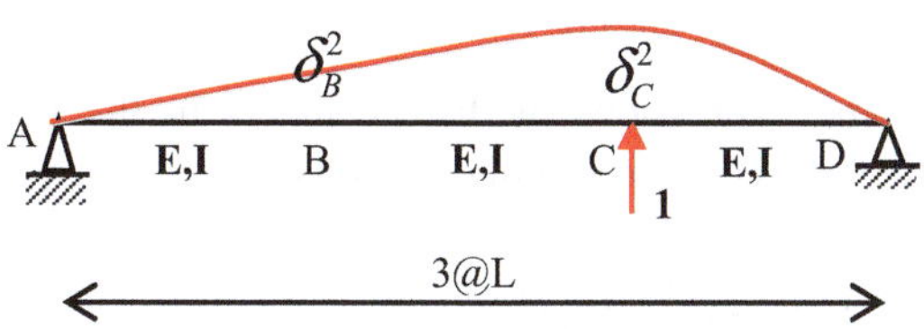

Figure 3.18. *Second unit system*

$$m_2(x) = -\frac{1}{3}x \qquad 0 \le x \le 2L$$

$$m_2(x) = \frac{x}{3} - 2L \qquad 2L \le x \le 3L$$

Displacement calculation:

$$\delta_B^0 = \frac{1}{EI}\int_0^{3L} \mu(x) m_1(x) dx = \frac{1}{EI}\left[\int_0^L (-\frac{2}{3}x)(-\frac{M_0}{3L}x)dx + \int_L^{3L} (\frac{1}{3}x - L)(-\frac{M_0}{3L}x)dx\right] = \frac{10}{27}\frac{M_0 L^2}{EI}$$

$$\delta_C^0 = \frac{1}{EI}\int_0^{3L} \mu(x) m_2(x) dx = \frac{1}{EI}\left[\int_0^L (-\frac{1}{3}x)(-\frac{M_0}{3L}x)dx + \int_L^{3L} (\frac{1}{3}x - L)(-\frac{M_0}{3L}x)dx\right] = \frac{52}{27}\frac{M_0 L^2}{EI}$$

$$\delta_B^1 = \frac{1}{EI}\int_0^{3L} m_1^2(x) dx = \frac{1}{EI}\left[\int_0^L (-\frac{2}{3}x)^2 dx + \int_L^{3L} (\frac{1}{3}x - L)^2 dx\right] = \frac{41}{27}\frac{L^3}{EI}$$

$$\delta_C^2 = \frac{1}{EI}\int_0^{3L} m_2^2(x) dx = \frac{1}{EI}\left[\int_0^L (-\frac{1}{3}x)^2 dx + \int_L^{3L} (\frac{1}{3}x - 2L)^2 dx\right] = \frac{1}{3}\frac{L^3}{EI}$$

$$\delta_B^2 = \delta_C^1 = \frac{1}{EI}\int_0^{3L} m_1(x).m_2(x) dx = \frac{1}{EI}\left[\int_0^L (-\frac{2}{3}x)(-\frac{1}{3}x)dx + \int_L^{2L} (\frac{1}{3}x - L)(-\frac{1}{3}x)dx + \int_{2L}^{3L} (\frac{1}{3}x - L)(\frac{1}{3}x - 2L)dx\right] = \frac{14}{27}\frac{L^3}{EI}$$

The kinematic conditions at the initial beam are given by:

$$\delta_B^0 + V_B \delta_B^1 + V_C \delta_C^1 = 0 \qquad [3.16a]$$

$$\delta_C^0 + V_B \delta_B^2 + V_C \delta_C^2 = 0 \qquad [3.16b]$$

By substituting the expressions of displacements in the system of equations [3.16], we obtain the following system of equations:

$$\frac{41}{27}L^3 V_B + \frac{14}{27}L^3 V_C = -\frac{10}{27}M_0 L^2$$

$$\frac{14}{27}L^3 V_B + \frac{9}{27}L^3 V_C = -\frac{52}{27}M_0 L^2$$

The support reactions are given as:

$$V_B = 3.68\frac{M_0}{L}$$

$$V_C = -11.50\frac{M_0}{L}$$

The moments at the joints can thus be calculated using the superposition principle of the bending moments of the fundamental and unit systems.

$$M_{AB} = 0$$

$$M_{BA} = \mu_{BA} + V_B m_1(L) + V_C m_2(L) = 1\,\text{kN.m}$$
$$M_{BC} = \mu_{BC} + V_B m_1(L) + V_C m_2(L) = -1\,\text{kN.m}$$

$$M_{CB} = \mu_{CB} + V_B m_1(2L) + V_C m_2(2L) = 4\,\text{kN.m}$$
$$M_{CD} = \mu_{CD} + V_B m_1(2L) + V_C m_2(2L) = -4\,\text{kN.m}$$

$$M_{DC} = -15\,\text{kN.m}$$

In the same way, the bending moment and shear force diagrams are shown in Figures 2.41 and 2.42.

3.6. Analysis of frames

3.6.1. *Frame with two degrees of static indeterminacy*

We analyze the simple frame in Figure 3.19. The method of forces also applies to the analysis of plane frames by performing the steps mentioned in simple and continuous beam analysis. We must analyze the frame while assuming that the flexural rigidity of the column is EI and that of the crossbar is 2EI.

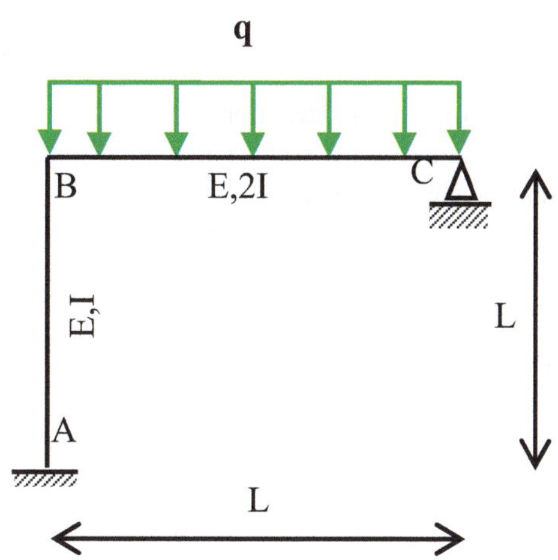

Figure 3.19. *Given frame*

The degree of static indeterminacy of the system is given as:

$$f = (3.b + r) - (3n + k) = (3.3 + 5) - (3.4 + 0) = 2$$

The system is twice statically indeterminate.

– *Equivalent system*

We substitute support C with a system of equivalent forces (Figure 3.20).

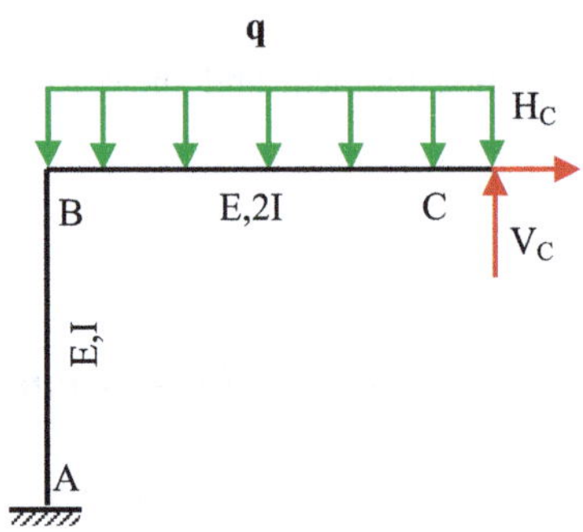

Figure 3.20. *Equivalent frame*

– *Fundamental system*

We ignore the effect of support reactions H_C and V_c, and thus the system becomes statically determinate (Figure 3.21). The expressions of bending moment of the statically determinate system are given as:

Bar (BC)

$$\mu(x) = -\frac{1}{2}q.x^2$$

Bar (AB)

$$\mu(x) = -\frac{1}{2}q.L^2$$

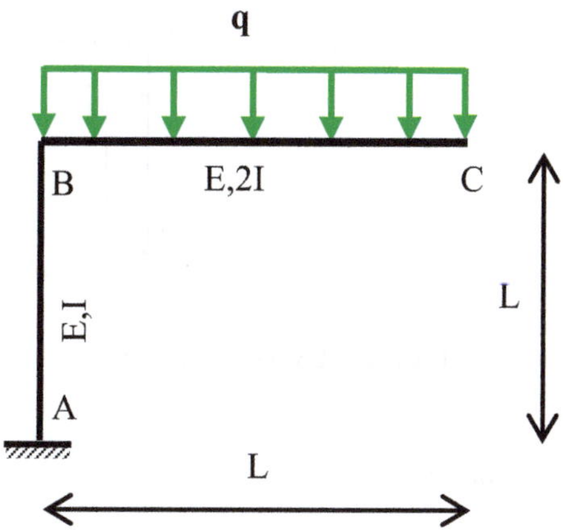

Figure 3.21. *Statically determinate system*

The bending moment diagram of the fundamental system is shown in Figure 3.22.

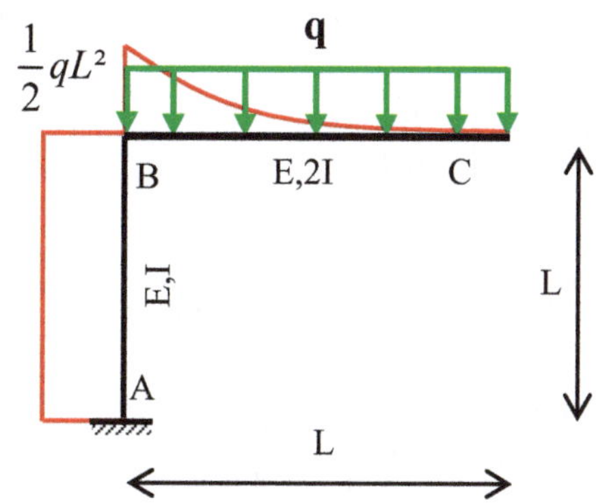

Figure 3.22. *Bending moment diagram of the fundamental system*

– *Unit systems*

Figure 3.23 shows bending moments diagrams due to horizontal and vertical unit forces applied at point C along both directions.

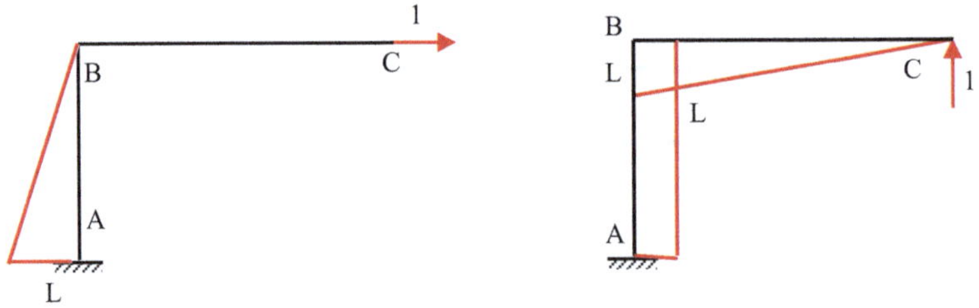

Figure 3.23. *Unit systems*

– *Displacement calculation*

Calculating displacements can be done using the graphical method.

$$\delta^0_{C,H} = \frac{1}{EI}\left[L(-\frac{1}{2}q.L^2)(-\frac{L}{2})\right] = \frac{qL^4}{4EI}$$

$$\delta^0_{C,V} = \frac{1}{2EI}\left[(-\frac{1}{6}q.L^3)(\frac{3}{4}L)\right] + \frac{1}{EI}\left[(-\frac{1}{2}q.L^2).L.L)\right] = -\frac{9qL^4}{16EI}$$

$$\delta^1_{C,H} = \frac{1}{EI}\left[\frac{1}{2}L.L.\frac{2}{3}L\right] = \frac{L^3}{3EI}$$

$$\delta^2_{C,V} = \frac{1}{2EI}\left[\frac{1}{2}L.L(\frac{2}{3}L)\right] + \frac{1}{EI}[L.L.L] = \frac{7L^3}{6EI}$$

$$\delta^1_{C,V} = \delta^2_{C,H} = \frac{1}{EI}\left[.\frac{1}{2}L.(-L).L\right] = -\frac{L^3}{2EI}$$

The displacements of joint C in the horizontal direction and the vertical direction make it possible to write:

$$\delta^0_{C,H} + \delta^1_{C,H}H_C + \delta^2_{C,H}V_C = 0$$

$$\delta^0_{C,V} + \delta^1_{C,V}H_C + \delta^2_{C,V}V_C = 0$$

The system of equations is constructed by substituting the values of the displacements above.

$$-\frac{qL}{4}+\frac{1}{3}H_C-\frac{1}{2}V_C=0$$

$$-\frac{9qL}{16}-\frac{1}{2}H_C+\frac{7}{6}V_C=0$$

Resolving the system of equations gives the reactions at support A, which are:

$$H_C=-\frac{3}{40}qL$$

$$V_C=\frac{9}{20}qL$$

The moments at the joints can be deduced by using the principle of superposition:

$$M_{ij}=M_{ij}^0+\sum_k m_{ij}^k X_k$$

Bar (AB)

$$M_{AB}=M_{AB}^0+\sum_k m_A^k X_k=$$

$$\frac{1}{2}qL^2+(-\frac{3}{40}qL)(L)+\frac{9}{20}qL(-L)=-\frac{1}{40}qL^2$$

$$M_{BA}=(-\frac{1}{2}qL^2)+\frac{9}{20}qL\ .L=-\frac{1}{20}qL^2$$

Bar (BC)

$$M_{BC}=\frac{1}{2}qL^2+\frac{9}{20}qL(-L)=\frac{1}{20}qL^2$$

$$M_{CB}=0.$$

The bending moment variations for each bar are given as:

Bar (AB)

$$M_{AB}(x) = \mu_{AB}(x) - M_{AB}(1-\frac{x}{L}) + M_{BA}(\frac{x}{L}) = \frac{1}{40}qL^2(1-\frac{x}{L}) - \frac{1}{20}qL^2(\frac{x}{L})$$

$$M_{AB}(x) = \frac{1}{40}qL^2 - \frac{3}{40}qLx$$

$$M_{AB}(0) = \frac{1}{40}qL^2$$

$$M_{AB}(L) = -\frac{1}{20}qL^2$$

Bar (BC)

$$M_{Bc}(x) = \mu_{BC}(x) - M_{BC}(1-\frac{x}{L}) + M_{BC}(\frac{x}{L}) = \frac{1}{2}qLx - \frac{1}{2}qx^2 - \frac{1}{20}qL^2(1-\frac{x}{L})$$

$$M_{Bc}(x) = \frac{11}{20}qLx - \frac{1}{2}qx^2 - \frac{1}{20}qL^2$$

$$M_{BC}(0) = -\frac{1}{20}qL^2$$

$$M_{BC}(L) = 0.$$

Similarly, the shear force and bending moment variations at each bar are given as:

Bar (AB)

$$T_{AB}(x) = \frac{dT(x)}{dx} = -\frac{3}{40}qL$$

Bar (BC)

$$T_{BC}(x) = \frac{11}{20}qL - qx$$

$$T_{BC}(0) = \frac{11}{20}qL$$

$$T_{BC}(L) = -\frac{9}{20}qL$$

Figures 3.24 and 3.25 show the bending moment diagram and the shear force diagram, respectively.

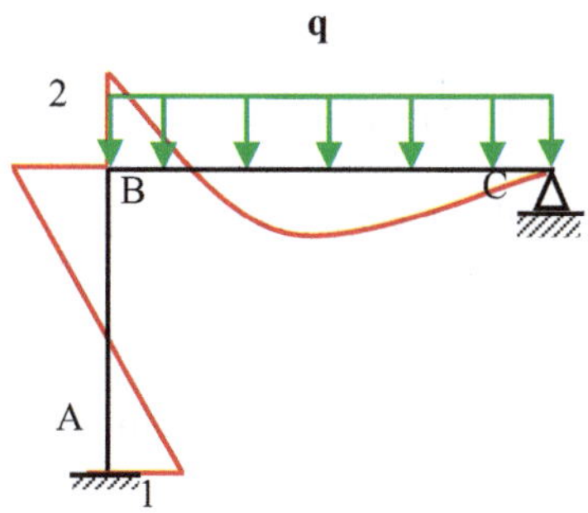

Figure 3.24. *Bending moment diagram* $*(\frac{qL^2}{40})$

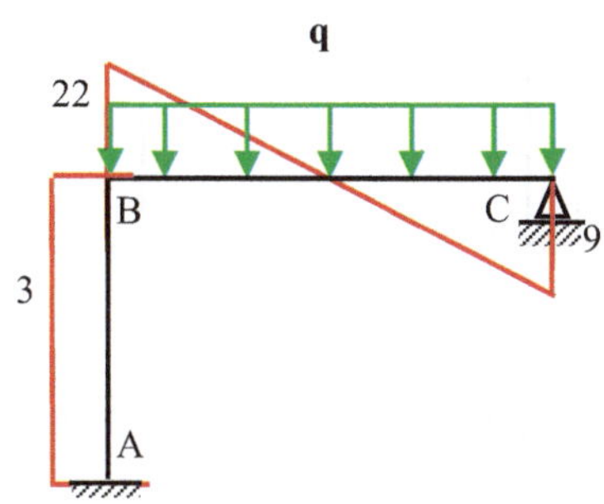

Figure 3.25. *Shear force diagram* $*(\frac{qL^2}{40})$

3.6.2. *Frame with cantilever*

We analyze the frame in Figure 3.26.

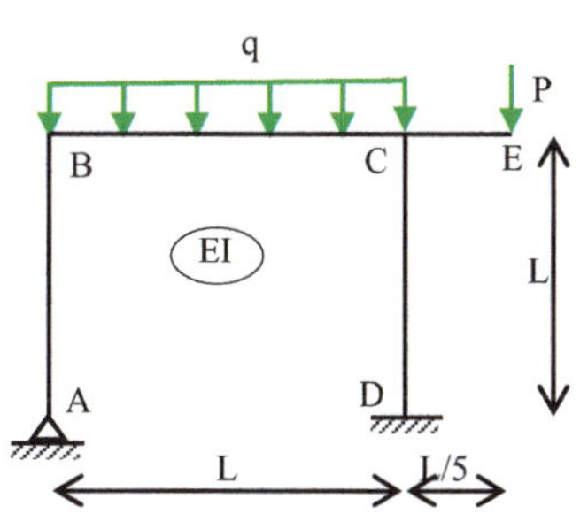

Figure 3.26. *Frame with cantilever*

We consider P = q.L for the sake of simplification.

Degree of static indeterminacy of the system:

$$f = (3.b + r) - (3n + k) = (3.3 + 5) - (3.4 + 0) = 2$$

The system is thus twice statically indeterminate.

– *Equivalent system*

We substitute support C with a system of equivalent forces (Figure 3.27).

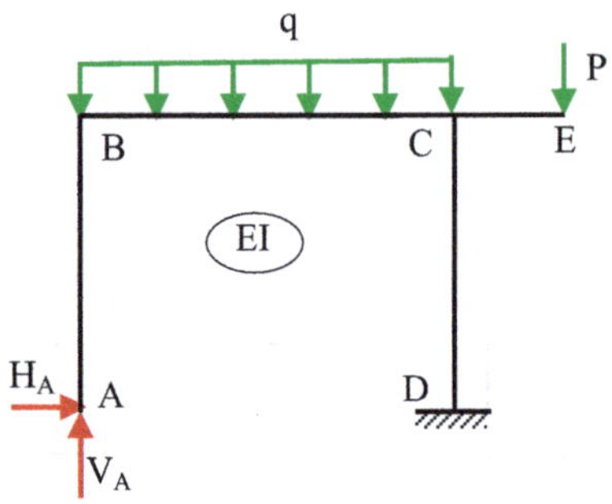

Figure 3.27. *Equivalent frame*

– *Basic system*

In this case, we neglect the effect of support reactions H_A and V_A, and the system becomes fundamental (Figure 3.28).

The expressions of bending moment of the statically determinate system are given as:

Bar (AB)

$$\mu(x) = 0$$

Bar (BC)

$$\mu(x) = -\frac{1}{2}qx^2$$

Bar (CD)

$$\mu(x) = Px - \frac{3}{10}PL$$

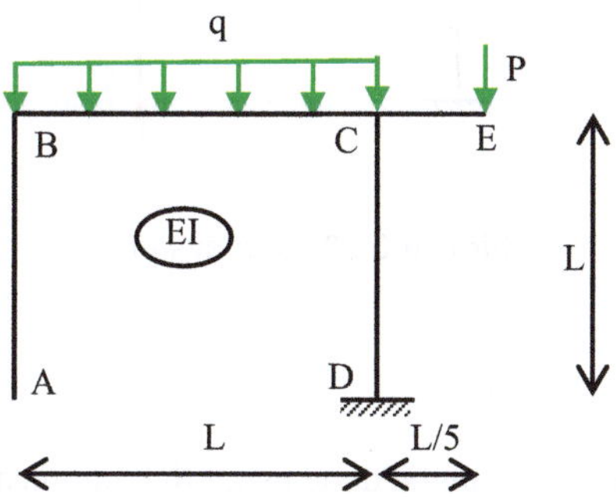

Figure 3.28. *Fundamental system*

The bending moment diagram of the fundamental system is shown in Figure 3.29.

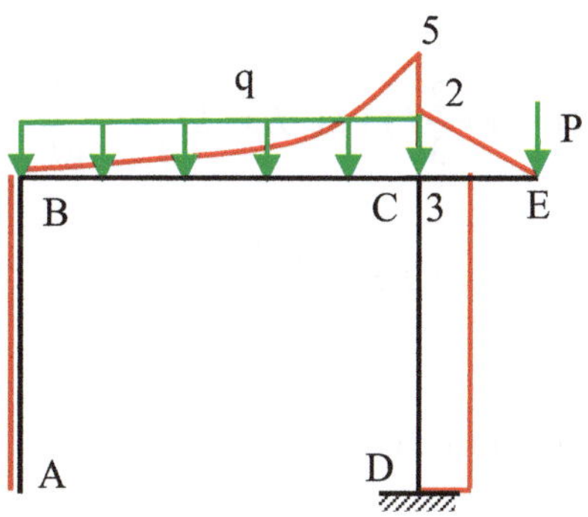

Figure 3.29. *Bending moment diagram of the fundamental system* *$(\frac{1}{10}qL^2)$

– *Unit systems*

Figure 3.30 shows bending moment diagrams due to horizontal and vertical unit actions.

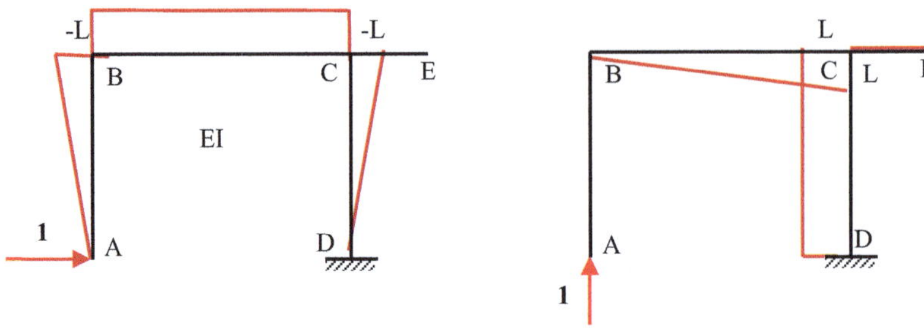

Figure 3.30. *Unit systems*

– *Displacement calculation*

Calculating displacements can be done using the graphical method.

$$\delta^0_{A,H} = \frac{1}{EI}\left[(-\frac{1}{6}qL^3)(-L) + (-\frac{3}{10}qL^2)(L)(-\frac{L}{2})\right] = \frac{19}{60}\frac{qL^4}{EI}$$

$$\delta^0_{A,V} = \frac{1}{EI}\left[(-\frac{1}{6}qL^3)(\frac{3}{4}L) + (-\frac{3}{10}qL^2)(L)(L)\right] = -\frac{17}{40}\frac{qL^4}{EI}$$

$$\delta^1_{A,H} = \frac{1}{EI}\left[2\frac{1}{2}L(-L)(-\frac{2}{3}L) + L(-L)(-L)\right] = \frac{5}{3}\frac{L^3}{EI}$$

$$\delta^2_{A,V} = \frac{1}{EI}\left[\frac{1}{2}L.L(\frac{2}{3}L) + L.L.L\right] = \frac{4}{3}\frac{L^3}{EI}$$

$$\delta^1_{A,V} = \frac{1}{EI}\left[\frac{1}{2}L.L(-L) + L.L(-\frac{1}{2}L)\right] = -\frac{L^3}{EI}$$

$$\delta^2_{A,H} = \frac{1}{EI}\left[L(-L)(\frac{1}{2}L) + \frac{1}{2}L(-L)L\right] = -\frac{L^3}{EI}$$

The displacements relative to joint A in the horizontal direction and the vertical direction make it possible to write:

$$\delta^0_{A,H} + \delta^1_{A,H}H_A + \delta^2_{A,H}V_A = 0$$

$$\delta^0_{A,V} + \delta^1_{A,V}H_A + \delta^2_{A,V}V_A = 0$$

The system of equations is constructed by substituting the values of the displacements obtained.

$$\frac{19}{20}P-\frac{5}{3}H_A-V_A=0$$

$$-\frac{17}{40}P-H_A+\frac{4}{3}V_A=0$$

Resolving the system of equations gives the reactions at support A.

$$H_A=\frac{1}{440}P$$

$$V_A=\frac{141}{440}P$$

The moments at the joints can be deduced by using the principle of superposition.

$$\mathrm{M_{ij}}=M_{ij}^0+\sum_k m_{ij}^k X_k$$

Bar (AB)

$$\mathrm{M_{AB}}=M_{AB}^0+\sum_k m_A^k X_k=0+0+0=0$$

$$\mathrm{M_{BA}}=0+(\text{-L})\,(\frac{1}{440}P)+0=-\frac{1}{440}PL$$

Bar (BC)

$$\mathrm{M_{BC}}=0+(\mathrm{L})\,(\frac{1}{440}P)+0=\frac{1}{440}PL$$

$$\mathrm{M_{CB}}=-\frac{5}{10}qL^2+(-L)(\frac{1}{440}P)+L(\frac{141}{440}P)=-\frac{80}{440}PL$$

Bar (CD)

$$M_{CD} = \frac{3}{10}PL + (L)(\frac{1}{440}P) + (-L)(\frac{141}{440}P) = -\frac{8}{440}PL$$

$$M_{CE} = \frac{2}{10}PL + 0 + 0 = \frac{2}{10}PL = \frac{88}{440}PL$$

$$M_{DC} = -\frac{3}{10}PL + 0 + (L)(\frac{141}{440}P) = \frac{9}{440}PL$$

The bending moment expressions of the bars are given as:

Bar (AB)

$$M_{AB}(x) = \mu_{AB}(x) + M_{BA}(\frac{x}{L}) = 0 + 0 - \frac{1}{440}PL(\frac{x}{L}) = -\frac{1}{440}Px$$

Bar (BC)

$$\begin{aligned} M_{BC}(x) &= \frac{qL}{2}x - \frac{qx^2}{2} - \frac{1}{440}PL(1-\frac{x}{L}) - \frac{80}{440}PL(\frac{x}{L}) \\ &= \frac{141}{440}Px - \frac{1}{2}qx^2 - \frac{1}{440}.PL \end{aligned}$$

Bar (CD)

$$M_{CD}(x) = -\frac{8}{440}PL(1-\frac{x}{L}) + \frac{9}{440}PL.(\frac{x}{L}) = -\frac{8}{440}PL + \frac{17}{440}Px$$

Bar (CE)

$$M_{CE}(x) = -\frac{1}{5}PL + Px$$

The diagrams of the bending moment and the shear force are shown in Figures 3.31 and 3.32.

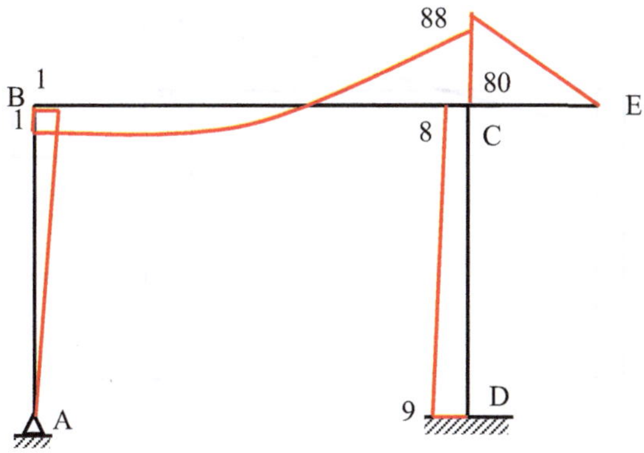

Figure 3.31. *Bending moment diagram * ($\frac{1}{440}PL$)*

Similarly, the expressions which give the shear force of the beam are given as:

Bar (AB)

$$T_{AB}(x) = -\frac{1}{440}P$$

Bar (BC)

$$T_{BC}(x) = \frac{141}{440}P - qx$$

Bar (CD)

$$T_{CD}(x) = \frac{17}{440}P$$

Bar (CE)

$$T_{CE}(x) = P$$

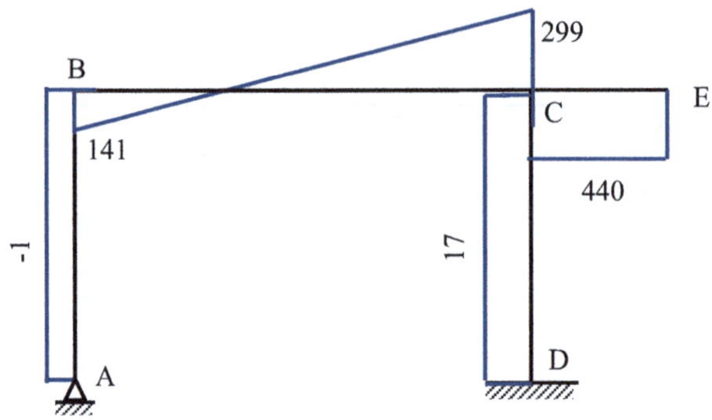

Figure 3.32. *Shear force diagram* $*\ (\frac{1}{440}P)$

3.6.3. *Frame with many degrees of static indeterminacy*

We analyze the frame in Figure 3.33 with columns of different lengths. The flexural rigidity of the members is assumed to be constant and the frame is stressed by a uniformly distributed load of intensity q applied along the horizontal bar.

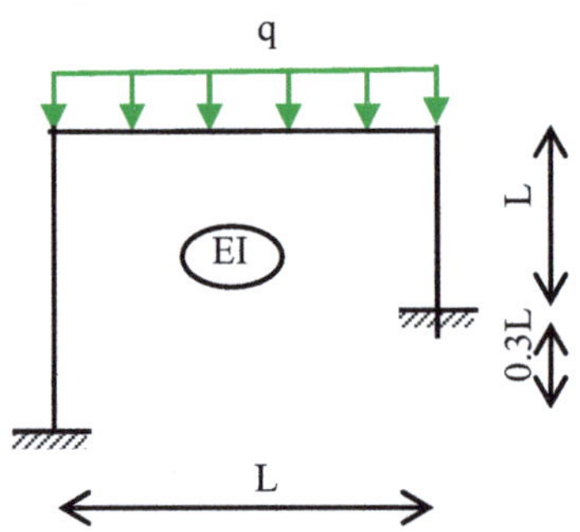

Figure 3.33. *Given frame*

Degree of static indeterminacy of the system

$$f = (3.b + r) - (3n + k) = (3.3 + 6) - (3.4 + 0) = 3$$

The system is three times statically indeterminate.

– *Equivalent system*

We substitute support D with a system of equivalent forces (Figure 3.34).

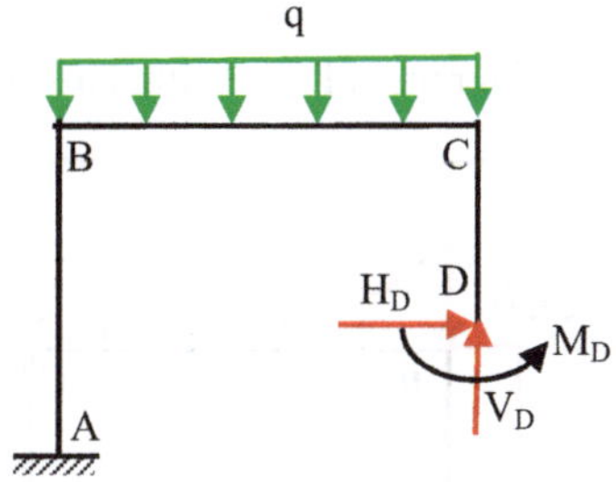

Figure 3.34. *Equivalent frame*

– *Fundamental system*

We ignore the effect of support reactions H_D and V_D and the effect of bending moment MD; the system becomes statically determinate (Figure 3.35).

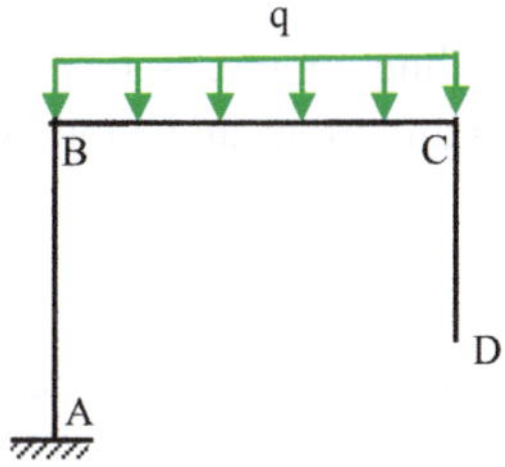

Figure 3.35. *Statically determinate system*

The expressions of bending moment of the statically determinate system are given as:

Bar (DC)

$$\mu(x) = 0$$

Bar (CB)

$$\mu(x) = -\frac{1}{2} q.x^2$$

Bar (BA)

$$\mu(x) = \frac{1}{2} qL^2$$

The bending moment diagram of the fundamental system is shown in Figure 3.36.

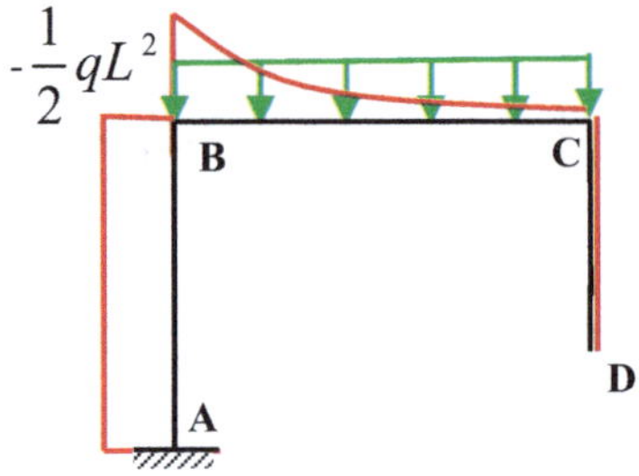

Figure 3.36. *Bending moment diagram of the fundamental system*

– *Unit systems*

Figure 3.37 shows the bending moments diagrams due to the application of forces in horizontal and vertical directions and a moment in rotational direction.

– *Displacement calculation*

Calculating displacements is carried out using the graphical method.

$$\delta^0_{D,H} = \frac{1}{EI}\left[(-\frac{1}{6}qL^3)(\frac{L}{2}) + (-\frac{1}{2}qL^2)(\frac{5L}{4})(\frac{L}{3} + \frac{L}{5})\right] = -\frac{5}{12}\frac{qL^4}{EI}$$

$$\delta^0_{D,V} = \frac{1}{EI}\left[(-\frac{1}{6}qL^3)(\frac{3}{4}L) + (-\frac{1}{2}qL^2)(\frac{4L}{5})(L)\right] = -\frac{21}{40}\frac{qL^4}{EI}$$

$$\omega^0_D = \frac{1}{EI}\left[(-\frac{1}{6}qL^3).1 + (-\frac{1}{2}qL^2)(\frac{4L}{5}).1\right] = -\frac{17}{40}\frac{qL^4}{EI}$$

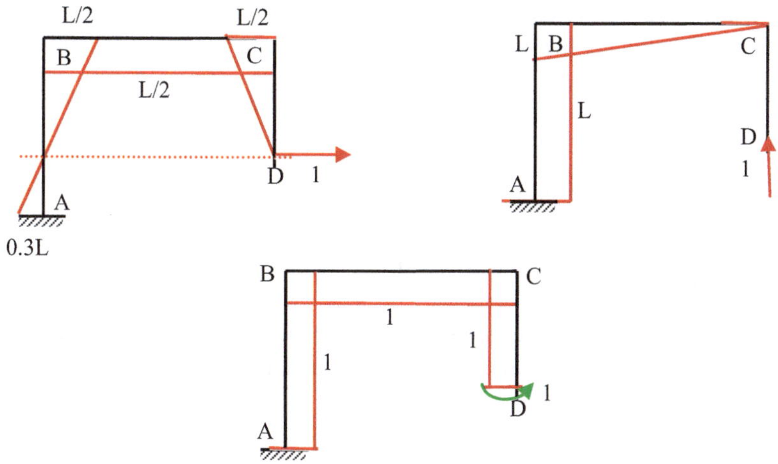

Figure 3.37. *Unit systems*

$$\delta_{D,H}^{1}=\frac{1}{EI}\left[\frac{1}{2}\frac{L}{2}\frac{L}{2}(\frac{L}{3})+\frac{L}{2}L\frac{L}{2}+\frac{1}{2}\frac{L}{2}\frac{L}{2}(\frac{L}{3})+\frac{1}{2}\frac{3L}{10}\frac{3L}{10}(\frac{L}{5})\right]=\frac{1027}{3000}\frac{qL^4}{EI}$$

$$\delta_{A,V}^{2}=\frac{1}{EI}\left[\frac{1}{2}L.L(\frac{2L}{3})+L\frac{4L}{5}(L)\right]=\frac{17}{15}\frac{L^3}{EI}$$

$$\omega_{D}^{1}=\frac{1}{EI}\left[1.\frac{1}{2}L.1+L.1.1+\frac{4L}{5}.1.1\right]=\frac{23}{10}\frac{L}{EI}$$

$$\delta_{D,H}^{2}=\frac{1}{EI}\left[\frac{1}{2}L.L(\frac{L}{2})+\frac{L}{2}\frac{L}{2}(\frac{L}{4})+\frac{1}{2}\frac{3L}{10}\frac{3L}{10}(\frac{3L}{20})\right]=\frac{57}{100}\frac{L^3}{EI}$$

$$\delta_{D,V}^{2}=\frac{1}{EI}\left[\frac{1}{2}L.L(1)+L.L(1)\right]=\frac{13}{10}\frac{L^3}{EI}$$

$$\omega_{D}^{2}=\frac{1}{EI}\left[\frac{1}{2}\frac{L}{2}(\frac{L}{2})+L.L+\frac{1}{2}\frac{L}{2}\frac{L}{2}+\frac{1}{2}\frac{3L}{10}\frac{3L}{10}\right]=\frac{259}{200}\frac{L^2}{EI}$$

The displacements relative to joint D in the horizontal direction, in the vertical direction and following the rotation allow us to write the following equations:

$$\delta^0_{D,H} + \delta^1_{D,H} H_D + \delta^2_{D,H} V_D + \omega^1_D M_D = 0$$

$$\delta^0_{D,H} + \delta^1_{D,H} H_D + \delta^2_{D,H} V_D + \omega^2_D M_D = 0$$

$$\omega^0_D + \omega^1_D H_D + \omega^2_D V_D + \omega^3_D M_D = 0$$

By substituting the deflection expressions in the preceding equations and after their resolution, we can determine the support reactions at joint D.

$$H_D = -0.133qL$$

$$V_D = 0.50qL$$

$$M_D = 0.0096qL^2$$

The moments at the joints can be deduced by using the principle of superposition.

$$\mathrm{M_{ij}} = M^0_{ij} + \sum_k m^k_{ij}.X_k$$

Bar (AB)

$$\mathrm{M_{AB}} = M^0_{AB} + \sum_k m^k_A.X_k = (-\frac{1}{2}qL^2) + \frac{3}{10}L(-0.125qL) + L(0.488qL)$$
$$+1.(0.0096qL^2) = -0.040qL^2$$

$$\mathrm{M_{BA}} = (-\frac{1}{2}qL^2) + \frac{L}{2}(-0.125qL) + L(0.488qL) + 1.(0.0096qL^2) = -0.065qL^2$$

Bar (BC)

$$\mathrm{M_{BC}} = (\frac{1}{2}qL^2) + (-\frac{L}{2})(-0.125qL) + (-L)(0.488qL) + (-1)(0.0096qL^2) = 0.065qL^2$$

$$M_{CB} = 0+(+\frac{L}{2})(-0.125qL)+0+(+1)(0.0096qL^2) = -0.053qL^2$$

Bar (CD)

$$M_{CD} = 0+(-\frac{L}{2})(-0.125qL)+0+(-1)(0.0096qL^2) = 0.053qL^2$$

$$M_{DC} = 0+0+0+(+1)(0.0096qL^2) = 0.0096qL^2$$

The variation of the bending moment in each bar is:

Bar (AB)

$$M_{AB}(x) = 0.040qL^2(1-\frac{5x}{4L}) - 0.065qL^2(\frac{5x}{4L}) = 0.040qL^2 - 0.131qLx$$

Bar (BC)

$$M_{BC}(x) = \frac{qLx}{2} - \frac{1}{2}qx^2 - 0.065qL^2(1-\frac{x}{L}) - 0.053qL^2(\frac{x}{L}) =$$
$$0.512\,\mathrm{qL}\,x - 0.5qx^2 - 0.065qL^2$$

Bar (CD)

$$M_{CD}(x) = -0.053qL^2(1-\frac{2x}{L}) + 0.0096qL^2(\frac{2x}{L}) = -0.053qL^2 + 0.125\,\mathrm{qL}\,x$$

Figures 3.38 and 3.39 show the bending moment and the shear force diagrams.

Similarly, the expressions of shear force are given as:

Bar (AB)

$$T_{AB}(x) = -0.131qL$$

Bar (BC)

$$T_{BC}(x) = 0.512\,\mathrm{qL} - qx$$

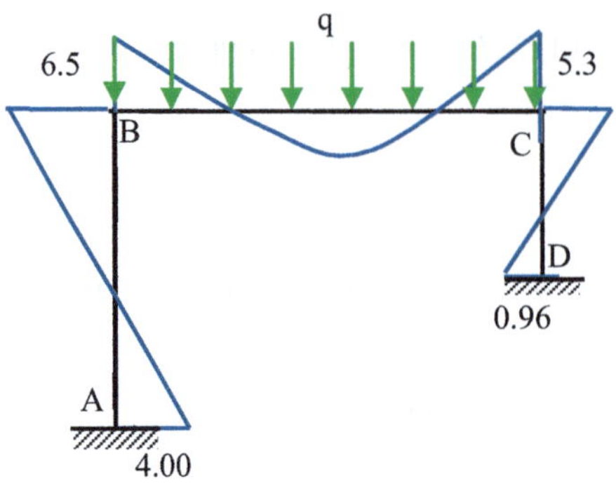

Figure 3.38. *Bending moment diagram* * $(\frac{1}{100}qL^2)$

Bar (CD)

$$T_{CD}(x) = 0.125\,\mathrm{qL}$$

Similarly, the diagram of shear force is given in Figure 3.39.

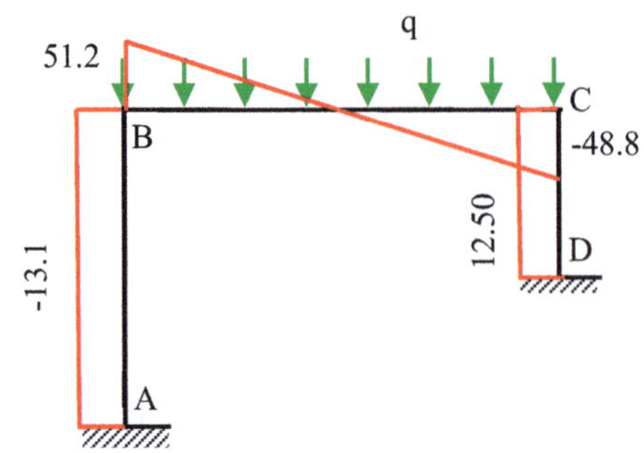

Figure 3.39. *Shear force diagram* * $(\frac{1}{100}qL)$

3.6.4. *Frame with oblique bars*

We analyze the frame in Figure 3.40 with columns of different lengths. The flexural rigidity of the members is assumed to be constant and the frame is stressed by a uniformly distributed load of intensity q applied on the horizontal bar.

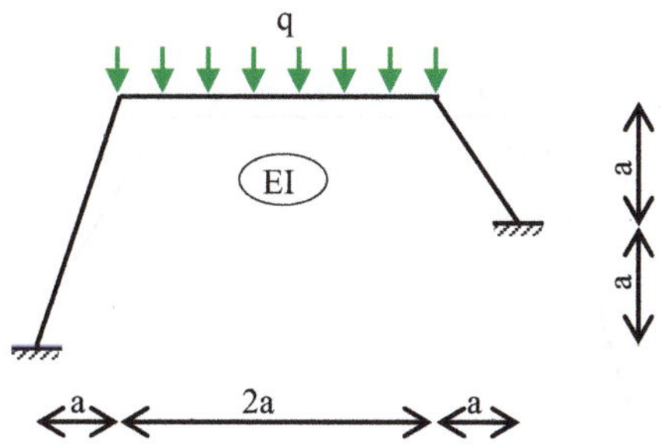

Figure 3.40. *Given frame*

Degree of static indeterminacy of the system:

f = (3.b + r) – (3n + k) = (3.3 + 6) – (3.4 + 0) = 3

The system is three times statically indeterminate.

– *Equivalent system*

We substitute support A with a system of equivalent forces (Figure 3.41).

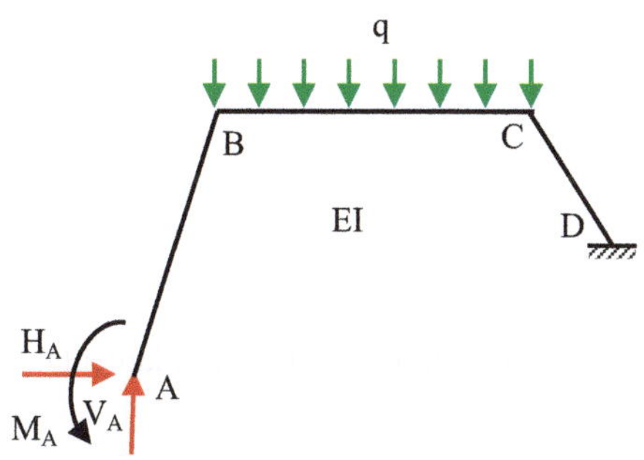

Figure 3.41. *Equivalent frame*

– *Fundamental system*

We ignore the effect of support reactions H_A and V_A and bending moment M_A; the system is statically determinate (Figure 3.42).

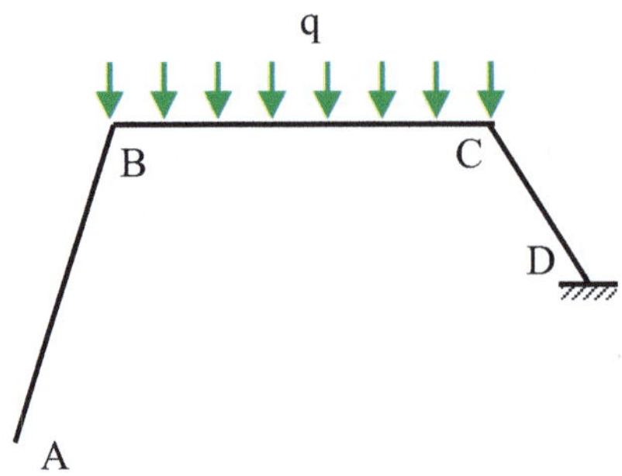

Figure 3.42. *Statically determinate system*

The bending moment expressions of the statically determinate system are given as:

Bar (AB)

$$\mu(x) = 0$$

Bar (BC)

$$\mu(x) = -\frac{1}{2}qx^2$$

Bar (CD)

$$\mu(x) = -2qa^2$$

The bending moment diagram of the fundamental system is shown in Figure 3.43.

– *Unit systems*

The systems stressed by unit actions applied to the joint A are shown in Figure 3.43.

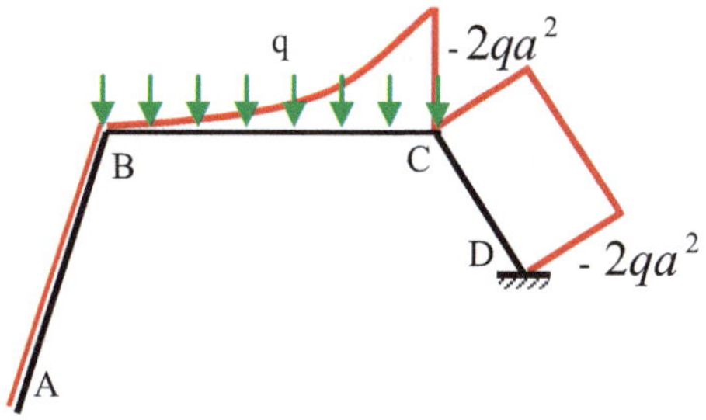

Figure 3.43. *Bending moment diagram of the fundamental system*

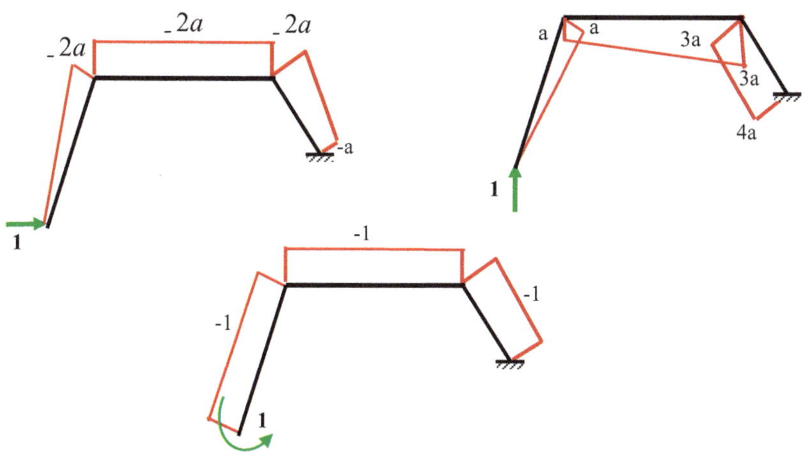

Figure 3.44. *Unit systems*

– *Displacement calculation*

The displacement calculation is done by:

$$\delta^0_{A,H} = \frac{1}{EI}\left[(-\frac{4}{3}qa^3)(-2a) + (-\frac{1}{2}qa^2)(\sqrt{2}a)(-\frac{3a}{2})\right] = 7\frac{qa^4}{EI}$$

$$\delta^0_{A,V} = \frac{1}{EI}\left[(-\frac{4}{3}qa^3)(a) + (-2qa^2)(\sqrt{2}a)(3a + \frac{2}{3}a)\right] = -16.90\frac{qa^4}{EI}$$

$$\omega^0_A = \frac{1}{EI}\left[(-\frac{4}{3}qa^3)(-1) + (-2qa^2)(\sqrt{2}a)(-1)\right] = 4.16\frac{qa^3}{EI}$$

$$\delta^1_{A,H} = \frac{1}{EI}\left[\frac{1}{2}\sqrt{2}a(-2a)(-\frac{2}{3}2a) + 2a.(-2a)(-2a) + \sqrt{2}a(-a)(-a) + \frac{1}{2}\sqrt{2}a(-a)(-a)\right] = 1.58\frac{a^3}{EI}$$

$$\delta^2_{A,V} = \frac{1}{EI}\left[\frac{1}{2}\sqrt{5}a(a)(\frac{2}{3}a) + a(2a)(a) + \frac{1}{2}2a(2a)(\frac{2}{3}2a) + 3a\sqrt{2}a(3a) + \frac{1}{2}\sqrt{2}a(a)(\frac{2}{3}a)\right]$$

$$= 18.610\frac{a^3}{EI}$$

$$\omega^3_A = \frac{1}{EI}\left[\sqrt{5}a(-1)(-1) + 2a(-1)(-1) + \sqrt{2}a(-1)(-1)\right] = 5.65\frac{a}{EI}$$

$$\omega^3_A = \frac{1}{EI}\left[\sqrt{5}a(-1)(-1) + 2a.(-1)(-1) + \sqrt{2}a(-1)(-1)\right] = 5.65\frac{a}{EI}$$

$$\delta^2_{A,H} = \frac{1}{EI}\left[\frac{1}{2}\sqrt{2}a(-2a)(\frac{2}{3}a) + 2a.(a)(-2a) + \sqrt{2}a(-a)(a) + \frac{1}{2}\sqrt{2}a(-a)(a)\right]$$

$$= -7.612\frac{a^3}{EI}$$

$$\delta^3_{A,V} = \frac{1}{EI}\left[\frac{1}{2}\sqrt{5}a(a)(-1) + a.(2a)(-1) + \frac{1}{2}2a(2a)(-1) + \sqrt{2}a(3a)(-1) + \frac{1}{2}\sqrt{2}a(a)(-1)\right] =$$

$$-10.07\frac{a^2}{EI}$$

$$\delta^3_{A,H} = \frac{1}{EI}\left[\frac{1}{2}\sqrt{5}a(-2a)(-1) + 2a.(-2a)(-1) + \sqrt{2}a(-a)(-1) + \frac{1}{2}\sqrt{2}a(-a)(-1)\right] =$$

$$8.357\frac{a^2}{EI}$$

The deflections at joint D in horizontal, vertical and rotational directions are written as:

$$\delta^0_{A,H} + \delta^1_{A,H}H_A + \delta^2_{A,H}V_A + \delta^3_{A,H}M_A = 0$$

$$\delta^0_{A,H} + \delta^2_{A,H}H_A + \delta^2_{A,V}V_A + \delta^3_{A,V}M_A = 0$$

$$\omega^0_A + \delta^3_{A,H}H_A + \delta^3_{A,V}V_A + \omega^3_A M_A = 0$$

By substituting the deflection expressions, resolving the system of equations leads to the determination of actions at joint A.

$$H_D = -0.133qL$$

$$V_D = 0.50qL$$

$$M_D = 0.0096qL^2$$

The moments at the joints can be deduced by using the principle of superposition.

$$M_{ij} = M_{ij}^0 + \sum_k m_{ij}^k . X_k$$

Bar (AB)

$$M_{AB} = M_{AB}^0 + \sum_k m_A^k . X_k = (-\frac{1}{2}qL^2) + \frac{3}{10}L(-0.125qL)$$
$$+L(0.488qL) + 1.(0.0096qL^2) = -0.216qL^2$$

$$M_{BA} = -\frac{1}{2}qL^2 + \frac{L}{2}(-0.125qL) + L.(0.488qL) + 1.(0.0096qL^2) = -0.335qL^2$$

Bar (BC)

$$M_{BC} = \frac{1}{2}qL^2 + (-\frac{L}{2})(-0.125qL) + (-L)(0.488qL) + (-1)(0.0096qL^2) = 0.335qL^2$$

$$M_{CB} = 0 + (+\frac{L}{2})(-0.125qL) + 0 + (+1)(0.0096qL^2) = -0.073qL^2$$

Bar (CD)

$$M_{CD} = 0 + (-\frac{L}{2})(-0.125qL) + 0 + (-1)(0.0096qL^2) = 0.073qL^2$$

$$M_{DC} = 0 + 0 + 0 + (+1)(0.105qL^2) = 0.105qL^2$$

The variation of the bending moment in each bar is:

Bar (AB)

$$M_{AB}(x) = 0.216qa^2(1 - \frac{x}{\sqrt{5}.a}) - 0.335qa^2(\frac{x}{\sqrt{5}.a})$$

Bar (BC)

$$M_{BC}(x) = qax - \frac{1}{2}qx^2 - 0.335qa^2(1 - \frac{x}{2.a}) - 0.073qa^2(\frac{x}{2.a})$$

Bar (CD)

$$M_{CD}(x) = -0.073qa^2(1 - \frac{x}{\sqrt{2}.a}) - 0.105qa^2(\frac{x}{\sqrt{2}.a})$$

The diagrams of the bending moment and the shear force are shown in Figures 3.45 and 3.46, respectively.

In the same way, the shear force expression is:

Bar (AB)

$$T_{AB}(x) = -\frac{0.551}{\sqrt{5}}qa$$

Bar (BC)

$$T_{BC}(x) = 1.131.qa - qx$$

Bar (CD)

$$T_{CD}(x) = -\frac{0.032}{\sqrt{2}}qa$$

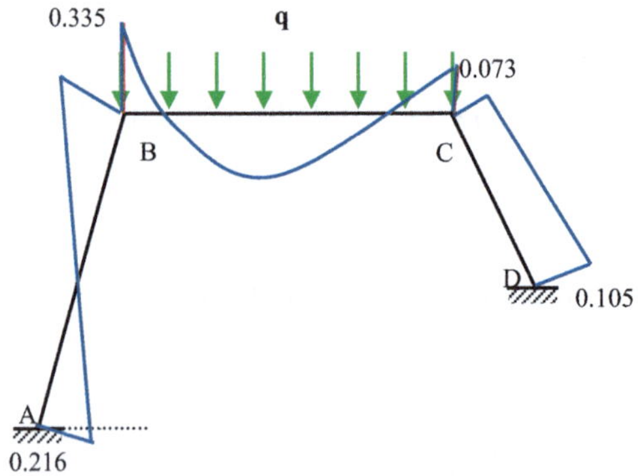

Figure 3.45. *Bending moment diagram *(qa²)*

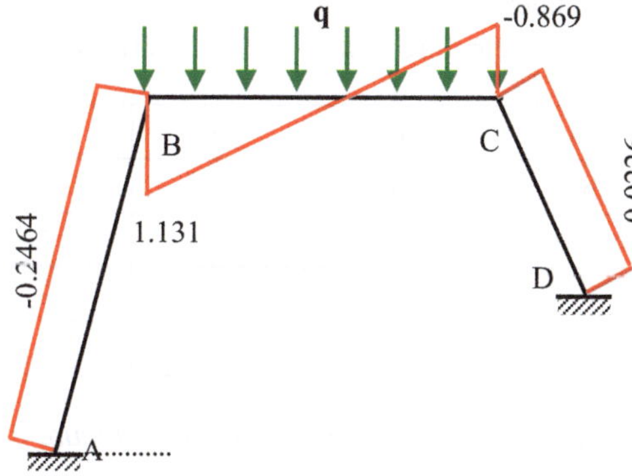

Figure 3.46. *Shear force diagram *(qa)*

3.7. Analysis of truss

Trusses can be grouped into three different classes: (1) internally statically indeterminate truss systems, (2) externally statically indeterminate truss systems and (3) internally and externally statically indeterminate truss systems (Figure 3.47).

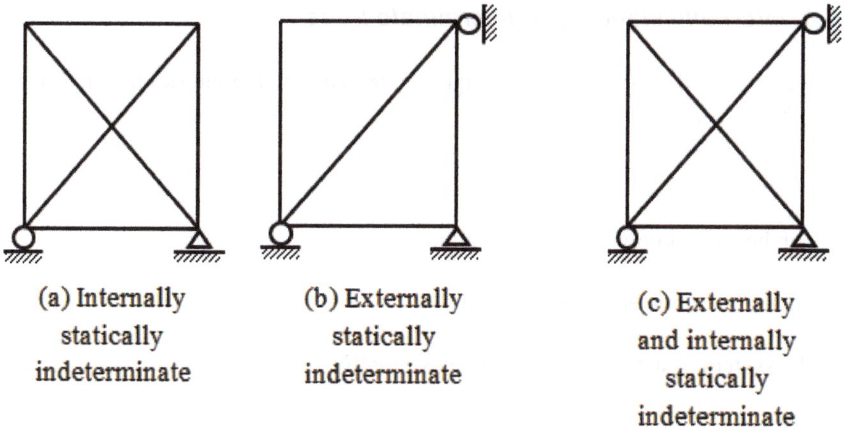

Figure 3.47. *Different types of truss structures*

Calculation of the degree of static indeterminacy of a truss is given by the following formula:

$$f = b - (2.n - 3)$$

The degree of static indeterminacy of each system (Figure 3.47) is presented in Table 3.1. Table 3.2 shows the nature of the static indeterminacy of truss systems.

System	(a)	(b)	(c)
b	6	5	6
2n – 3	5	5	5
r	3	4	3
f	1	0	1

Table 3.1. *Degree of static indeterminacy (Figure 3.47)*

System	Internal analysis	External analysis
(a)	Once statically indeterminate	Statically determinate
(b)	Statically determinate	Once statically indeterminate
(c)	Once statically indeterminate	Once statically indeterminate

Table 3.2. *Nature of static indeterminacy*

3.7.1. *Internally statically indeterminate truss*

We analyze the truss system in Figure 3.48. The membrane rigidity ($E\Omega$) of all of the bars is assumed to be constant.

– *External analysis*

Number of unknowns (support reactions) = 3

Number of static equations = 3

The system is statically determinate externally.

– *Internal analysis:*

Number of bars: b = 6.

Number of possible equations: 2n – 3 = 5

Therefore, the system is once statically indeterminate internally.

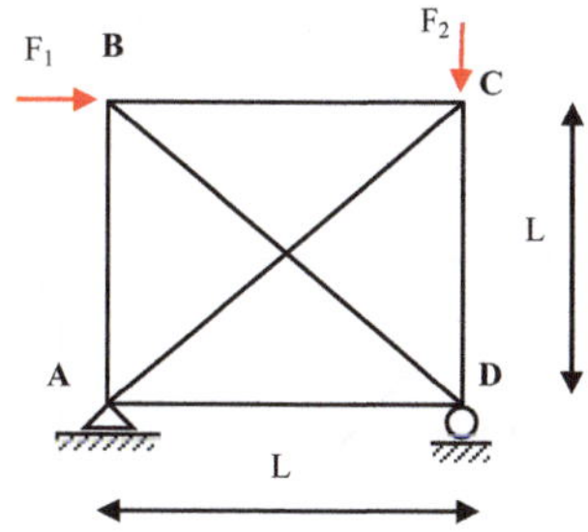

Figure 3.48. *Given truss structure*

To analyze the truss (Figure 3.48), it is necessary to use the same method presented during the analysis of beams and frames.

– *Equivalent system*

The given truss is once statically indeterminate internally. It is necessary to replace a bar with an equivalent system of force (Figure 3.49), for example bar (AC).

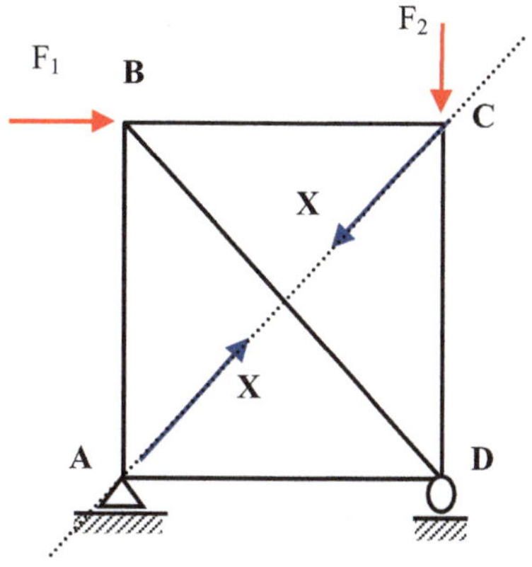

Figure 3.49. *Equivalent truss system*

– *Fundamental system*

We ignore the existence of bar (AC) in order to obtain the fundamental system (Figure 3.50). The system obtained is internally and externally statically determinate; the degree of static indeterminacy of the corresponding system can be verified by:

$$f = b - (2n - 3) = 5 - (2 \times 4 - 3) = 0$$

To calculate the internal forces in the bars of this structure, we use the method of joint equilibrium.

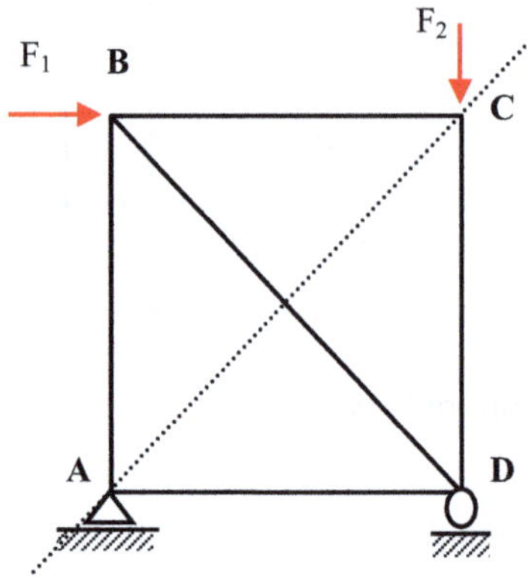

Figure 3.50. *Basic truss system*

– *Joint C*

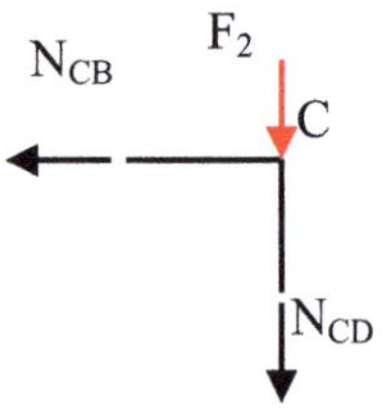

The equilibrium of joint C allows us to write:

$$\sum \vec{F} = \vec{0}$$

$$F_2 + N_{CB} + N_{CD} = 0$$

$$N_{CB} = 0$$

$$N_{CD} = -F_2$$

– *Joint B*

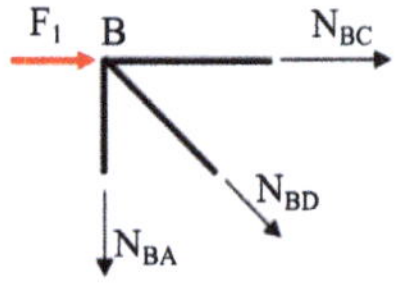

The equilibrium of joint B allows us to write:

$$\sum F = 0$$

$$F_1 + N_{BC} + N_{BA} + N_{BD} = 0$$

The projection of forces and internal forces along the horizontal and vertical axes is:

$$N_{BD} = -\sqrt{2}F_1$$

$$N_{BA} = F1$$

– *Joint A*

The support reactions can be deduced using equilibrium equations of the fundamental system.

HA = - F1

VA = - F1

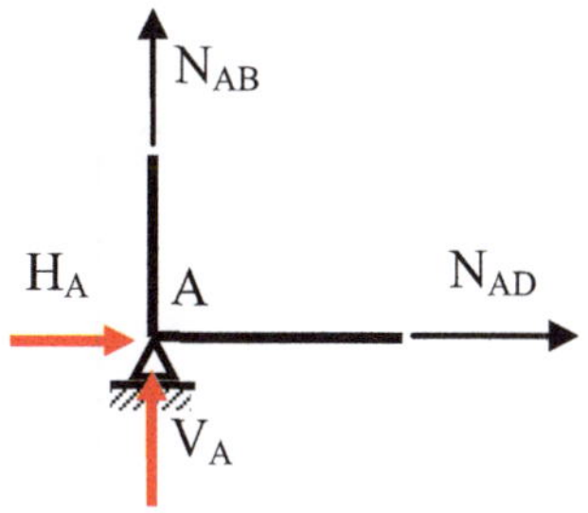

The equilibrium of joint A leads to:

$$\sum \vec{F} = \vec{0}$$

$$H_A + V_A + N_{AD} + N_{AB} = 0$$

The projection of forces along the horizontal and vertical axes gives:

$$N_{AD} = F_1$$

The internal forces on the bars of the truss structure are grouped in Table 3.3.

Bar	N^0	Nature of force
AB	F_1	Traction
AD	F_1	Traction
BC	0	Neutral
BD	$-\sqrt{2}\,F_1$	Compression
CD	$-F_2$	Compression
AC	0	Neutral

Table 3.3. *Forces on the bars of the fundamental system*

– *Unit system*

Instead of the eliminated force (Figure 3.50), a unit force is applied (Figure 3.51). The forces on the bars of the fundamental truss structure are grouped in Table 3.4.

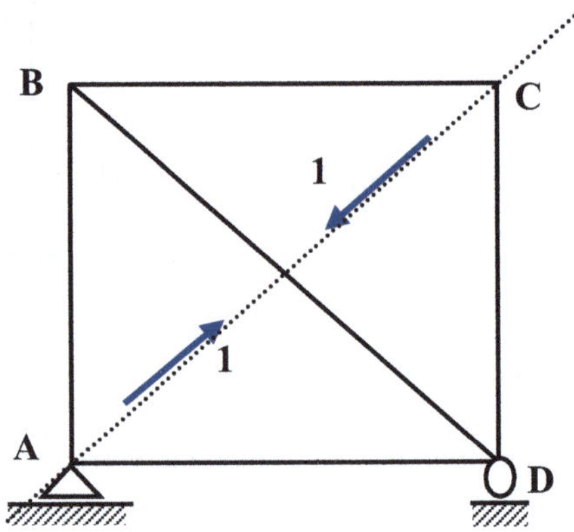

Figure 3.51. *Unit truss system*

The internal forces due to a unit action applied in the direction of the eliminated bar are grouped in Table 3.4.

Bar	n	L	n.L	N°	$\delta^1_{A,AC} = \sum \frac{n_i^2.L_i}{E\Omega}$	$\delta^0_{A,AC} = \sum \frac{N_i^0.n_i.L_i}{E\Omega}$
AB	$-\sqrt{2}/2$	L	$-\sqrt{2}L/2$	F_1		
AD	$-\sqrt{2}/2$	L	$-\sqrt{2}L/2$	F_1		
BC	$-\sqrt{2}/2$	L	$-\sqrt{2}L/2$	0		
BD	1	$\sqrt{2}L$	$\sqrt{2}L$	$-\sqrt{2}F_1$	$(2+2\sqrt{2})\frac{L}{E\Omega}$	$-(2+\sqrt{2})\frac{F_1 L}{E\Omega}$ $+\sqrt{2}\frac{F_2 L}{E\Omega}$
CD	$-\sqrt{2}/2$	L	$-\sqrt{2}L/2$	$-F_2$		
AC		$\sqrt{2}L$	$\sqrt{2}L$	0		

Table 3.4. *Forces on the bars of the unit system and displacement along the diagonal (AC)*

We apply the compatibility condition of the displacement along the diagonal of bar (AC); we thus obtain:

$$\delta^0_{A,AC} + \delta^1_{A,AC} X = 0$$

The internal force in the bar (AC) can be deduced by:

$$X = -\frac{\delta^0_{AC}}{\delta^1_{AC}} = \frac{2+\sqrt{2}}{2+2\sqrt{2}} F_1 - \frac{\sqrt{2}}{2(2+2\sqrt{2})} F_2$$

The forces on the bars are determined using the superposition of effects principle (Table 3.5).

$$N_i = N_i^0 + Xn_i$$

Bar	N^0	n	$N = N^0 + X\,n$ $*(\frac{1}{4(1+\sqrt{2})})$
AB	F_1	$-\sqrt{2}/2$	$(2+2\sqrt{2})\,F_1 + F_2$
AD	F_1	$-\sqrt{2}/2$	$(2+2\sqrt{2})\,F_1 + F_2$
BC	0	$-\sqrt{2}/2$	$-(2+2\sqrt{2})\,F_1 + F_2$
BD	$-\sqrt{2}\,F1$	1	$-(4+2\sqrt{2})\,F_1 - \sqrt{2}\,F_2$
CD	$-F_2$	$-\sqrt{2}/2$	$-(2+2\sqrt{2})\,F_1 - (3+4\sqrt{2})\,F_2$
AC	0	1	$(4+2\sqrt{2})\,F_1 - \sqrt{2}\,F_2$

Table 3.5. *Forces on the bars of the given structure*

In particular, for $F_1 = F_2 = F$, in this case we calculate the normal force and its nature for each bar (Table 3.6).

Bar	$N = N^0 + X.n$ $*(\frac{1}{4(1+\sqrt{2})})$	N	Nature of force
AB	$(3+2\sqrt{2})$ F	0.603 F	Traction
AD	$(3+2\sqrt{2})$ F	0.603 F	Traction
BC	$-(1+2\sqrt{2})$ F	0.396 F	Compression
BD	$-(4+3\sqrt{2})$ F	0.853 F	Compression
CD	$-(5+6\sqrt{2})$ F	1.396 F	Compression
AC	$(4+\sqrt{2})$ F	0.560 F	Traction

Table 3.6. *Forces on the bars of the given structure for $F = F_1 = F_2$*

The internal forces acting on the bars of the truss structure are shown schematically in Figure 3.52.

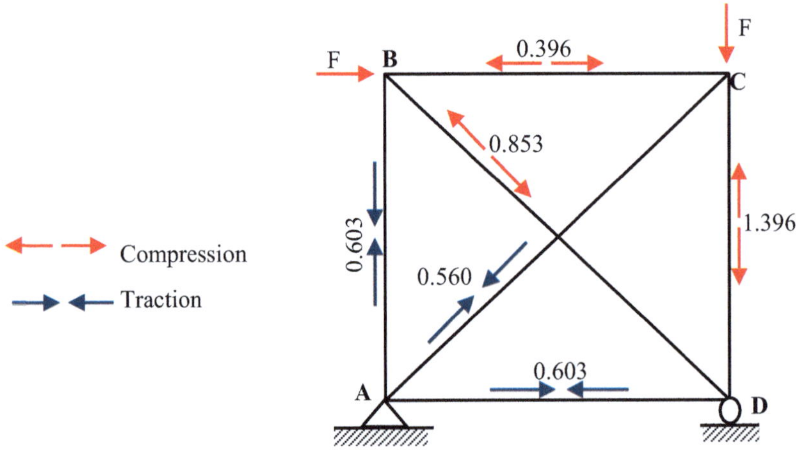

Figure 3.52. *Normal force diagram *(F)*

3.7.2. Externally statically indeterminate truss

The system (Figure 3.53) is externally statically indeterminate. We assume that the membrane rigidity ($E\Omega$) is constant for all bars.

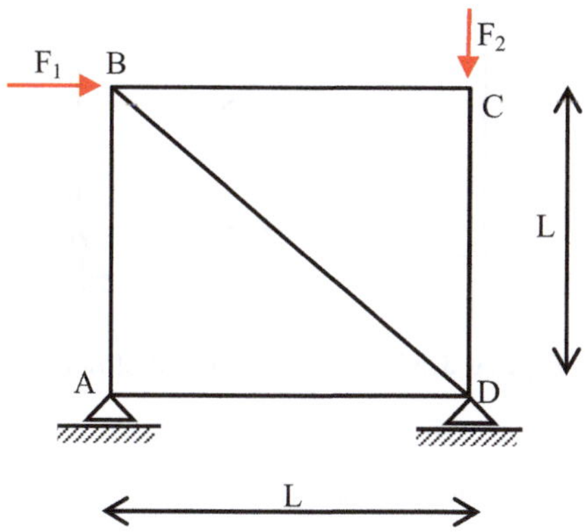

Figure 3.53. *Given truss structure*

– *Internal analysis*

Number of bars: b = 5

Number of equations: 2.n – 3 = 5

The system is internally statically determinate.

– *External analysis*

Number of reactions: r = 4

Number of static equations: 3

The system is statically determinate internally and once statically indeterminate externally.

– *Equivalent system*

In the equivalent system, the support reactions are replaced by forces of a number equal to that of the degree of static indeterminacy (Figure 3.54).

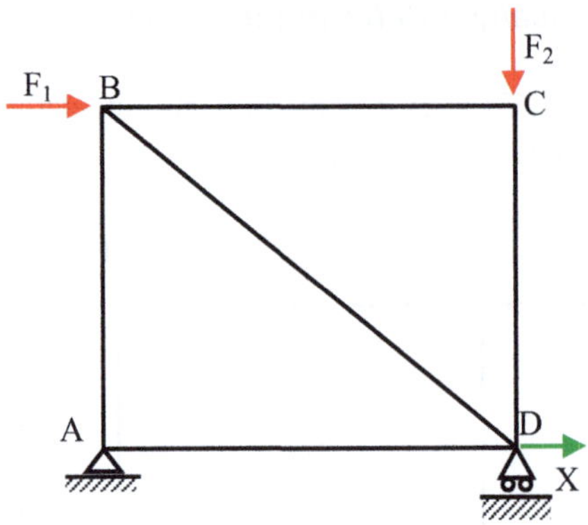

Figure 3.54. *Equivalent truss structure*

The construction of the fundamental system is represented in Figure 3.55.

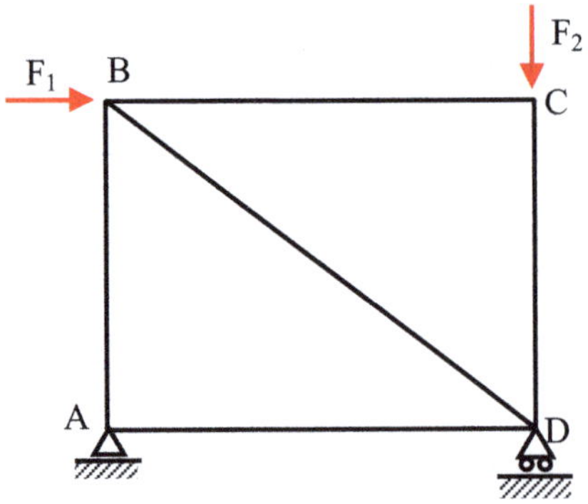

Figure 3.55. *Fundamental structure*

The forces on the bars have already been determined and grouped in (Table 3.3). The unit system is constructed by applying a unit force in the direction of the free body link (Figure 3.56).

The method of joint equilibrium makes it possible to determine internal forces in the bars, due to the application of a unit force, which are:

$$n_{AD} = 1$$

$$n_{AB} = n_{BC} = n_{BD} = n_{CD} = 0$$

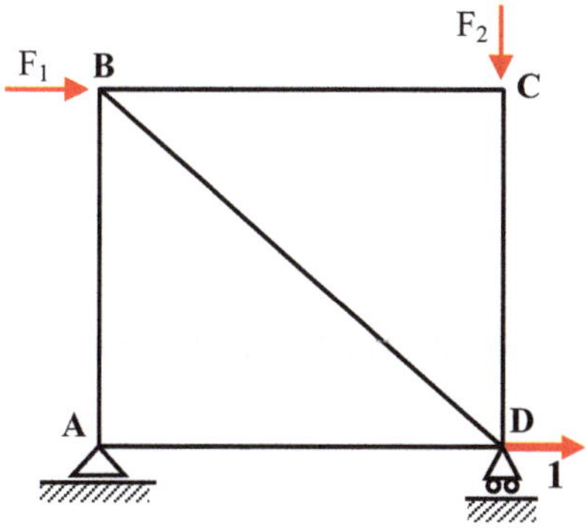

Figure 3.56. *Unit truss structure*

– *Displacement calculation*

The displacement of point D in the direction (xx) of the fundamental system has already been calculated (Table 3.3), which can be evaluated by:

$$\delta^0_{D,X} = \sum \frac{N^0 L}{E\Omega} = -\frac{F_2 L}{E\Omega}$$

The displacement of point D in the direction (xx) due to the effect of a unit force applied to joint D in the horizontal direction is given in Table 3.7.

Bar	N	L	n.L	$\delta^1_{D,xx} = \sum \frac{n\,.L}{E\Omega}$
AB	0	L	0	$\frac{L}{E\Omega}$
AD	1	L	L	
BC	0	L	0	
BD	0	$\sqrt{2}$ L	0	
CD	0	L	0	

Table 3.7. *Horizontal displacement of joint D*

The compatibility condition of the displacement in the direction (xx) of joint D must be satisfied.

$$\delta^0_{D,X} + \delta^1_{D,X} X = 0$$

The compatibility condition of the displacement makes it possible to determine the support reaction of point D.

$$X = -\frac{\delta^0_{D,X}}{\delta^1_{D,X}} = -F_2$$

The forces on the bars are calculated using the superposition of effects principle (Table 3.8).

$$N_i = N_i^0 + Xn_i^1$$

Bar	N^0	n	X.n	N	Nature of force
AB	F_1	0	0	F_1	Traction
AD	F_1	1	$-F_2$	$F_1 - F_2$	?
BC	0	0	0	0	Neutral
BD	$-\sqrt{2}\,F_1$	0	0	$-\sqrt{2}\,F_1$	Compression
CD	$-F_2$	0	0	$-F_2$	Compression

Table 3.8. *Internal forces on the bars of the given system*

The normal force diagram is represented in Figure 3.57 considering $F = F_1 = F_2$.

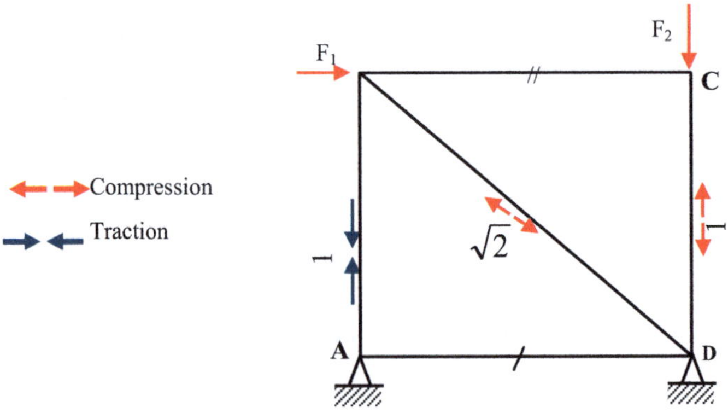

Figure 3.57. *Normal force diagram *(F)*

3.7.3. *Internally and externally statically indeterminate truss*

The superposition of the above systems makes it possible to obtain a truss that is once statically indeterminate internally and externally. It is assumed that all of the bars have the same membrane rigidity ($E\Omega$) and we must calculate the forces in the bars of the truss structure (Figure 3.58).

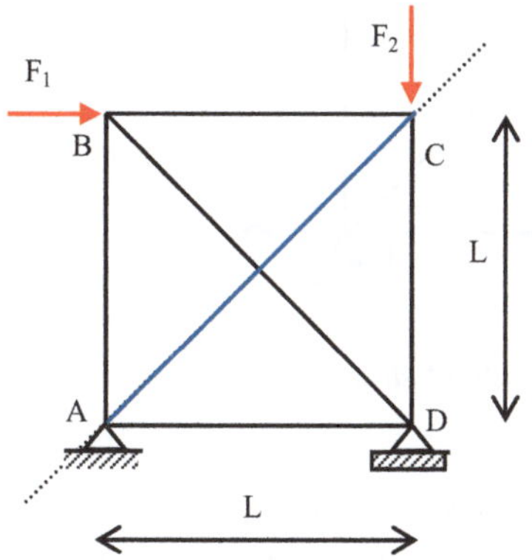

- Bar (AC) displays internal static indeterminacy.
- Support D presents external static indeterminacy.

Figure 3.58. *Given truss structure*

– *External analysis*

Number of unknowns (support reactions) = 4

Number of static equations = 3

The system is once externally statically indeterminate.

– *Internal analysis*

Number of bars: b = 6

Number of possible equations: 2n – 3 = 5

The system is once internally statically indeterminate.

Hence the system is statically indeterminate once internally and once externally.

– *Equivalent system (Figure 3.59)*

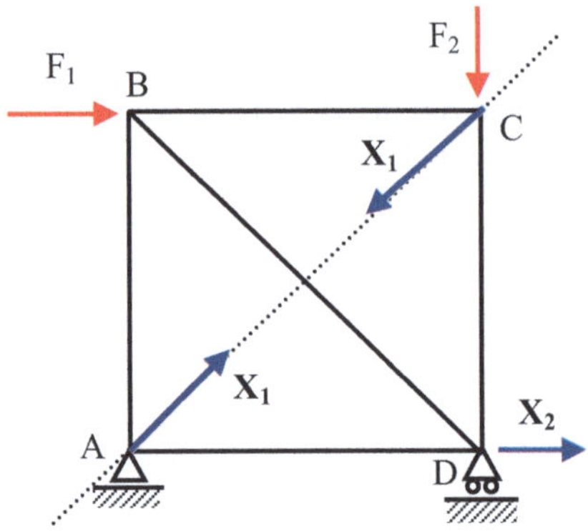

Figure 3.59. *Equivalent structure*

In the equivalent system, the internal force in bar (AC) and the horizontal reaction of support D are substituted by the equivalent forces X_1 and X_2 in sections 3.7.1 and 3.7.2.

– *Fundamental system (Figure 3.60)*

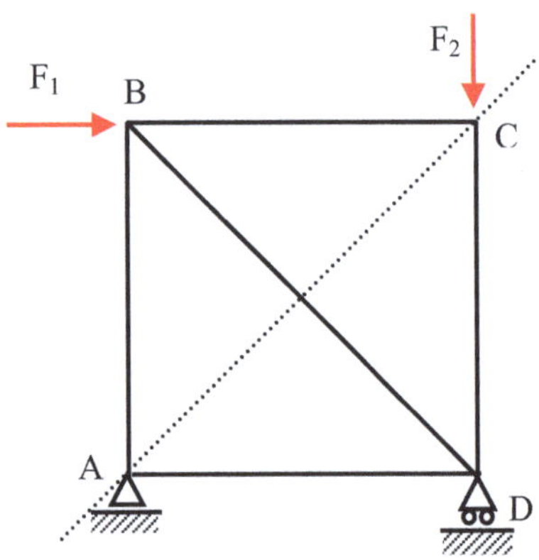

Figure 3.60. *Fundamental structure*

The forces on the bars of the truss have been determined and grouped in Table 3.3.

– *Unit systems (Figure 3.61)*

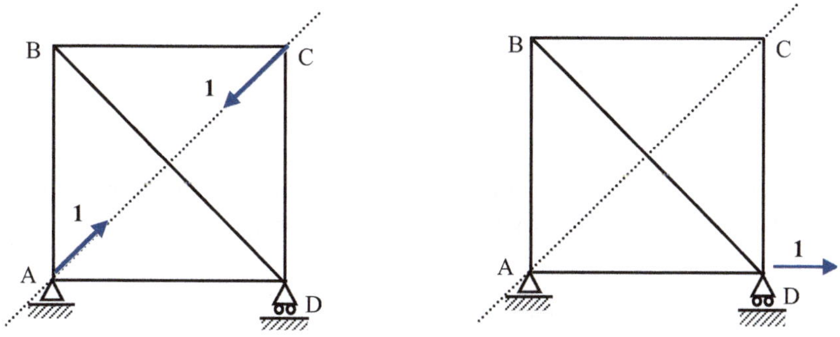

(a) Unit internal force (b) Unit external force

Figure 3.61. *Trusses under unit loads*

The compatibility equations for displacements are written as:

$$\delta^0_{A,AC} + \delta^1_{A,AC} X_1 + \delta^{12}_{A,AC} X_2 = 0$$

$$\delta^0_{D,xx} + \delta^{21}_{D,xx} X_1 + \delta^2_{D,xx} X_2 = 0$$

knowing that:

$$\delta^0_{A,AC} = -\frac{4+\sqrt{2}}{2}\frac{FL}{E\Omega}$$

$$\delta^1_{A,AC} = (2+2\sqrt{2})\frac{L}{E\Omega}$$

$$\delta^0_{D,xx} = \frac{FL}{E\Omega}$$

$$\delta^1_{D,xx} = \frac{L}{E\Omega}$$

Displacements $\delta^{12}_{A,AC}$ and $\delta^{21}_{D,xx}$ have been evaluated using the virtual work method.

$$\delta^{12}_{A,AC} = \delta^{21}_{D,xx} = \sum_{i=1}^{5} \frac{n_i^1 . n_i^2}{E\Omega} L_i = -\frac{\sqrt{2}}{2}\frac{L}{E\Omega}$$

Resolving the system of equations leads to the determination of the normal force on bar (AC) as well as the horizontal reaction to support D.

$$X_1 = \frac{4}{3+4\sqrt{2}} F = 0.462F$$

$$X_2 = 0.673F$$

The forces on the bars of the unit systems due to $X_1 = 1$ (Figure 3.61(a)) and $X_2 = 1$ (Figure 3.61(b)), respectively, are grouped in Table 3.9.

The internal stresses in the bars of the given truss structure can be deduced using the principle of superpostition of effects.

$$N_i = N_i^{(0)} + X_1 n_i^{(1)} + X_2 n_i^{(2)}$$

Bar	N^0	$X_1 n_1$	$X_2 n_2$	N	Nature of force
AB	F	–0.326 F	0	0.674 F	Traction
AD	F	–0.326 F	–0.673 F	0	Neutral
BC	0	–0.326 F	0	–0.326 F	Compression
BD	$-\sqrt{2}$ F	0.462 F	0	–0.952 F	Compression
CD	–F	–0.326F	0	–1.327 F	Compression
AC	0	0.462 F	/	0.426 F	Traction

Table 3.9. *Internal forces on the bars*

The internal force diagram is shown in Figure 3.62.

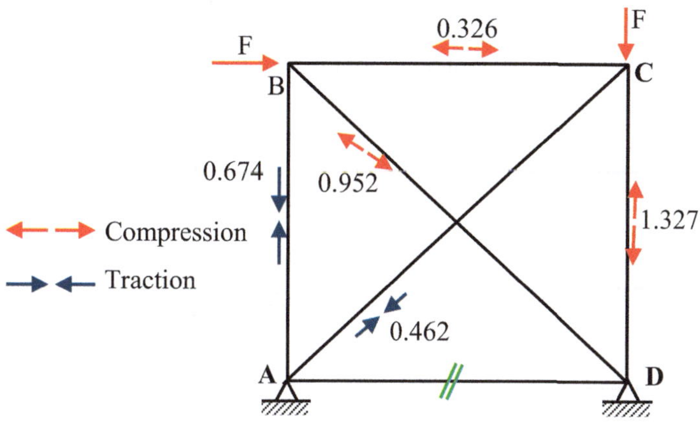

Figure 3.62. *Internal force diagram * (F)*

3.8. Conclusion

In this chapter, we presented the formulation and application of the method of forces to analyze the different types of statically indeterminate structures: trusses, beams and plane frames.

The method of forces is based on applying the deflection compatibility method, which leads to determining the redundant links whose number must be equal to the number of degrees of static indeterminacy of the given structure.

In the first place, it is necessary to construct the equivalent system in which the support reactions or the internal actions must be schematized. Redundant links are the unknowns of the problem, which can be determined using the kinematic compatibility conditions. Once the actions of redundant links have been evaluated, the internal actions can be determined using the superposition of effects principle of the statically determinate system on the one hand, and the unit systems on the other hand.

3.9. Problems

Exercise 1

Determine the degree of static indeterminacy of the following structures:

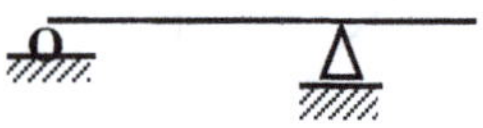
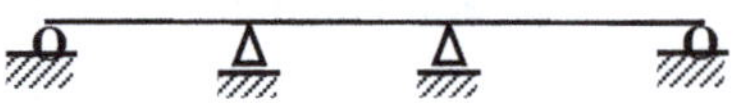
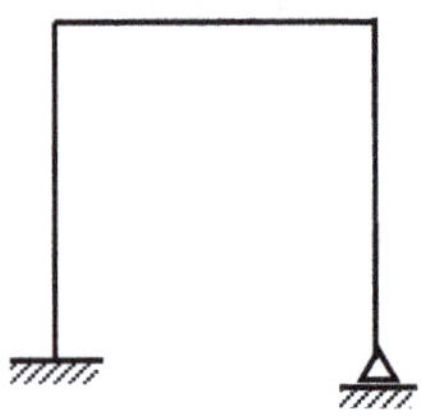
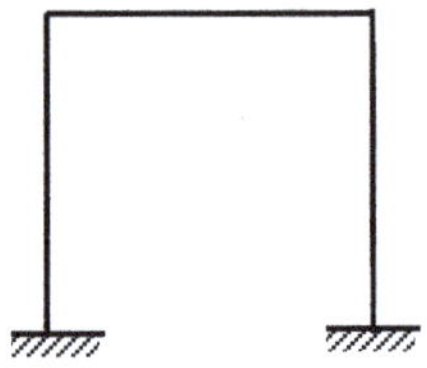
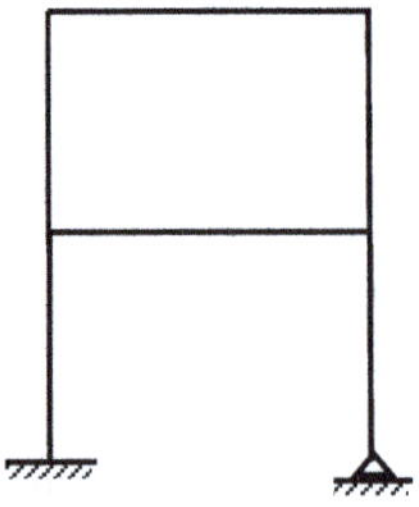
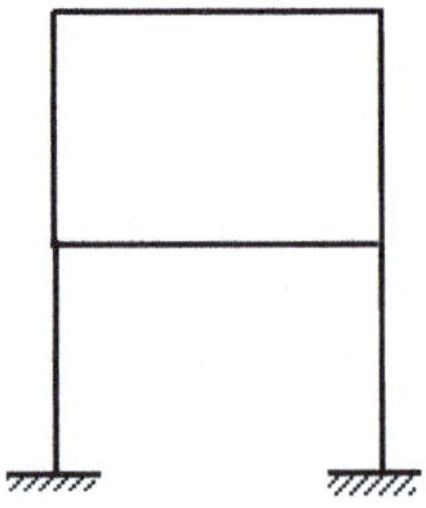
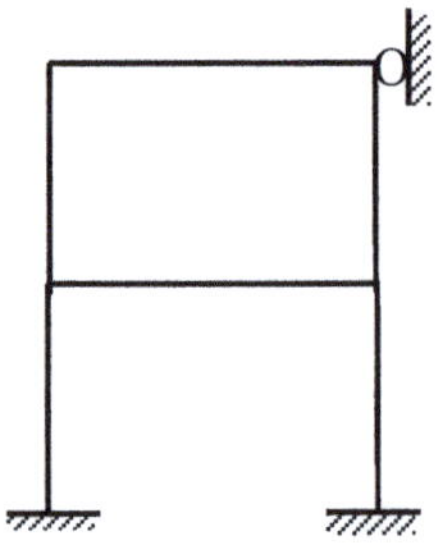
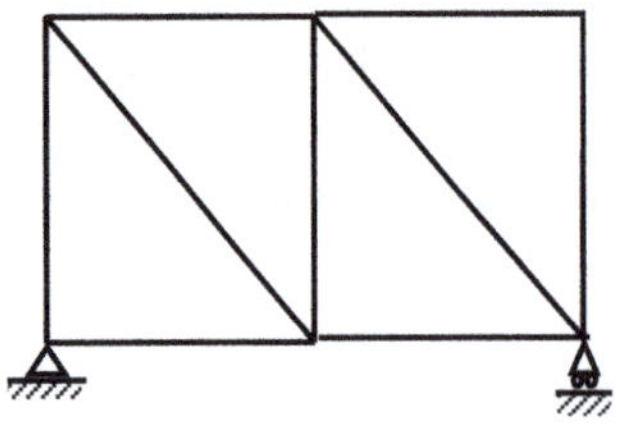
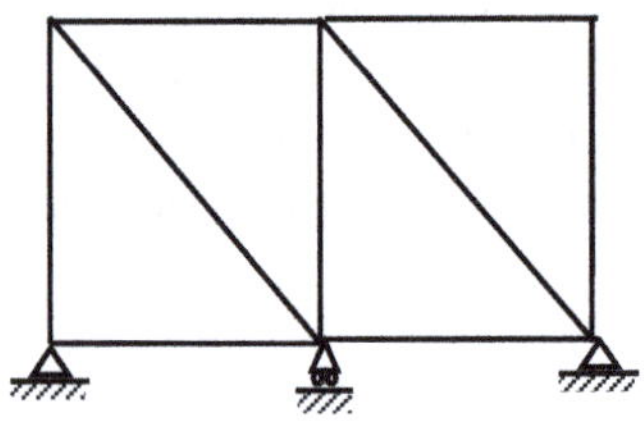

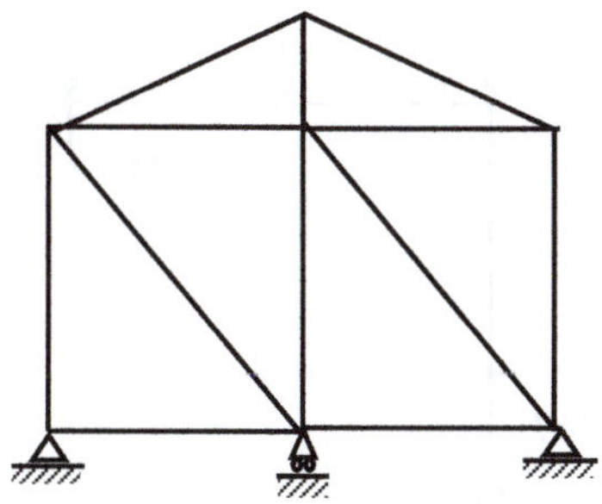

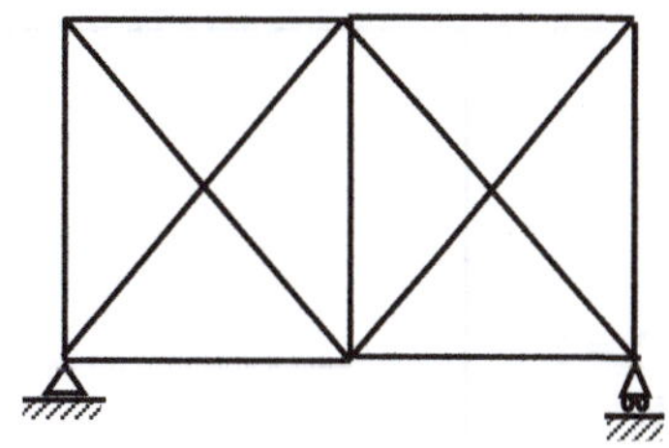

Exercise 2

Determine the unknown indicated for each structure.

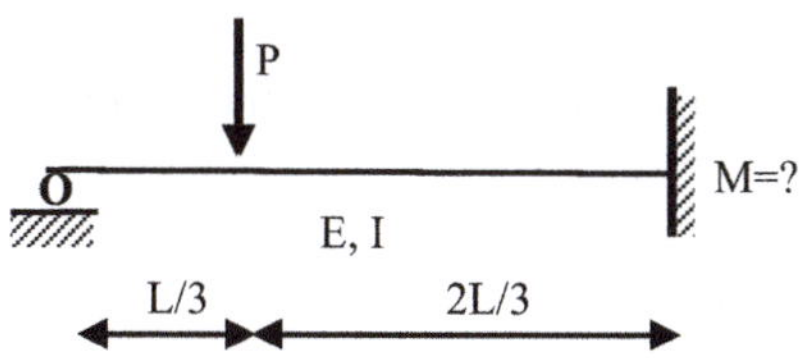

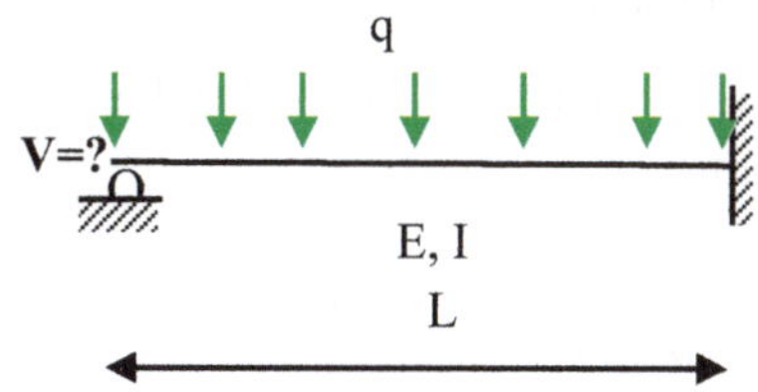

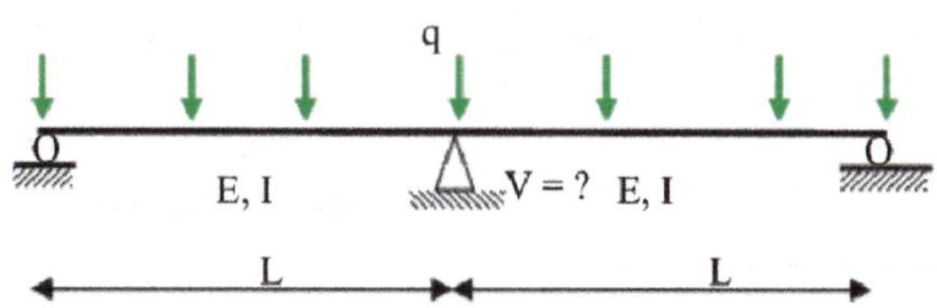

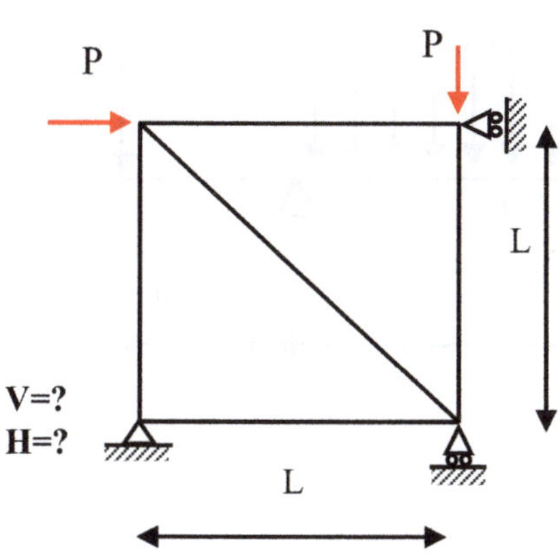

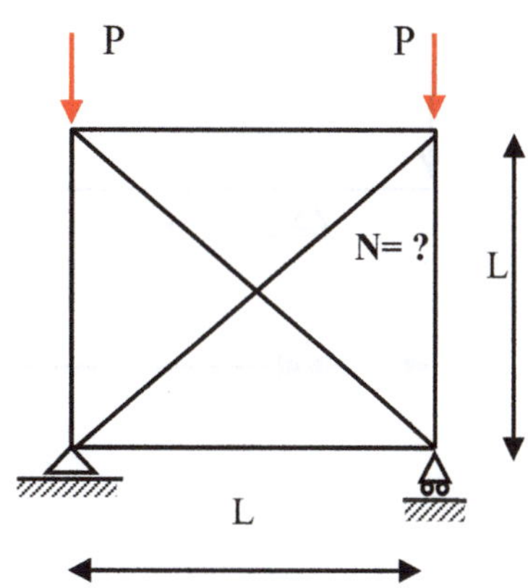

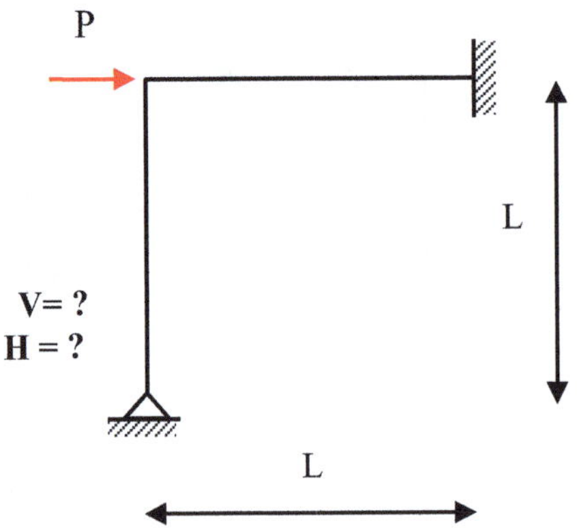

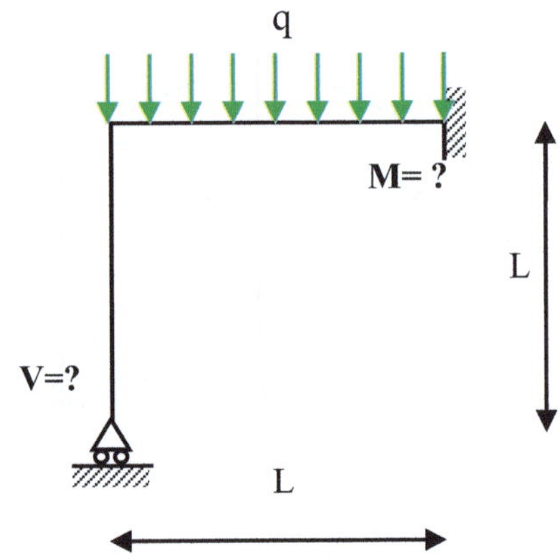

Exercise 3

Draw diagrams of the bending moment and the shear force for the following beams:

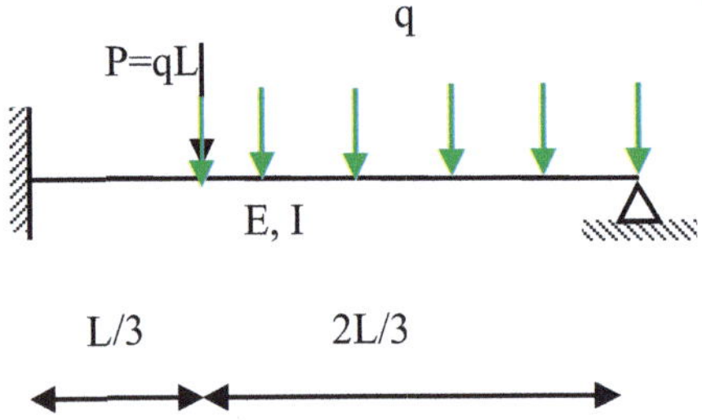

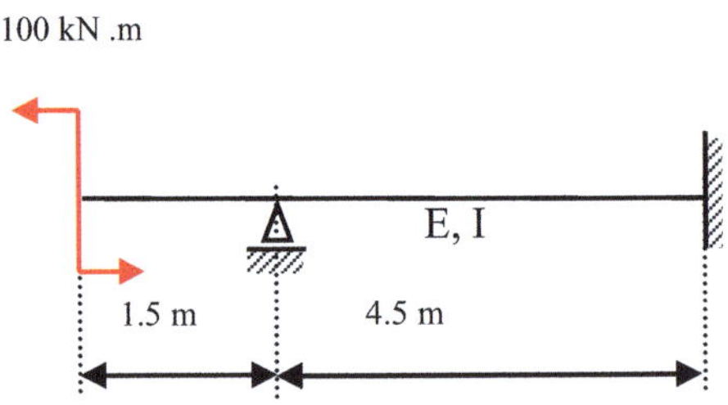

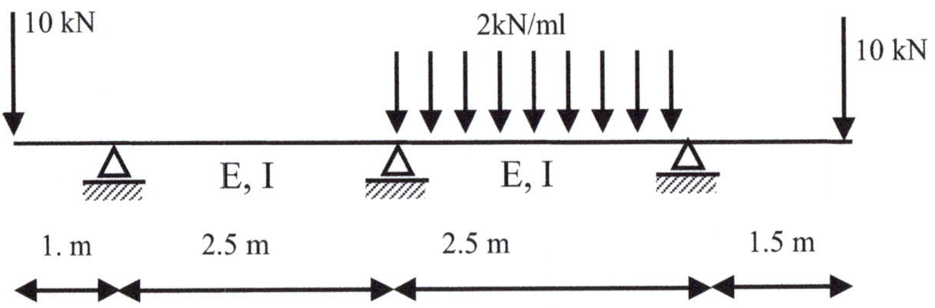

Exercise 4

Calculate the moments at the joints of the following frames.

– Draw diagrams of the bending moment and the shear force.

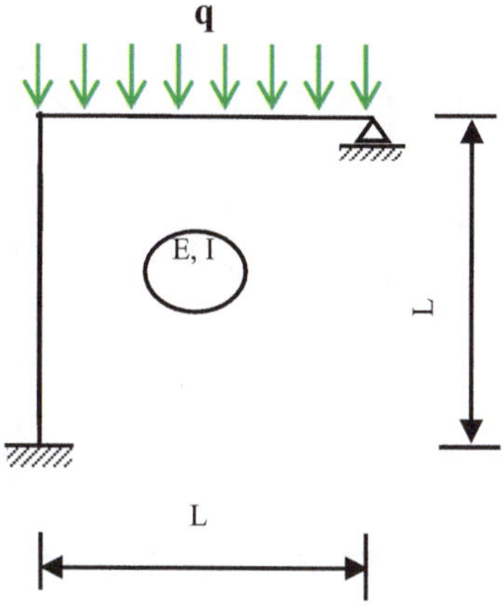

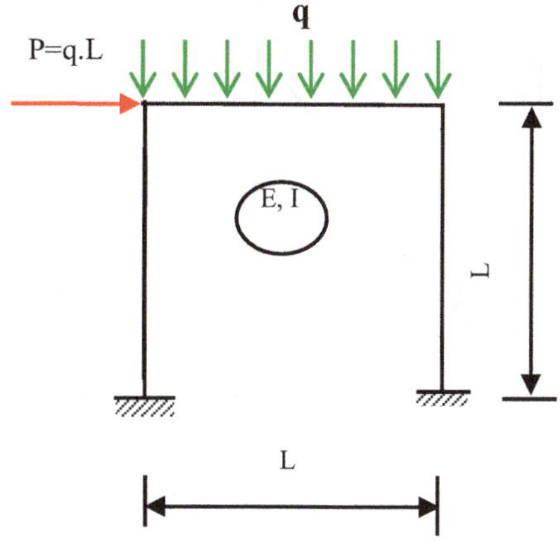

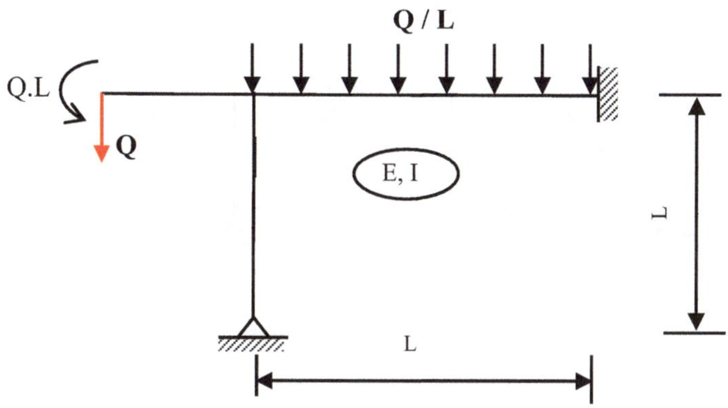

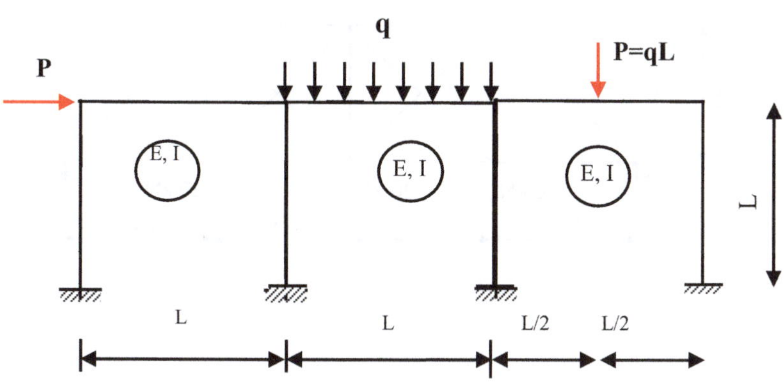

Exercise 5

Calculate the forces of the following truss systems (E Ω = constant):

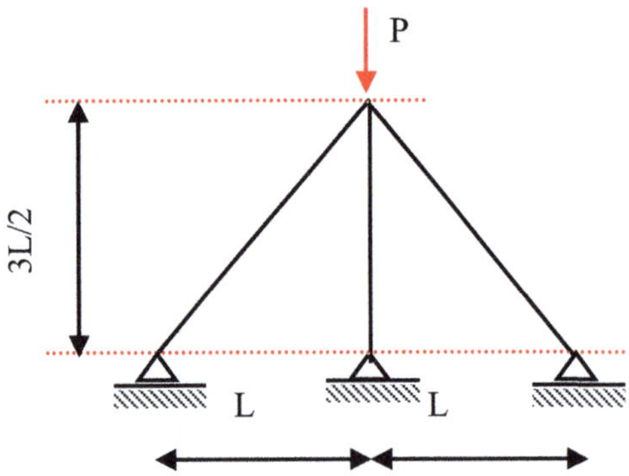

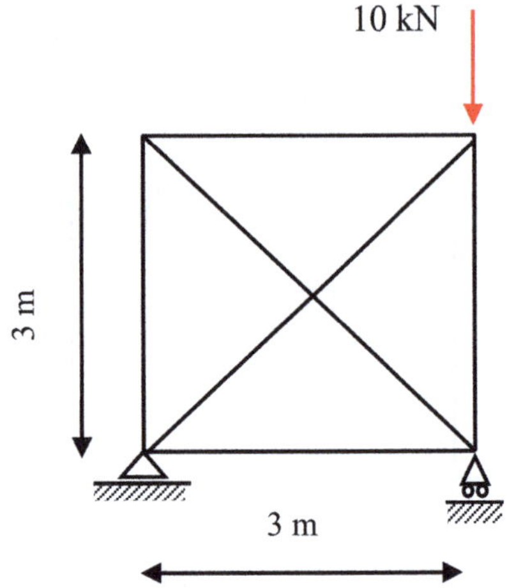

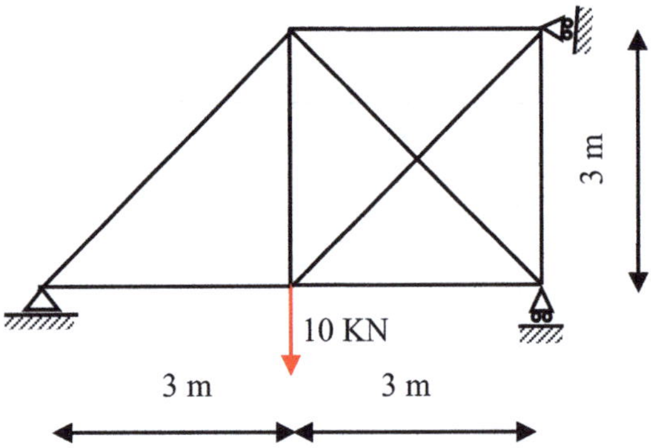

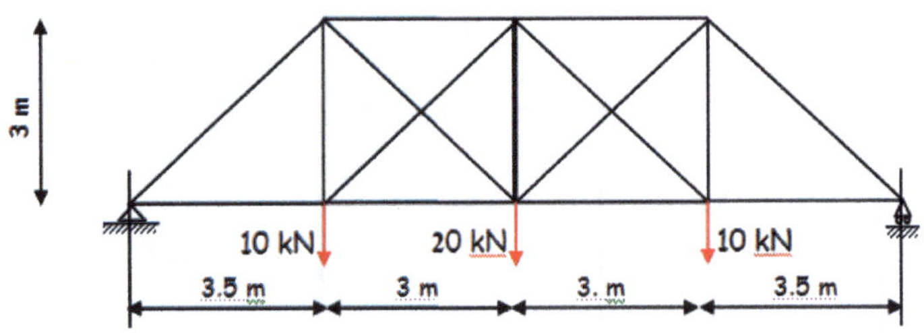

Exercise 6

Evaluate the displacement at point A of the following structures:

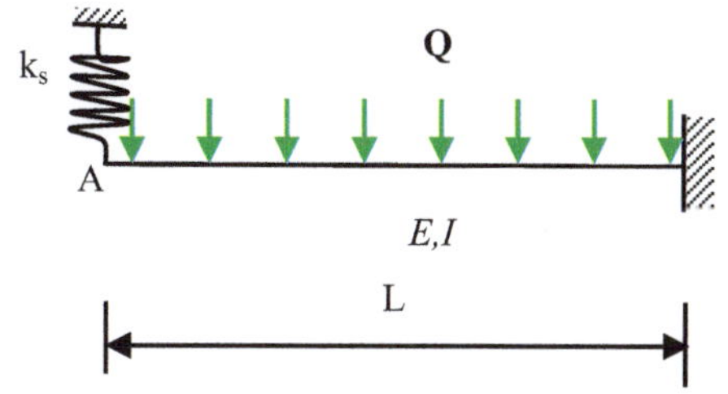

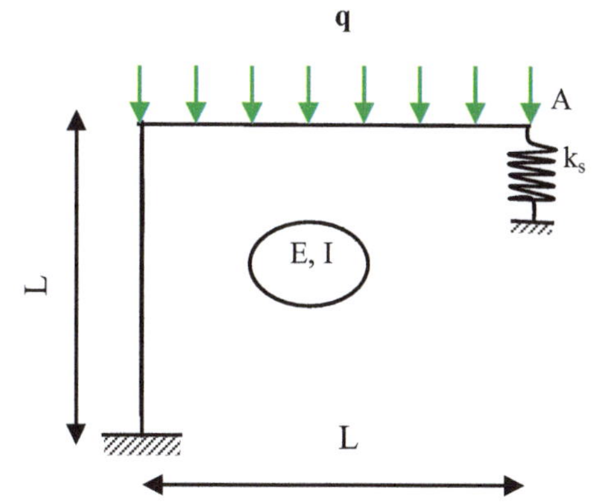

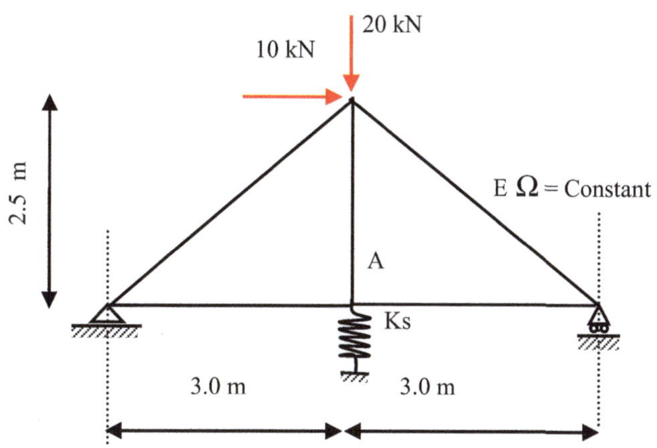

4

Slope-Deflection Method

This chapter hopes to impart the following knowledge:

– formulating the equations of the slope-deflection method of bars with and without support settlements;

– evaluating fixed-end moments;

– the notion of the rigidity factor and the transmission coefficient of a bar according to the kinematic conditions;

– implementing the slope-deflection method when analyzing continuous beams with and without support settlements, frames with sidesway, frames without sidesway and multistory frames.

First, we present the foundations of the slope-deflection method and then we apply them to the analysis of different types of structures such as continuous beams, frames with sidesway and frames without sidesway, and with or without support settlements.

4.1. Relationship between deflections and transmitted moments

The slope-deflection method is based on the description of moments at the ends of an element of a structure according to the slopes at the joints and the applied loads. To derive the formulas of this method, consider a structural element (AB) of a continuous beam (Figure 4.1). When the beam is stressed by an external load or subjected to a support settlement, the element (AB) deforms and generates moments M_{AB} and M_{BA} at the ends of the bar (Figure 4.2). It is assumed that the statically indeterminate beam has constant flexural rigidity (EI) and length (L). Without considering the connection of the beam at its ends and using the superposition of effects principle, we can write:

$$\omega_A = \omega_A^{'} + \theta_A(M_{AB}) + \theta_A(M_{BA}) \quad [4.1a]$$

$$\omega_B = \omega_B^{''} + \theta_B(M_{AB}) + \theta_B(M_{BA}) \quad [4.1b]$$

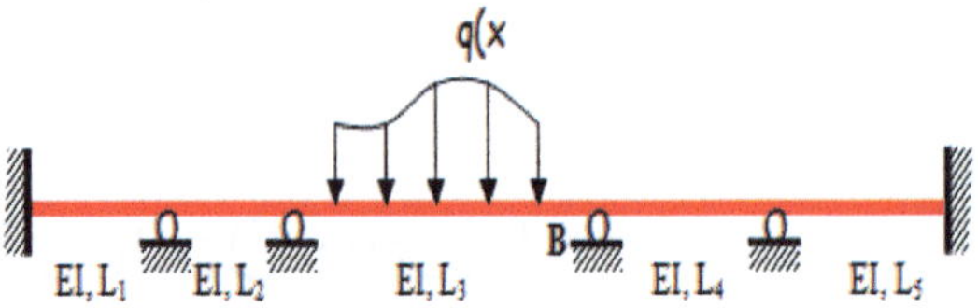

Figure 4.1. *Statically indeterminate beam*[1]

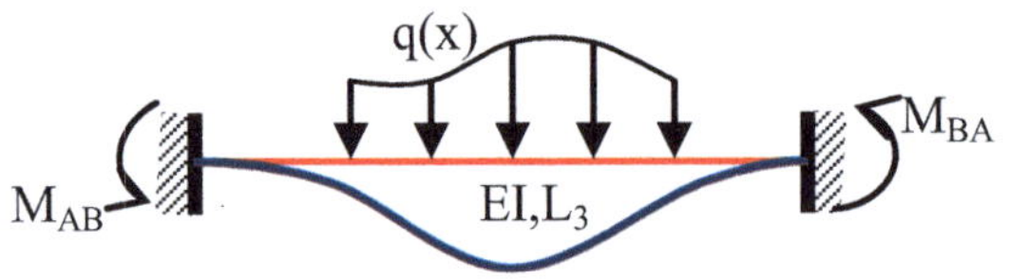

Figure 4.2. *The moments at the ends of the studied beam*

If the support settlement is taken into consideration (Figure 4.3), the system of equations [4.1] is written as

$$\omega_A = \omega_A^{'} + \theta_A(M_{AB}) + \theta_A(M_{BA}) + \beta_A \quad [4.2a]$$

$$\omega_B = \omega_B^{''} + \theta_B(M_{AB}) + \theta_B(M_{BA}) + \beta_B \quad [4.2b]$$

From Figure 4.3, we can say that

$$\beta = \beta_A = \beta_B \quad [4.2c]$$

The relationships of [4.2] become

$$\omega_A = \omega_A^{'} + \theta_A(M_{AB}) + \theta_A(M_{BA}) + \beta \quad [4.3a]$$

$$\omega_B = \omega_B^{''} + \theta_B(M_{AB}) + \theta_B(M_{BA}) + \beta \quad [4.3b]$$

1 All of the figures in this chapter are available to view in full color at www.iste.co.uk/khalfallah/analysis2.zip.

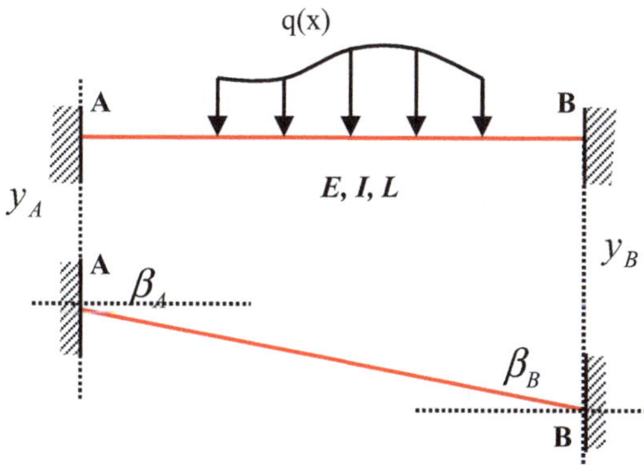

Figure 4.3. *Deflection due to support settlement*

The loading applied to element (AB) generates at ends A and B bending moments called “fixed-end moments”, denoted as γ (Figure 4.4) whose expressions depend on the applied loading. They can be determined using the method of three moments (Chapter 2). In the case of a beam with constant inertia, determining the fixed-end moments γ_{AB} and γ_{BA} is written as

$$\omega_A' + a\gamma_{AB} - b\gamma_{BA} = 0 \quad [4.4a]$$

$$\omega_B'' - b\gamma_{AB} + c\gamma_{BA} = 0 \quad [4.4b]$$

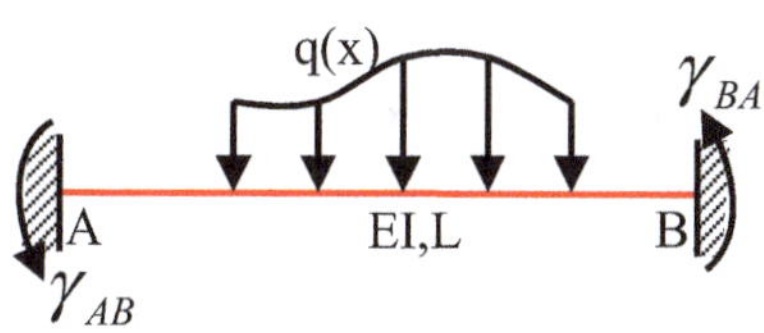

Figure 4.4. *Fixed-end moment*

The expressions of the slopes at the ends of the fundamental system due to the applied load are given as

$$\omega_A' = -a\gamma_{AB} + b\gamma_{BA} \quad [4.5a]$$

$$\omega_B'' = b\gamma_{AB} - c\gamma_{BA} \quad [4.5b]$$

The slopes at joints A and B due to moments M_{AB} and M_{BA} are given as

$$\omega_A = \omega_A' + aM_{AB} - bM_{BA} + \beta \quad [4.6a]$$

$$\omega_B = \omega_B'' - bM_{AB} + cM_{BA} + \beta \quad [4.6b]$$

We substitute the system of equation [4.5] with [4.6]; we obtain the slope expressions of the given system.

$$\omega_A = -a\gamma_{AB} + b\gamma_{BA} + aM_{AB} - bM_{BA} + \beta \quad [4.7a]$$

$$\omega_B = b\gamma_{AB} - c\gamma_{BA} - bM_{AB} + cM_{BA} + \beta \quad [4.7b]$$

The system of equations [4.7] is written in the following form:

$$aM_{AB} - bM_{BA} = \omega_A + a\gamma_{AB} - b\gamma_{BA} - \beta \quad [4.8a]$$

$$-bM_{AB} + cM_{BA} = \omega_B - b\gamma_{AB} + c\gamma_{BA} - \beta \quad [4.8b]$$

Resolving the system of equations [4.8] allows us to determine the moments at the end joints of the bar (AB).

$$M_{AB} = \gamma_{AB} + \frac{c}{ac - b^2}\omega_A + \frac{b}{ac - b^2}\omega_B - \frac{b + c}{ac - b^2}\beta \quad [4.9a]$$

$$M_{BA} = \gamma_{BA} + \frac{b}{b^2 - ac}\omega_A + \frac{a}{b^2 - ac}\omega_B - \frac{b + a}{b^2 - ac}\beta \quad [4.9b]$$

The bending moments at the ends of the bar (AB) are expressed according to the fixed-end moments, the slopes at joints ω_A, ω_B and the support settlement β.

4.2. Fixed-end moments

4.2.1. *Bi-hinged beam*

The formulas of the fixed-end moments due to external loading can be deduced using the method of three moments (Chapter 2). In this case, we consider a bi-fixed support beam subjected to any load (Figure 4.4). The equation of three moments makes it possible to write

$$\omega_A^{'} + a\gamma_{AB} - b\gamma_{BA} = 0$$

$$\omega_B^{''} - b\gamma_{AB} + c\gamma_{BA} = 0$$

The fixed-end moment expressions are given as

$$\gamma_{AB} = \frac{c\omega_A^{'} + b\omega_B^{''}}{b^2 - ac}$$

$$\gamma_{BA} = \frac{b\omega_A^{'} + a\omega_B^{''}}{b^2 - ac}$$

For example, a beam of length L and constant flexural rigidity EI are stressed by a concentrated load applied at distance a from support A (Figure 4.5). The slopes at the ends of a statically determinate beam are

$$\omega_A^{'} = -\frac{Pab(L+b)}{6EIL}$$

$$\omega_B^{''} = \frac{Pab(L+a)}{6EIL}$$

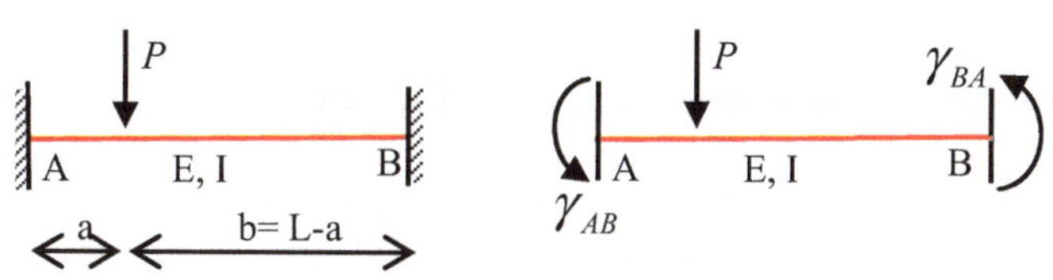

Figure 4.5. *Bi-hinged beam*

Hence the fixed-end moments at A and B are given as

$$\gamma_{AB} = \frac{Pab^2}{L^2}$$

$$\gamma_{BA} = -\frac{Pa^2b}{L^2}$$

4.2.2. *Simply supported beam*

The slope-deflection method equations [4.9] depend on the connection between the bar and the external environment. When this connection is a roller or hinge (Figure 4.6), which allows for a slope, moment M_{AB} is therefore zero. Equation [4.9b] becomes

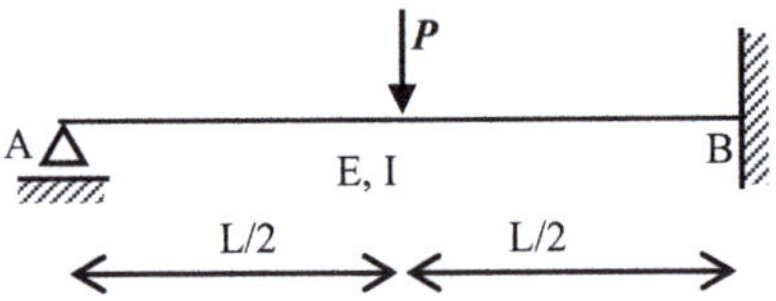

Figure 4.6. *Simply supported beam*

The fixed-end moment at joint B is

$$\gamma_{BA} = -\frac{\omega_B^{"}}{c} = -\frac{3}{16}PL$$

The fixed-end moments of some beams according to types of loading are given in the Appendix.

4.3 Rigidity factor and transmission coefficient

Beam (AB) of constant rigidity (EI) and length L is subject to the effect of couple M_{AB} (Figure 4.7). The beam is not loaded and the support settlement is neglected. We apply the slope-deflection method formula [4.9a].

$$M_{AB} = \frac{c}{ac - b^2}\omega_A \qquad [4.10]$$

Equation [4.10] describes the relationship between the applied couple M_{AB} and the slope of joint A. The connection parameter is defined by the rigidity factor and written as *k*.

$$k_{AB} = \frac{c}{ac - b^2} \quad [4.11]$$

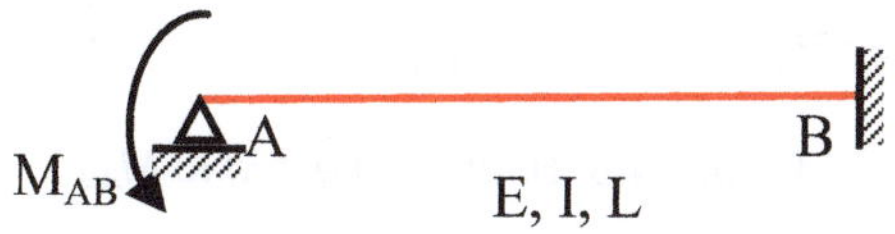

Figure 4.7. *Beam subjected to couple M_{AB}*

Substituting the relationship [4.11] with [4.10], we obtain

$$M_{AB} = k_{AB}\omega_A \quad [4.12]$$

In the case of a beam with constant inertia, we have

$$a = 2\,b = c = \frac{L}{3EI} \quad [4.13]$$

For a bi-fixed support element, the rigidity factor is given as

$$k_{AB} = \frac{c}{ac - b^2} = \frac{2b}{2b.2b - b^2} = \frac{2}{3b} = \frac{4EI}{L} \quad [4.14]$$

The rigidity factor *k* represents the distribution of flexural rigidity (EI) along the studied element.

So we can establish the expression of moment M_{BA} due to the effect of the transmission of moment M_{AB} using equation [4.9b] (Figure 4.8).

$$M_{BA} = \frac{b}{b^2 - ac}\omega_A \quad [4.15]$$

In another way:

$$M_{BA} = \frac{c}{b^2 - ac}\frac{b}{c}\omega_A \qquad [4.16]$$

Figure 4.8. *Beam subjected to M_{AB}*

Substituting the expression [4.10] with [4.16], we obtain

$$M_{BA} = \frac{b}{c}M_{AB} \qquad [4.17]$$

We conclude that moment M_{AB} transmits to the adjacent end a moment $M_{BA} = \frac{b}{c}M_{AB}$

We define the transmission factor by the quantity λ_{AB} which has the expression

$$\lambda_{AB} = \frac{b}{c} \qquad [4.18]$$

For a beam with constant inertia, the transmission factor is

$$\lambda_{AB} = \frac{b}{2b} = \frac{1}{2} \qquad [4.19]$$

Generally, we can write the expressions of moments at the ends of the joints [4.9] by introducing the rigidity factors and the transmission coefficients by

$$M_{ij} = \gamma_{ij} + k_{ij}\omega_i + k_{ji}\lambda_{ji}\omega_j - k_{ij}(1+\lambda_{ij})\beta \qquad [4.20a]$$

$$M_{ji} = \gamma_{ji} + k_{ij}\lambda_{ij}\omega_i + k_{ji}\omega_j - k_{ji}(1+\lambda_{ji})\beta \qquad [4.20b]$$

Particular cases

– *Support B is fixed (Figure 4.9)*

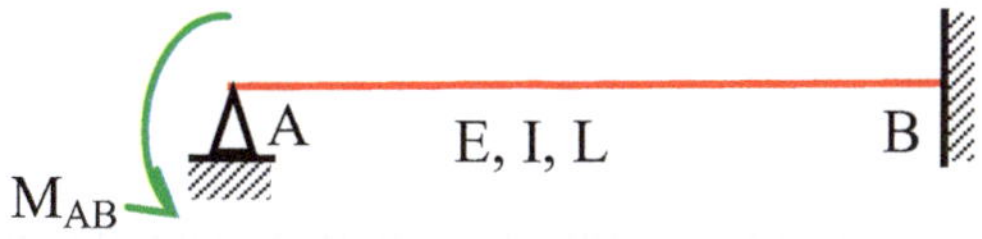

Figure 4.9. *Beam fixed in B*

Using relationships [4.14] and [4.19], the rigidity factor and the transmission coefficient of bar (AB) are given as

$$k_{AB} = \frac{4EI}{L}$$

$$\lambda_{AB} = \frac{1}{2}$$

– *Support B is a roller (Figure 4.10)*

$$M_{AB} = k'_{AB}\omega_A + k'_{AB}\lambda'_{AB}.\omega_B \qquad [4.21]$$

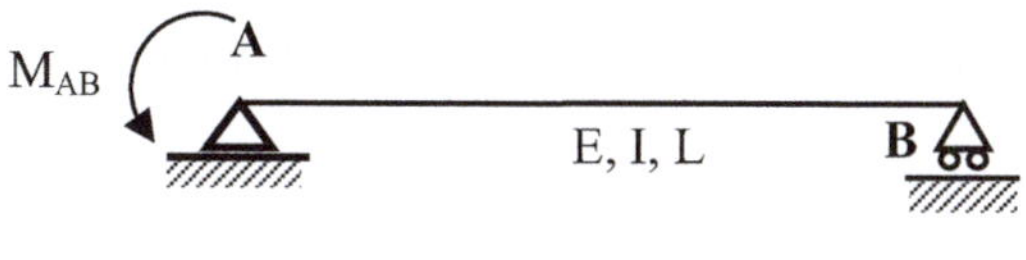

Figure 4.10. *Simply supported beam in B*

The index (') is introduced to describe the nature of the beam's connection with the external environment at point B.

The quantity $k'_{AB}.\lambda'_{AB}.\omega_B$ represents the moment transmitted due to the effect of M_{AB}. Knowing that this moment is zero means that the effect of the transmission did not take place. This allows us to write

$$\lambda'_{AB} = 0$$

The relationship [4.21] is written as

$$M_{AB} = k'_{AB}.\omega_A \tag{4.22}$$

$$M_{BA} = 0 \tag{4.23}$$

It is necessary to evaluate the slope at support A using relationship [4.22] under the effect of the applied moment M_{AB} (Figure 4.11).

Figure 4.11. *Deformed beam*

The expression of the bending moment becomes

$$M(x) = - M_{AB}(1-\frac{x}{L}) \tag{4.24}$$

Applying a unit moment at support A, we obtain the bending moment variation.

$$m(x) = -(1-\frac{x}{L}) \tag{4.25}$$

The expression of the slope ω_A using the method of virtual work can be deduced by

$$\omega_A = \int_0^L \frac{M(x)m(x)}{EI}dx = \int_0^L \frac{M_{AB}(1-\frac{x}{L})(1-\frac{x}{L})}{EI}dx = M_{AB}a \tag{4.26}$$

Using relationships [4.22] and [4.26], we can say

$$M_{AB} = \frac{\omega_A}{a} = \frac{3EI}{L}\omega_A \tag{4.27}$$

Hence the rigidity factor of the bar (AB) is

$$k'_{AB} = \frac{3EI}{L} \tag{4.28}$$

Moment M_{BA} is zero, which means that no transmission $\lambda'_{AB} = 0$ of the moment can be made due to the effect of the applied moment M_{AB}.

– *Support settlement*

The effect of the support settlement can be shown in the formation of moments at the ends of the studied beam (Figure 4.12).

$$M_{AB} = -k_{AB}(1+\lambda_{AB})\beta \quad [4.29]$$

$$M_{BA} = -k_{BA}(1+\lambda_{BA})\beta \quad [4.30]$$

For small slopes, we can write

$$tg\beta \approx \beta = -\frac{y_B - y_A}{L} \quad [4.31]$$

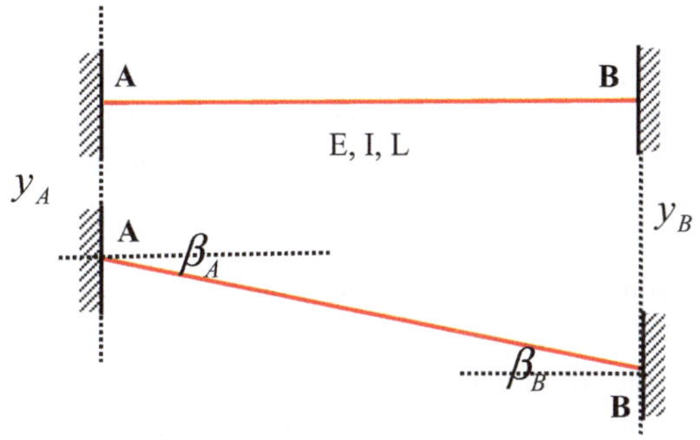

Figure 4.12. *Support settlement*

We put $\Delta = y_B - y_A$

$$M_{AB} = k_{AB}(1+\lambda_{AB})\frac{\Delta}{L} \quad [4.32]$$

$$M_{BA} = k_{BA}(1+\lambda_{BA})\frac{\Delta}{L} \quad [4.33]$$

4.4. Beam analysis

The steps for continuous beam analysis using the slope-deflection method can be grouped into:

– the identification of unknown joint slopes of a continuous beam, which will subsequently be the unknowns of the problem;

– we calculate the rigidity factors, the transmission coefficients and the fixed-end moments of each loaded span;

– if the support settlement is taken into account, the corresponding slopes are also calculated by dividing the relative translation over the length of each span;

– we write the expressions of moments at the ends of each bar of the continuous beam;

– then, the equilibrium conditions of the bending moments are applied to each joint that can be rotated;

– resolving the system of equations, which has been constructed, makes it possible to determine the slopes at the joints;

– the moments at the ends of each bar of the continuous beam are recalculated by substituting the slopes calculated in the previous step in the expression of the moment at each end. In this case, we can check the equilibrium of the joints by adding up the moments at each joint.

In general, we formulate the expressions of the bending moment, shear force and support reactions. The bending moment is established from the results obtained in the last step, while the shear force is evaluated using external loading and the moments at the ends of each bar. Applying the equilibrium equations for each bar leads to calculating the shear forces. The support reactions are determined using the equilibrium of the joints of the structure.

4.4.1. *Single span beam*

The single-span beam was analyzed by the method of three moments (Figure 2.8) and then by the method of forces (Figure 3.2).

– *Rigidity factor*

$$K'_{AB} = \frac{3EI}{L}$$

– *Transmission coefficient*

$$\lambda'_{AB} = 0.$$

– *Fixed-end moments*

To calculate the fixed-end moment γ_{AB}, we use the method of three moments.

$$\omega_A = \omega'_A + a\gamma_{AB} + b\gamma_{BA} = 0$$

Knowing that $\gamma_{BA} = 0.$ and $\omega'_A = -\frac{5}{81}\frac{PL^2}{EI}$

$$\gamma_{AB} = \frac{5}{27}PL$$

The expression of the bending moment is

$$M_{AB} = \gamma_{AB} + K'_{AB}\omega_A + K'_{AB}\lambda'_{AB}\omega_B$$

Substituting the rigidity factor values K'_{AB} and the transmission coefficient λ'_{AB}, we obtain

$$M_{AB} = \frac{5}{27}PL$$

$$M_{BA} = 0.$$

The equilibrium in relation to point A is

$$V_B L + M_{AB} - \frac{PL}{3} = 0.$$

Hence the support reaction at B is

$$V_B = \frac{P}{3} - \frac{M_{AB}}{L} = \frac{4}{27}P$$

The vertical equilibrium leads to

$$V_A = P - V_B = \frac{23}{27}P$$

Figure 4.14 shows support reactions.

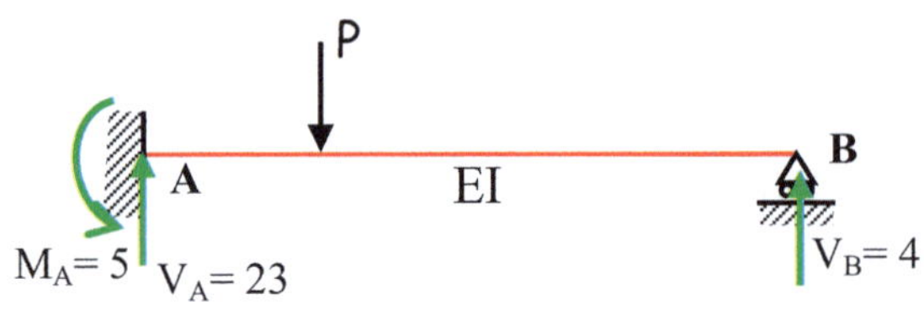

Figure 4.13. *Support reaction* $*(\frac{P}{27})$ *and moment* $*(\frac{PL}{27})$

The beam deflection is given by Figure 4.15.

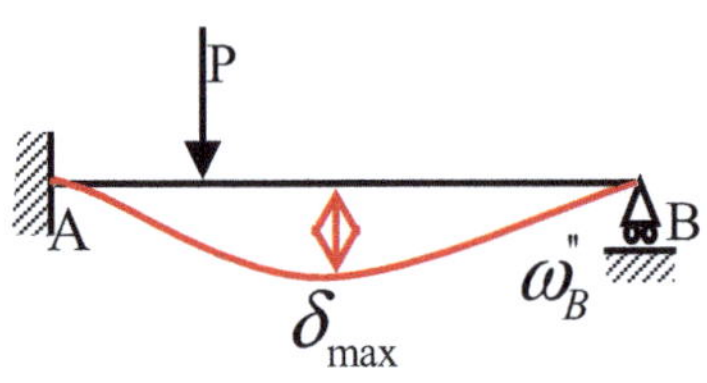

Figure 4.14. *Deformed beam*

4.4.2. Continuous beam

This involves analyzing the continuous beam (Figure 2.21). We assume that the flexural rigidity (EI) is constant. The beam is subjected to a uniformly distributed load q and by a concentrated force P. This example is examined using the method of three moments (Figure 2.21) and the method of forces (Figure 3.5).

– *Rigidity factors*

$$K_{AB} = K_{BA} = K_{BC} = K_{CB} = \frac{4EI}{L}$$

– *Transmission coefficients*

$$\lambda_{AB} = \lambda_{BA} = \lambda_{BC} = \lambda_{CB} = \frac{1}{2}$$

– *Fixed-end moments*

$$\gamma_{AB} = -\gamma_{BC} = \frac{1}{12}qL^2$$

$$\gamma_{BC} = \frac{4}{27}PL = \frac{4}{27}qL^2$$

$$\gamma_{CB} = -\frac{2}{27}PL = -\frac{2}{27}qL^2$$

– *Unknowns of problem*

In the given structure (Figure 2.21), the slopes at joints A and C are zero. The only unknown is the slope at joint B.

– *Equilibrium equations*

Resolving this problem requires one single equation.

We consider the equilibrium at joint B.

$$\text{MB} = 0:\ \text{M}_{\text{BA}} + \text{M}_{\text{BC}} = 0$$

– *Expressions of moments at the ends*

The formulas that give the moments at the joints of each span are

$$M_{AB} = \frac{qL^2}{12} + \frac{2EI}{L}\omega_B$$

$$M_{BA} = -\frac{qL^2}{12} + \frac{4EI}{L}\omega_B$$

$$M_{BC} = \frac{4}{27}qL^2 + \frac{4EI}{L}\omega_B$$

$$M_{CB} = -\frac{2}{27}qL^2 + \frac{2EI}{L}\omega_B$$

– *Calculation of the slope at joint B*

To calculate slope ω_B, we substitute the expressions of moments in the previous equilibrium equation. We obtain

$$-\frac{qL^2}{12} + \frac{4EI}{L}\omega_B + \frac{4qL^2}{27} + \frac{4EI}{L} = 0$$

Hence

$$\omega_B = -\frac{7}{864}\frac{qL^3}{EI}$$

– *Moments at the ends of bars*

Substituting the expressions of slopes ω_B with the expressions giving the bending moments, we obtain

$$M_{AB} = \frac{qL^2}{12} + \frac{2EI}{L}\omega_B = \frac{29}{432}qL^2$$

$$M_{BA} = -\frac{qL^2}{12} + \frac{4EI}{L}\omega_B = -\frac{50}{432}qL^2$$

$$M_{BC} = \frac{4}{27}qL^2 + \frac{4EI}{L}\omega_B = \frac{50}{432}qL^2$$

$$M_{CB} = -\frac{2}{27}qL^2 + \frac{2EI}{L}\omega_B = -\frac{39}{432}qL^2$$

The diagrams of the bending moment and the shear force are shown in Figures 2.21 and 2.22.

Applying the equilibrium equations at each span leads to

$$V_A = \frac{M_{BA} - M_{AB}}{L} + \frac{1}{2} qL = \frac{192}{432} qL$$

$$V_B^{(1)} = \frac{M_{BA} - M_{AB}}{L} + \frac{1}{2} qL = \frac{237}{432} qL$$

$$V_B^{(2)} = \frac{M_{CB} - M_{BC}}{L} + \frac{2P}{3} = \frac{299}{432} qL$$

$$V_B = V_B^{(1)} + V_B^{(2)} = \frac{536}{432} qL$$

$$V_C = \frac{M_{CB} - M_{BC}}{L} + \frac{P}{3} = \frac{133}{432} qL$$

Figure 4.15 shows the deflection and the support reactions.

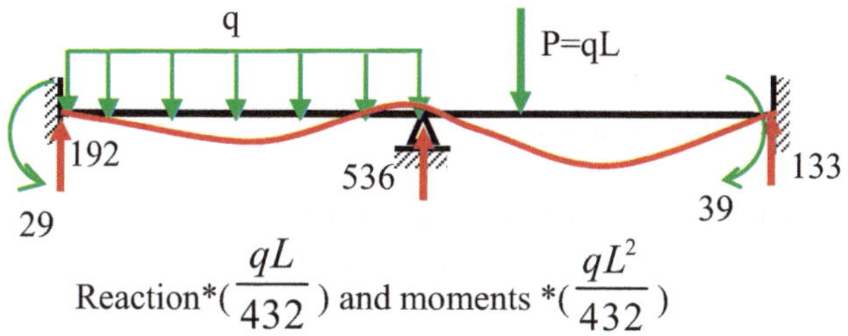

Figure 4.15. *Deflection of the beam and support reaction*

4.4.3. Continuous beam with cantilever

We analyze the continuous beam with a cantilever as shown in Figure 2.27. Flexural rigidity (EI) is constant for both spans.

– *Rigidity factors*

$$K_{12} = K_{21} = K_{23} = K_{32} = \frac{4EI}{L}$$

– *Transmission coefficients*

$$\lambda_{12} = \lambda_{21} = \lambda_{23} = \lambda_{32} = \frac{1}{2}$$

– *Fixed-end moments*

For simplification reasons, we put P = qL.

$$\gamma_{23} = -\gamma_{32} = \frac{1}{12} q.L^2$$

– *Equilibrium equations*

To analyze the given beam (Figure 2.27), it is necessary to build a system of equations which allows the slopes of joints ω_2 and ω_3 to be quantified.

The equilibrium of the beam allows us to write

$$M_{21} + M_{23} = 0$$

$$M_{32} + \frac{PL}{5} = 0.$$

– *Slope-deflection method equation*

The direct application of the slope-deflection method and the use of formulas, which give the moments at the joints of each span, result in

$$M_{12} = \frac{2EI}{L}\omega_2$$

$$M_{21} = \frac{4EI}{L}\omega_2$$

$$M_{23} = \frac{1}{12} q.L^2 + \frac{4EI}{L}\omega_2 + \frac{2EI}{L}\omega_3$$

$$M_{32} = -\frac{1}{12} q.L^2 + \frac{2EI}{L}\omega_2 + \frac{4EI}{L}\omega_3$$

– *Calculating slopes at joints*

The equilibrium equations established at joints 2 and 3 are written as

$$8\omega_2 + 2\omega_3 = -\frac{qL^3}{12EI}$$

$$2\omega_2 + 4\omega_3 = \frac{qL^3}{12EI} - \frac{PL^2}{5EI}$$

The resolution of the equation leads to calculating the slopes

$$\omega_2 = -\frac{1}{280}\frac{qL^3}{EI}$$

$$\omega_3 = -\frac{23}{840}\frac{qL^3}{EI}$$

– *Moments at the ends of bars*

Substituting the expressions of slopes ω_2 and ω_3 with the expressions of moments, we obtain

$$M_{12} = -\frac{1}{140}q.L^2$$

$$M_{21} = -\frac{1}{70}q.L^2$$

$$M_{23} = \frac{1}{12}q.L^2 - \frac{4}{280}q.L^2 - \frac{46}{840}q.L^2 = \frac{1}{70}q.L^2$$

$$M_{32} = -\frac{1}{12}q.L^2 - \frac{2}{280}q.L^2 - \frac{96}{840}q.L^2 = -\frac{1}{5}q.L^2$$

The results obtained are identical to those obtained in Chapters 2 and 3.

From the shear force diagram (Figure 2.29), we can evaluate the support reactions of the beam, which are:

$$\text{R1} = \frac{-3}{140}qL$$

$$R2 = \frac{47}{140}qL$$

$$R3 = \frac{236}{140}qL$$

The diagram of the support reactions and deflection is represented in Figure 4.16.

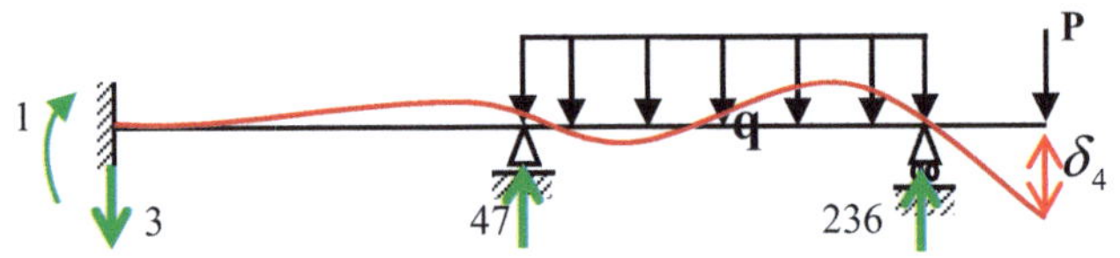

Figure 4.16. *Support reactions: load (* $\frac{1}{140}qL$) and moment (* $\frac{1}{140}qL^2$)*

4.4.4. *Beam with support settlements*

Determine the support reactions and calculate the moments at the ends of each bar of a continuous beam subjected to support settlements B and C of Δ_1 and Δ_2 using the slope-deflection method (Figure 2.24).

Draw the bending moment diagram for $\Delta_1 = 2\Delta_2$.

The unknowns of the problem are effectively the slopes ω_B and ω_C.

– *Rigidity factors*

$$K_{AB} = K_{BA} = K_{BC} = K_{CB} = \frac{4EI}{L}$$

$$K_{CD} = \frac{3EI}{L}$$

– *Transmission coefficients*

$$\lambda_{AB} = \lambda_{BA} = \lambda_{CD} = \lambda_{CD} = \frac{1}{2}$$

$$\lambda'_{CD} = 0$$

– *Fixed-end moments*

No external load is applied along the beam, so fixed-end moments are zero.

– *Equilibrium equations*

The equilibrium of the beam is written as

MBA + MBC = 0

MCB + MCD = 0

– *Slope-deflection method equation*

Applying the slope-deflection method equations for each span, we obtain

$$M_{AB} = \frac{2EI}{L}\omega_B - \frac{6EI}{L}\beta_1$$

$$M_{BA} = \frac{4EI}{L}\omega_B - \frac{6EI}{L}\beta_1$$

$$M_{BC} = \frac{4EI}{L}\omega_B + \frac{2EI}{L}\omega_C - \frac{6EI}{L}\beta_2$$

$$M_{CB} = \frac{2EI}{L}\omega_B + \frac{4EI}{L}\omega_C - \frac{6EI}{L}\beta_2$$

$$M_{CD} = \frac{3EI}{L}\omega_C + \frac{3EI}{L}\beta_3$$

$$M_{DC} = 0$$

Knowing that $\beta_1 = -\frac{\Delta_1}{L} = -\frac{2\Delta_2}{L}$, $\beta_2 = \frac{\Delta_1 - \Delta_2}{L} = \frac{\Delta_2}{L}$ and $\beta_3 = \frac{\Delta_2}{L}$, the new expressions of the bending moments at the ends of each bar are given as

$$M_{AB} = \frac{2EI}{L}\omega_B + \frac{12EI}{L^2}\Delta_2$$

$$M_{BA}=\frac{4EI}{L}\omega_B+\frac{12EI}{L^2}\Delta_2$$

$$M_{BC}=\frac{4EI}{L}\omega_B+\frac{2EI}{L}\omega_C-\frac{6EI}{L^2}\Delta_2$$

$$M_{CB}=\frac{2EI}{L}\omega_B+\frac{4EI}{L}\omega_C-\frac{6EI}{L^2}\Delta_2$$

$$M_{CD}=\frac{3EI}{L}\omega_C-\frac{3EI}{L^2}\Delta_2$$

– *Calculating slopes at joints*

Applying the equilibrium equations to joints B and C, we deduce the following system of equations:

$$8\omega_B+2\omega_C=6\frac{\Delta_2}{L}$$

$$2\omega_B+7\omega_C=-9\frac{\Delta_2}{L}$$

Resolving the system of equations leads to evaluating the slopes at joints:

$$\omega_B=-\frac{15}{13}\frac{\Delta_2}{L}$$

$$\omega_C=\frac{21}{13}\frac{\Delta_2}{L}$$

– *Moments at the ends of bars*

The moments at the ends of each bar are as follows

$$M_{AB}=\frac{126}{13}\frac{EI}{L^2}\Delta_2$$

$$M_{BA}=\frac{96}{13}\frac{EI}{L^2}\Delta_2$$

$$M_{BC} = -\frac{96}{13}\frac{EI}{L^2}\Delta_2$$

$$M_{CB} = -\frac{24}{13}\frac{EI}{L^2}\Delta_2$$

$$M_{CD} = \frac{24}{13}\frac{EI}{L^2}\Delta_2$$

The diagrams of the bending moment and the shear force are shown, respectively, in Figures 2.25 and 2.26.

From the shear force diagram (Figure 2.26), we can evaluate the support reactions.

$$R_A = \frac{222}{13}\frac{EI}{L^3}\Delta_2$$

$$R_B = -\frac{222}{13}\frac{EI}{L^3}\Delta_2 + (-\frac{120}{13}\frac{EI}{L^3}\Delta_2) = -\frac{342}{13}\frac{EI}{L^3}\Delta_2$$

$$R_C = \frac{144}{13}\frac{EI}{L^3}\Delta_2$$

$$R_D = -\frac{24}{13}\frac{EI}{L^3}\Delta$$

The bending moment diagram is shown in Figure 4.17.

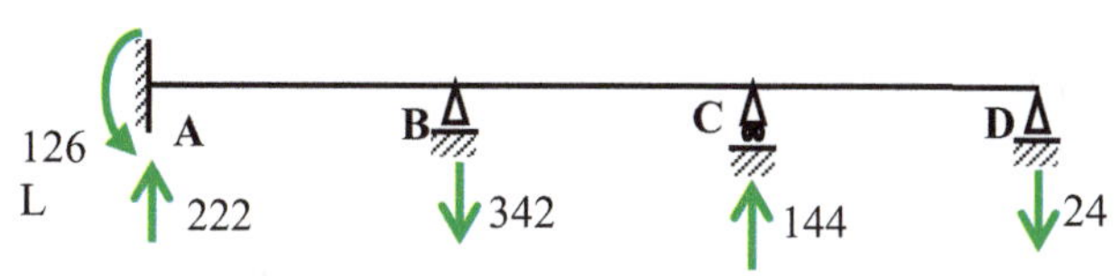

Figure 4.17. *Support actions: reactions* * $(\frac{1}{13}\frac{EI}{L^3}\Delta_2)$ *and moments* * $(\frac{1}{13}\frac{EI}{L^2}\Delta_2)$

4.4.5. *Beam subjected to a moment*

We analyze the beam in Figure 2.30 using the slope-deflection method. We assume that the flexural rigidity EI and the length of the spans are constant. The purpose of this example is to present the distribution of moment M_4 = 15 kN·m along the continuous beam using the slope-deflection method.

– *Rigidity factors*

$$K_{21} = K_{34} = \frac{3EI}{L}$$

$$K_{23} = K_{32} = \frac{4EI}{L}$$

– *Transmission coefficients*

$$\lambda_{21} = \lambda_{34} = \frac{3EI}{L}$$

$$\lambda_{23} = \lambda_{32} = \frac{4EI}{L}$$

– *Calculation of moments*

The beam is extended (Figure 2.40) by length $L_5 = 0$ (Figure 4.18).

$$M_{12} = 0$$

$$M_{21} = \frac{3EI}{L}\omega_2$$

$$M_{23} = \frac{4EI}{L}\omega_2 + \frac{2EI}{L}\omega_3$$

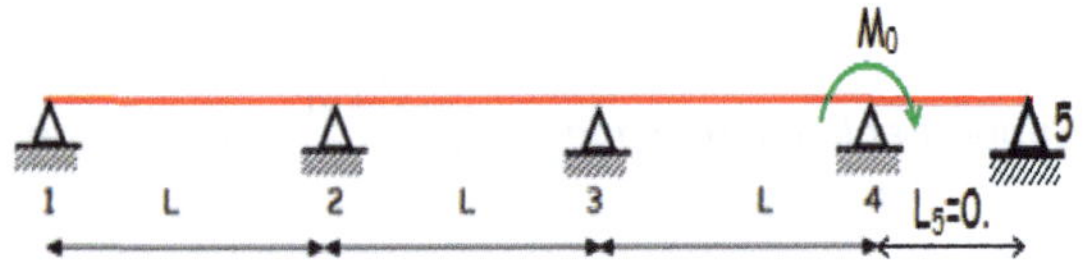

Figure 4.18. *Beam extension*

$$M_{32} = \frac{4EI}{L}\omega_3 + \frac{2EI}{L}\omega_2$$

$$M_{34} = \frac{4EI}{L}\omega_3 + \frac{2EI}{L}\omega_4$$

$$M_{43} = \frac{2EI}{L}\omega_3 + \frac{4EI}{L}\omega_4$$

$$M_{45} = 0.$$

– *Joint equilibrium*

Joint 2

$$M_{21} + M_{23} = 0 \; : \; 7\omega_2 + 2\omega_3 = 0.$$

$$M_{32} + M_{34} = 0 \; : \; 2\omega_2 + 8\omega_3 + 2\omega_4 = 0.$$

$$M_{43} + M_{45} = -M_0 \; : \; 2\omega_3 + 4\omega_4 = \frac{M_0 L}{EI}$$

Resolving the previous equations leads to the determination of the slopes of joints 2–4.

Knowing that:

$$\omega_2 = \frac{2}{7}\omega_3$$

$$\omega_4 = \frac{M_0 L}{4EI} - \frac{\omega_3}{2}$$

Hence

$$\omega_3 = -\frac{7M_0 L}{90EI}$$

and consequently,

$$\omega_2 = \frac{M_0 L}{45EI}$$

$$\omega_4 = \frac{13 M_0 L}{45EI}$$

$$M_{12} = 0$$

$$M_{21} = \frac{3EI}{L}\omega_2 = \frac{M_0}{15} = 1\,\text{kN.m}$$

$$M_{23} = \frac{4EI}{L}\omega_2 + \frac{2EI}{L}\omega_3 = -\frac{M_0}{15} = -1\,\text{kN.m}$$

$$M_{32} = \frac{4EI}{L}\omega_3 + \frac{2EI}{L}\omega_2 = -\frac{4M_0}{15} = -4\,\text{kN.m}$$

$$M_{34} = \frac{4EI}{L}\omega_3 + \frac{2EI}{L}\omega_4 = \frac{4}{15}M_0 = 4\,\text{kN.m}$$

$$M_{43} = \frac{2EI}{L}\omega_3 + \frac{4EI}{L}\omega_4 = M_0 = 15\,\text{kN.m}$$

$$M_{45} = 0.$$

The diagrams of the bending moment and the shear force are shown in Figures 2.41 and 2.42, respectively. We associate the moments at the ends of each bar (Figure 4.19).

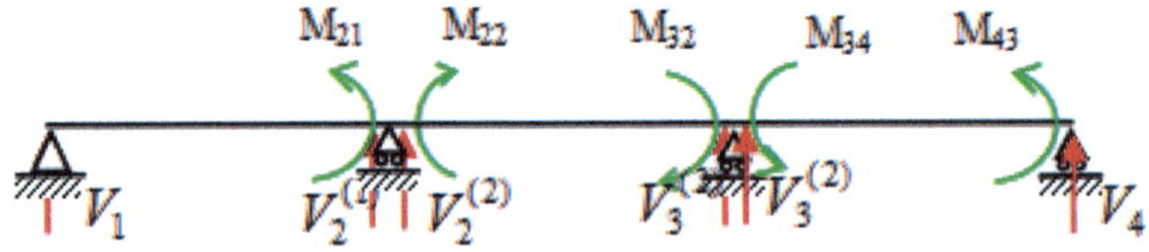

Figure 4.19. *The moments at the ends of each beam*

Applying the equilibrium equations at each span leads to:

$$V_1 = \frac{M_{21}}{L} = \frac{1}{L}$$

$$V_2^{(1)} = \frac{-M_{21}}{L} = -\frac{1}{L}$$

$$V_2^{(2)} = -\frac{M_{23} + M_{32}}{L} = -\frac{5}{L}$$

$$V_2 = V_2^{(1)} + V_2^{(2)} = -\frac{6}{L}$$

$$V_3^{(2)} = \frac{M_{23} + M_{32}}{L} = \frac{5}{L}$$

$$V_3^{(3)} = \frac{M_{34} + M_{34}}{L} = \frac{19}{L}$$

$$V_3 = V_3^{(2)} + V_3^{(3)} = \frac{24}{L}$$

$$V_4 = -\frac{M_{34} + M_{34}}{L} = -\frac{19}{L}$$

Figure 4.20 shows the nature of the deflection and the support reactions.

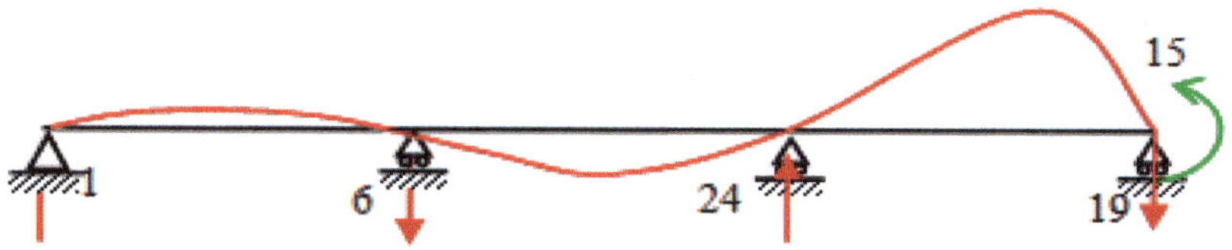

Figure 4.20. *Deflection of the beam and support reaction*($\frac{1}{L}$)*

4.5. Analysis of frames

The slope-deflection method is a widely used tool in indeterminate structure analysis. For structures subjected to vertical loads, the axial deflections have a weak or very weak influence. For this reason, they are neglected in the calculation.

In literature, frames are grouped into two categories: (1) frames without sidesway and (2) frames with sidesway.

4.5.1. *Frame without sidesway*

4.5.1.1. *Simple frame*

We analyze the simple frame in Figure 3.25 using the slope-deflection method. Flexural rigidities and geometric dimensions are shown in the figure.

– *Rigidity factors*

$$K_{AB} = K_{BA} = \frac{4EI}{L}$$

$$K_{BC} = \frac{6EI}{L}$$

– *Transmission coefficients*

$$\lambda_{AB} = \lambda_{BA} = \frac{1}{2}$$

$$\lambda_{BC} = 0$$

– *Fixed-end moments*

$$\gamma_{BC} = \frac{1}{8}qL^2$$

– *Moments at the ends of bars*

$$M_{AB} = \frac{2EI}{L}\omega_B$$

$$M_{BA} = \frac{4EI}{L}\omega_B$$

$$M_{BC} = \gamma_{BC} + \frac{6EI}{L}\omega_B = \frac{1}{8}qL^2 + \frac{6EI}{L}\omega_B$$

$$M_{CB} = 0$$

– *Joint equilibrium*

Joint B

$$M_{BA} + M_{BC} = 0$$

$$\frac{4EI}{L}\omega_B + \frac{6EI}{L}\omega_B + \frac{1}{8}qL^2 = 0$$

The slope at joint B is

$$\omega_B = -\frac{1}{80}\frac{qL^3}{EI}$$

The values of the moments at the ends of each bar are given as:

$$M_{AB} = \frac{2EI}{L}\omega_B = -\frac{1}{40}qL^2$$

$$M_{BA} = \frac{4EI}{L}\omega_B = -\frac{1}{20}qL^2$$

$$M_{BC} = \gamma_{BC} + \frac{6EI}{L}\omega_B = \frac{1}{8}qL^2 + \frac{6EI}{L}\omega_B = \frac{1}{20}qL^2$$

$$M_{CB} = 0$$

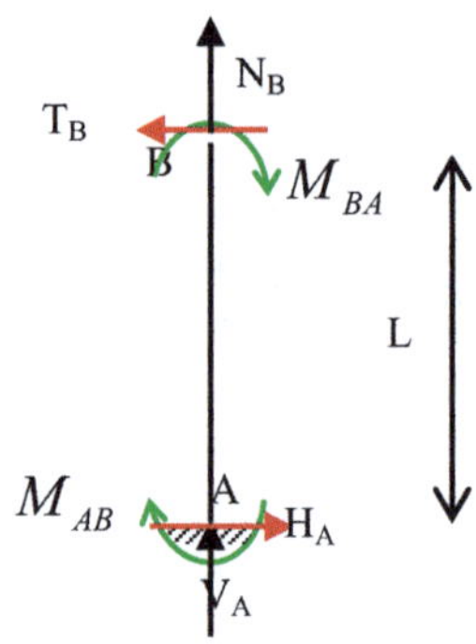

Figure 4.21. *The Column (AB)*

or for bar (AB):

$$\sum M_{/B} = 0 \ : \ T_A L - M_{AB} - M_{BA} = 0$$

$$H_A = \frac{3}{40} qL$$

$$T_B = -\frac{3}{40} qL$$

This time we consider bar (BC):

$$\sum M_{/B} = 0 \ : V_C L + M_{BA} - \frac{1}{2} qL^2 = 0$$

$$V_C = \frac{9}{20} qL$$

$$\sum M_{/C} = 0 \ : \ T_B L - M_{BA} - \frac{1}{2} qL^2 = 0$$

$$T_B = \frac{11}{20} qL$$

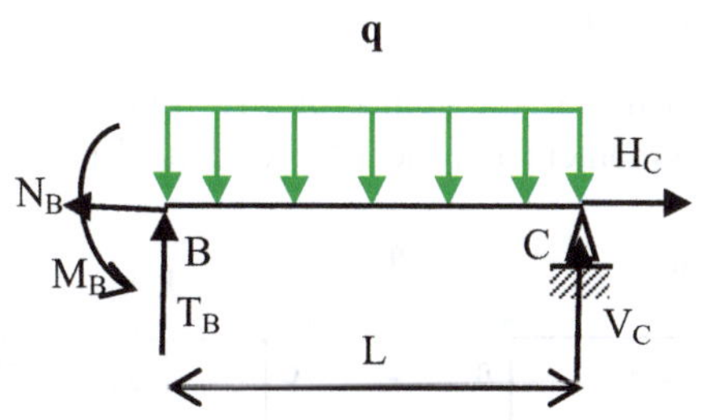

Figure 4.22. *The bar (BC)*

The diagrams of the bending moment and the shear force are shown in Figures 3.30 and 3.31.

These results allow us to deduce the support reactions.

$$V_B = T_B = \frac{11}{20} qL$$

$$H_C = -H_A = -\frac{3}{40} qL$$

In the same way, the deflection and the support reactions can be shown in Figure 4.21.

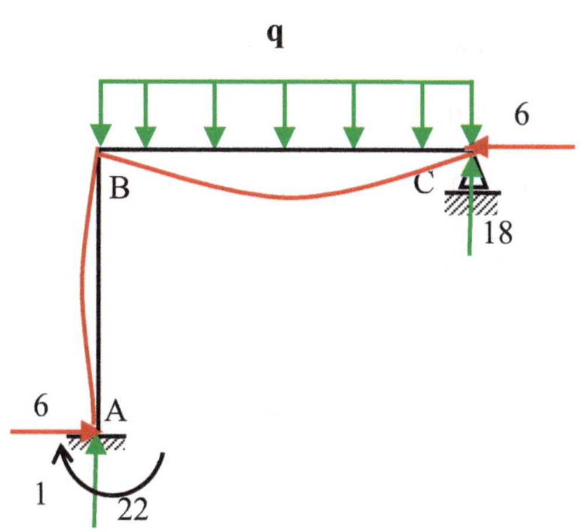

Figure 4.23. *Deflection of the beam, support reaction*($\frac{qL}{40}$) and moment ($\frac{qL^2}{40}$)*

4.5.1.2. *Multiple span frame*

Calculate the moments at the joints of the frame (Figure 4.24) and draw diagrams of the bending moment and shear force of the frame.

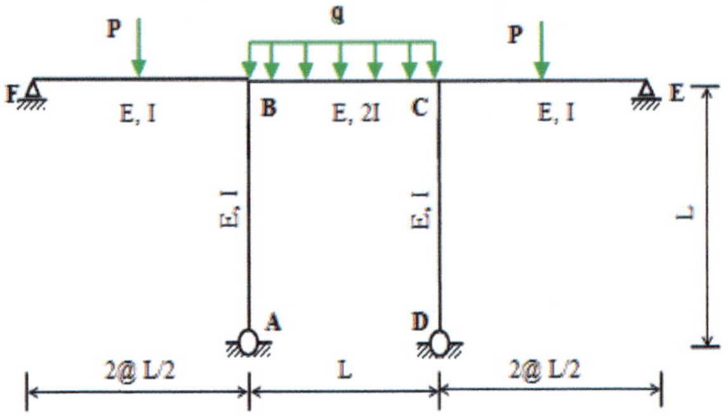

Figure 4.24. *Symmetrical frame*

– *Equilibrium equations*

The equilibrium equations of the joints, which can be rotated, are given as

$$M_B = 0$$

$$M_C = 0$$

In another way, we can write

$$M_{BA} + M_{BC} + M_{BF} = 0$$

$$M_{CB} + M_{CD} + M_{CE} = 0$$

– *Expressions of moments*

We have P = qL.

The fixed-end moments of each span are given by

$$\gamma_{BF} = -\frac{3}{16}PL = -\frac{3}{16}q.L^2$$

$$\gamma_{AB} = \gamma_{BA} = 0$$

$$\gamma_{BC} = -\gamma_{CB} = \frac{1}{12}qL^2$$

$$\gamma_{CE} = \frac{3}{16}PL = \frac{3}{16}qL^2, \quad \gamma_{EC} = 0$$

$$\gamma_{CD} = \gamma_{DC} = 0$$

The expressions of the moments at the ends of each bar can be formulated using the advantage of the structure's symmetry ($\omega_B = -\omega_C$).

$$M_{AB} = 0$$

$$M_{BA} = \frac{3EI}{L}\omega_B$$

$$M_{BC} = \frac{1}{12}qL^2 + \frac{8EI}{L}\omega_B + \frac{4EI}{L}\omega_C = \frac{1}{12}qL^2 + \frac{4EI}{L}\omega_B$$

$$M_{BF} = -\frac{3}{16}qL^2 + \frac{3EI}{L}\omega_B$$

$$M_{FB} = 0$$

– *The unknowns of the problem*

Substituting the expressions of the moments described above with equilibrium equations of the joints, we can obtain the slope of joint B.

$$\omega_B = \frac{qL^3}{96.EI}$$

– *Moments at the ends of bars*

Substituting the expressions of slopes ω_B and ω_C with the expressions of the moments at the ends of each bar, we get

$$M_{BA} = M_{CD} = \frac{3EI}{L}\omega_B = \frac{1}{32}qL^2$$

$$M_{BC} = M_{CB} = \frac{1}{12}qL^2 + \frac{4EI}{L}\omega_B = \frac{1}{8}qL^2$$

$$M_{BF} = M_{CE} = -\frac{3}{16}qL^2 + \frac{3EI}{L}\omega_B = -\frac{5}{32}qL^2$$

The bending moment and shear force expressions are given as

Bar (AB)

$$M_{AB}(x) = \mu_{AB}(x) - M_{AB}(1-\frac{x}{L}) + M_{BA}(\frac{x}{L}) = \frac{1}{32}qL^2(\frac{x}{L}) = \frac{1}{32}qLx$$

$$M_{AB}(0) = 0$$

$$M_{AB}(L) = \frac{1}{32}qL^2$$

Bar (BC)

$$M_{BC}(x) = \frac{1}{2}qLx - \frac{1}{2}qx^2 - \frac{1}{8}qL^2(1-\frac{x}{L}) - \frac{1}{8}qL^2(\frac{x}{L}) = \frac{1}{2}qLx - \frac{1}{2}qx^2 - \frac{1}{8}qL^2$$

$$M_{BC}(0) = -\frac{1}{8}qL^2$$

$$M_{BC}(L) = -\frac{1}{8}qL^2$$

Bar (BF)

$$0 \le x \le \frac{L}{2}$$

$$M_{BF}(x) = \frac{1}{2}qLx - \frac{5}{32}qL^2(\frac{x}{L}) = \frac{11}{32}qLx$$

$$M_{BF}(0) = 0$$

$$M_{BF}(\frac{L}{2}) = \frac{11}{62}qL^2$$

$$\frac{L}{2} \le x \le L$$

$$M_{FB}(x) = \frac{1}{2}qL^2 - \frac{1}{2}qLx - \frac{5}{32}qL^2(\frac{x}{L}) = \frac{1}{2}qL^2 - \frac{21}{32}qLx$$

$$M_{FB}(\frac{L}{2}) = \frac{11}{64}qL^2$$

$$M_{FB}(L) = -\frac{5}{32}qL^2$$

The bending moment diagram is shown in Figure 4.25.

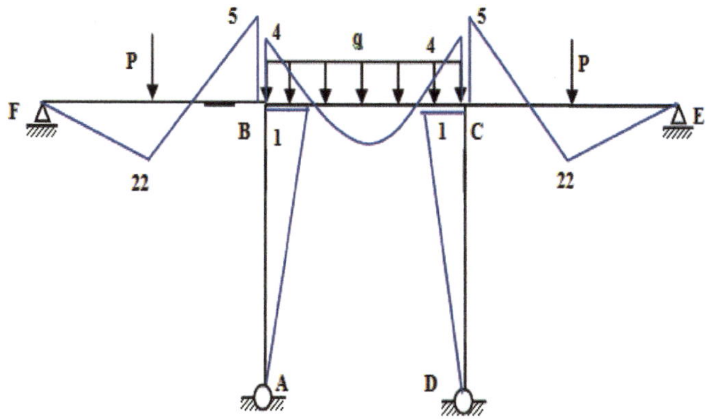

Figure 4.25. *Bending moment diagram* * $(\frac{1}{32}qL^2)$

In the same way, the distribution of the shear force is given by

Bar (AB)

$$T_{AB}(x) = \frac{dM(x)}{dx} = \frac{d\mu_{AB}(x)}{dx} + \frac{M_{AB}+M_{BA}}{L} = \frac{1}{32}qL$$

Bar (BC)

$$T_{BC}(x) = \frac{1}{2}q.L - q.x$$

$$T_{BC}(0) = \frac{1}{2}qL$$

$$T_{BC}(L) = -\frac{1}{2}qL$$

Bar (FB)

$$0 \leq x \leq \frac{L}{2}$$

$$T_{FB}(x) = \frac{11}{32}\ q.L$$

$$\frac{L}{2} \leq x \leq L$$

$$T_{FB}(x) = -\frac{21}{32}q.L$$

Bar (CD)

$$T_{CD}(x) = -\frac{1}{32}qL$$

Bar (CE)

$$0 \leq x \leq \frac{L}{2}$$

$$T_{CE}(x) = \frac{21}{32}qL$$

$$\frac{L}{2} \leq x \leq L$$

$$T_{CE}(x) = -\frac{21}{32}qL$$

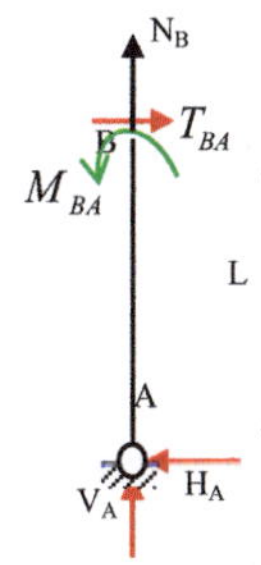

Figure 4.26. *The Column (AB)*

Bar (AB)

$$\sum M_{/B} = 0 : H_A L - M_{BA} = 0$$

$$H_A = \frac{1}{32} qL$$

$$T_{BA} = \frac{1}{32} qL$$

The shear force diagram is shown in Figure 4.27.

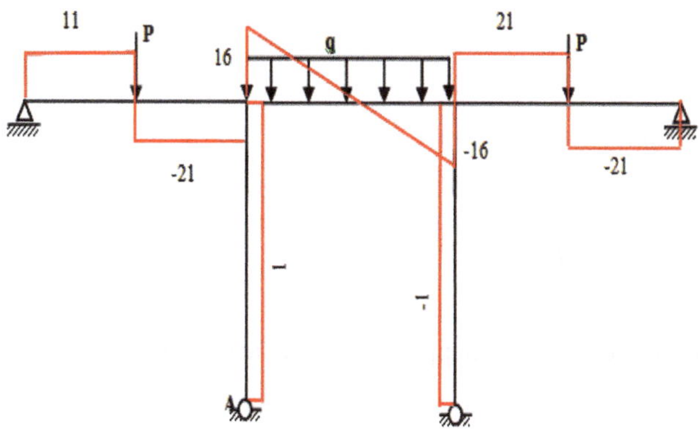

Figure 4.27. *Shear force diagram* * $(\frac{1}{32} qL)$

This time we consider bar (FB).

$$\sum M_{/B} = 0 \; : \; V_F L + M_{BF} - PL = 0$$

$$V_F = \frac{11}{32} qL$$

$$\sum F_{/yy} = 0 \; : \; T_{BF} + V_F - P = 0$$

$$T_{BF} = \frac{21}{32} P = \frac{21}{32} qL$$

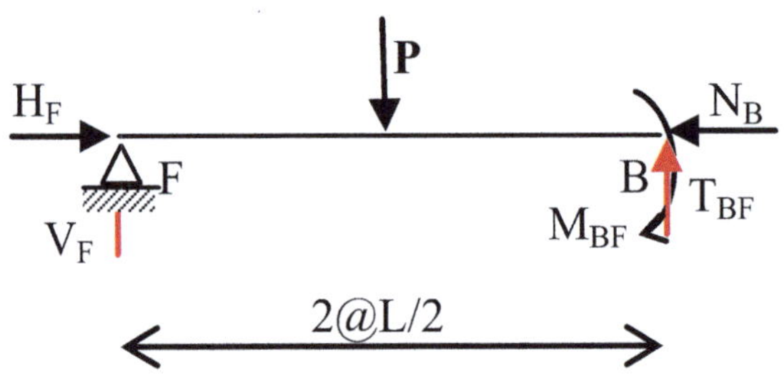

Figure 4.28. *Bar (FB)*

Hence

$$V_A = \frac{21}{32} qL$$

$$H_F = T_{BA} = \frac{1}{32} qL$$

Finally, the deflection and support reactions acting on the frame are plotted in Figure 4.29.

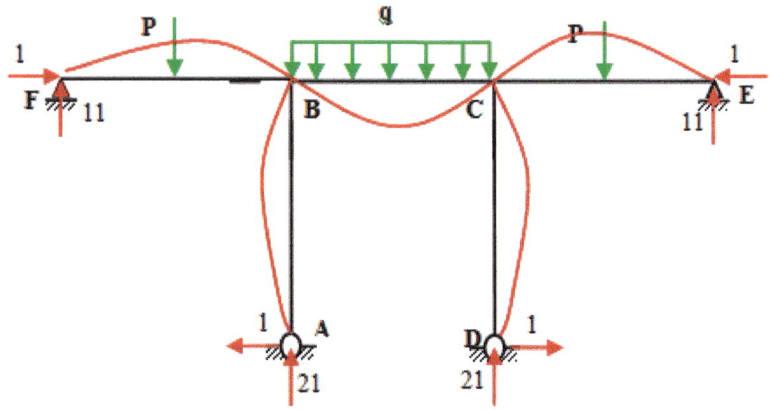

Figure 4.29. *Deflection of the beam and support reactions* $*(\frac{qL}{32})$

4.5.1.3. *Frames with support settlement*

Determine the moments at the joints and plot the diagrams of the bending moment and shear force due to a settlement of support B of a quantity Δ. It is assumed that the axial deformation of bar (BC) is neglected; the settlement Δ of support C causes a settlement of joint B of the same quantity (Figure 4.30). This settlement causes the bars to incline and AB and BD are AB' and DB' in the deformed configuration. This settlement generates a slope on the bars; this slope is negative and $\psi = -\frac{\Delta}{L}$.

Using the adopted sign convention, we associate a negative slope ($-\psi$) with bar AB and a positive slope (ψ') with bar BD (Chapter 1).

Applying the slope-deflection method means we can calculate the moments at the joints of the given frame.

$$M_{AB} = \frac{2EI}{L}\omega_B - \frac{6EI}{L}\psi$$

$$M_{BA} = \frac{4EI}{L}\omega_B - \frac{6EI}{L}\psi$$

$$M_{BD} = \frac{3EI}{L}\omega_B - \frac{3EI}{L}\psi'$$

$$M_{BC} = \frac{4EI}{L}\omega_B$$

$$M_{CB} = \frac{2EI}{L}\omega_B$$

$$M_{DB} = 0.$$

Knowing that $\psi = -\psi'$

$$M_{BD} = \frac{3EI}{L}\omega_B + \frac{3EI}{L}\psi$$

The equilibrium of joint B is written as

$$\sum M_B = 0: \frac{11EI}{L}\omega_B - \frac{3EI}{L}\psi = 0$$

This relationship allows us to deduce the slope at joint B, which is

$$\omega_B = +\frac{3}{11}\psi$$

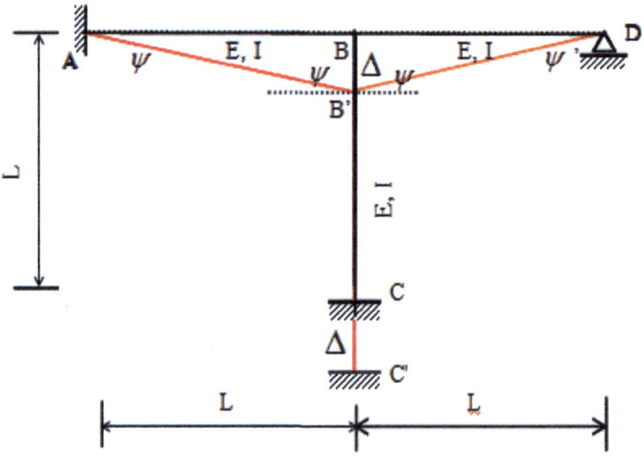

Figure 4.30. *Frame under support settlement*

The expressions of moments at the ends in relation to the settlement of support B are

$$M_{AB} = -\frac{60EI}{11L}\psi$$

$$M_{BA} = -\frac{54EI}{11L}\psi$$

$$M_{BD} = \frac{42EI}{11L}\psi$$

$$M_{BC} = \frac{12EI}{11L}\psi$$

$$M_{CB} = \frac{6EI}{11L}\psi$$

$$M_{DB} = 0$$

Substituting the value of the slope due to the settlement at support C, the moments at the ends of the bars become

$$M_{AB} = \frac{60EI}{11L^2}\Delta$$

$$M_{BA} = \frac{54EI}{11L^2}\Delta$$

$$M_{BD} = -\frac{42EI}{11L^2}\Delta$$

$$M_{BC} = -\frac{12EI}{11L^2}\Delta$$

$$M_{CB} = -\frac{6EI}{11L^2}\Delta$$

$$M_{DB} = 0$$

The bending moment expressions are given as

Bar (AB)

$$M_{AB}(x) = -\frac{60EI}{11L^2}\Delta(1-\frac{x}{L}) + \frac{54EI}{11L^2}\Delta(\frac{x}{L})$$

Bar (BC)

$$M_{BC}(x) = \frac{12EI}{11L^2}\Delta(1-\frac{x}{L}) - \frac{6EI}{11L^2}\Delta(\frac{x}{L})$$

Bar (BD)

$$M_{BD}(x) = \frac{42EI}{11L^2}\Delta(1-\frac{x}{L})$$

The shear force expression of each element is given by

$$T_{AB} = \frac{60EI}{11L^3}\Delta + \frac{54EI}{11L^3}\Delta = \frac{114}{11}\frac{EI}{L^3}\Delta$$

$$T_{BD} = -\frac{42EI}{11L^3}\Delta$$

$$T_{BC} = -\frac{12EI}{11L}\Delta - \frac{6EI}{11L}\Delta = -\frac{18EI}{11L}\Delta$$

Figure 4.31 plots the bending moment diagram.

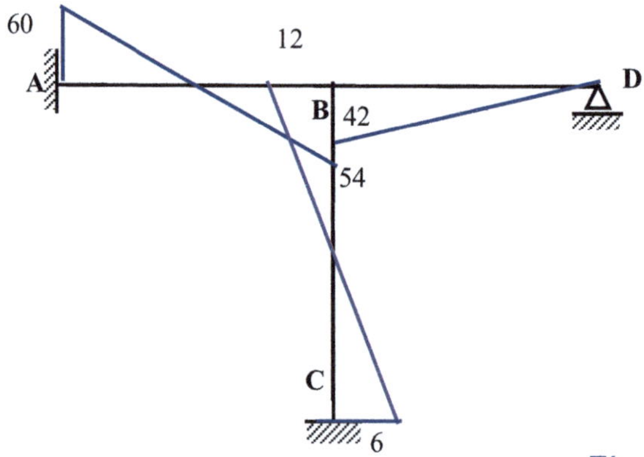

Figure 4.31. *Bending moment diagram* $*\frac{EI}{11L^2}\Delta$

Similarly, Figure 4.33 shows the variation of the shear force of the frame.

The equilibrium of bar (AB) allows us to write

$$\sum M_{/B}=0 \; : \; V_A L - M_{AB} - M_{BA} = 0$$

$$V_A = \frac{114}{11}\frac{EI}{L^3}\Delta$$

$$T_{BA} = -\frac{114}{11}\frac{EI}{L^3}\Delta$$

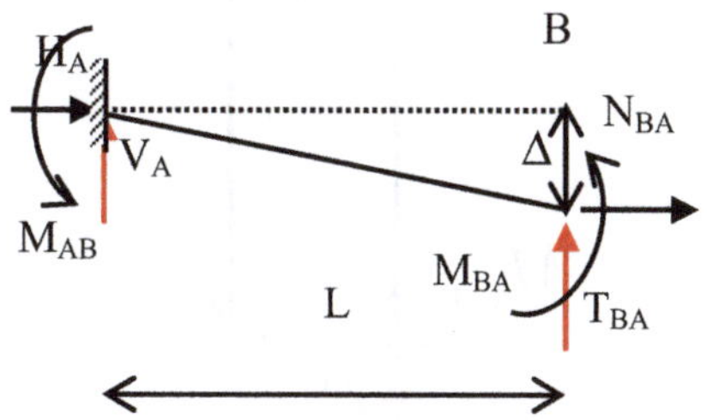

Figure 4.32. *Bar (AB)*

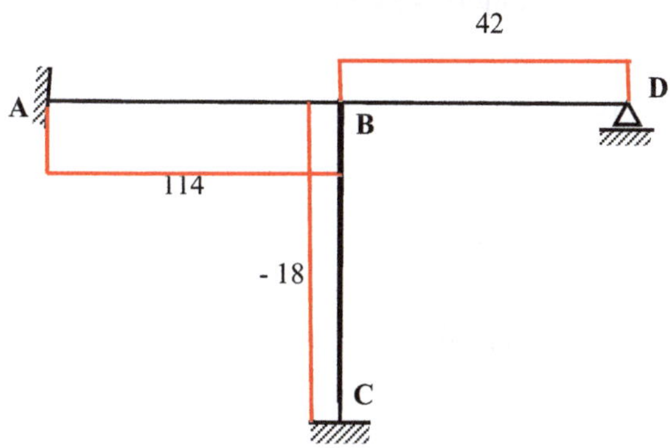

Figure 4.33. *Shear force diagram* $*(\frac{1}{11}\frac{EI}{L^3}\Delta)$

Now we consider bar (CB).

$$\sum M_{/B} = 0 : H_C L - M_{AB} - M_{BA} = 0$$

$$H_C = \frac{18}{11}\frac{EI}{L^3}\Delta$$

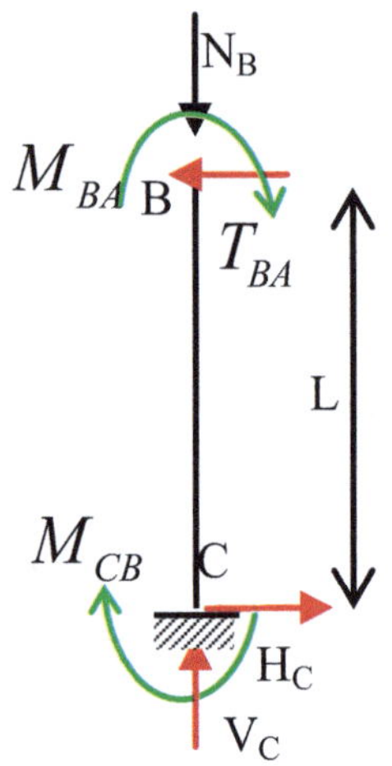

Figure 4.34. *Bar (CB)*

Hence, $T_{BA} = H_C = \frac{18}{11}\frac{EI}{L^3}\Delta$

Bar (BD)

$$\sum M_{/D} = 0 \; : \; T_{BD} L + M_{BD} = 0$$

$$T_{BD} = -\frac{42}{11}\frac{EI}{L^3}\Delta$$

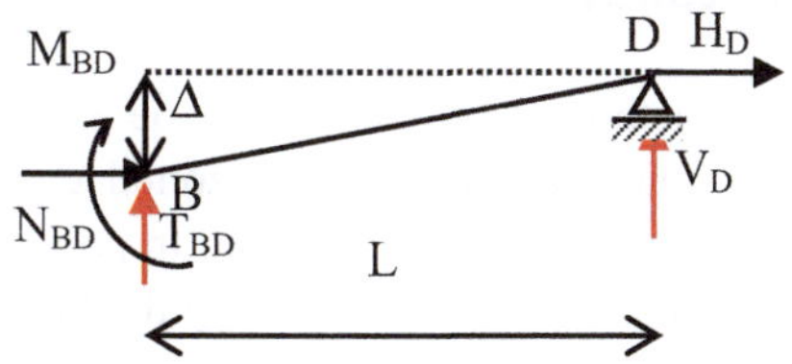

Figure 4.35. *Bar (BD)*

Hence; $V_D = -T_{BD} = \dfrac{42}{11}\dfrac{EI}{L^3}\Delta$

We can deduce that

$$V_C = -T_{BD} - T_{BA} = \frac{156}{11}\frac{EI}{L^3}\Delta$$

The deflection and support reactions acting on the frame at the joints are represented in Figure 4.36.

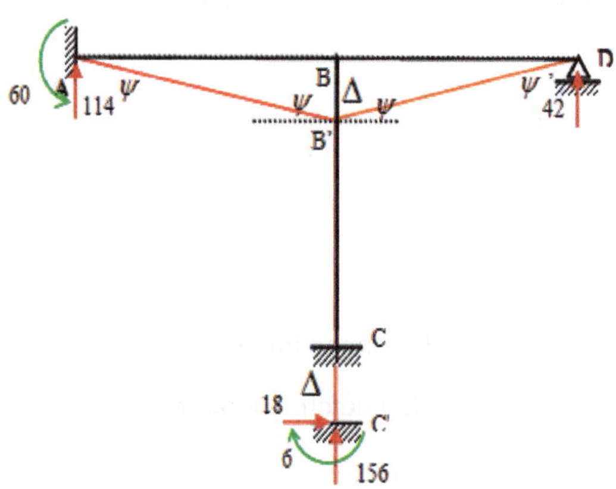

Figure 4.36. *Deflection, fixed moments * ($\frac{1}{11}\frac{EI}{L^2}\Delta$) and support reactions of frame and support reactions *($\frac{1}{11}\frac{EI}{L^3}\Delta$)*

4.5.2. *Frames with sidesway*

The slope-deflection method expressions can be applied to the analysis of statically indeterminate frames subjected to horizontal displacements. In general, axial deformations in frame elements are neglected. In previous sections, we studied frames without sidesway, which are characterized by the following:

– the joints of the frame are considered fixed;

– the frame's geometry and loading are symmetrical.

The horizontal displacement of the frames is due to the non-symmetry in geometry or the non-symmetry of the applied loading or the effect of a horizontal force.

To analyze frames with sidesway, consider frame (Figure 4.37). In this case, the horizontal load P causes a horizontal displacement Δ of the frame in the direction of force's application. In this case, we must take into account the angular rotation β of columns (AB) and (CD) due to horizontal displacement Δ . The horizontal bar (BC) can be translated without angular rotation.

The effect of the horizontal force can be expressed by neglecting the vertical actions, which generates moments at the ends of the vertical bars.

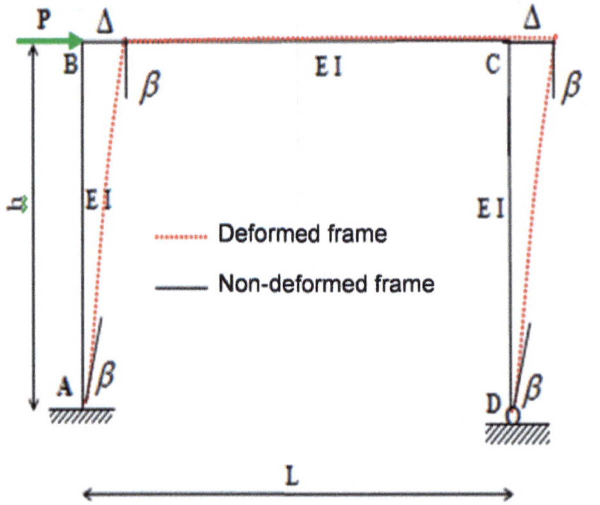

Figure 4.37. *Frame with sidesway*

Column (AB)

$$M_{AB} = -k_{AB}(1+\lambda_{AB})\beta = -\frac{4EI}{h}(1+\frac{1}{2})\beta = -\frac{6EI}{h}\beta \qquad [4.34a]$$

$$M_{BA} = -k_{BA}(1+\lambda_{BA})\beta = -\frac{4EI}{h}(1+\frac{1}{2})\beta = -\frac{6EI}{h}\beta \qquad [4.34b]$$

Column (CD)

$$M_{CD} = -k'_{CD}(1+\lambda'_{CD})\beta = -\frac{3EI}{h}\beta = -\frac{3EI}{h}\beta \qquad [4.35a]$$

$$M_{DC} = 0 \qquad [4.35b]$$

4.5.2.1. *Frames with cantilever*

We use the slope-deflection method to analyze the frame shown in Figure 3.26. We assume that EI is constant for each span and P = q.L.

– *Equilibrium equations*

The equilibrium of the frame allows us to write the following equilibrium equations:

$$M_B = 0$$

$$M_C = -\frac{1}{5}PL = -\frac{1}{5}qL^2$$

In another way, we can write

$$M_{BA} + M_{BC} = 0$$

$$M_{CB} + M_{CD} = -\frac{1}{5}qL^2$$

– *Fixed-end moments*

$$\gamma_{AB} = \gamma_{BA} = 0$$

$$\gamma_{BC} = -\gamma_{CB} = \frac{1}{12}qL^2$$

$$\gamma_{CD} = \gamma_{DC} = 0$$

– *Moments at the ends of bars*

$$M_{AB}=0$$

$$M_{BA}=\frac{3EI}{L}\omega_B-\frac{3EI}{L}\beta$$

$$M_{BC}=\frac{1}{12}qL^2+\frac{4EI}{L}\omega_B+\frac{2EI}{L}\omega_C$$

$$M_{CB}=-\frac{1}{12}qL^2+\frac{2EI}{L}\omega_B+\frac{4EI}{L}\omega_C$$

$$M_{CD}=\frac{4EI}{L}\omega_C-\frac{6EI}{L}\beta$$

$$M_{DC}=\frac{2EI}{L}\omega_C-\frac{6EI}{L}\beta$$

– *Rotation of displaced joints*

Substituting the expressions of moments at the ends of bars with the equilibrium equations, we get

$$7\omega_B+2\omega_C-3\beta=-\frac{qL^3}{12EI}$$

$$2\omega_B+8\omega_C-6\beta=-\frac{7qL^3}{60EI}$$

The equilibrium of displaced joints allows for the construction of two equations with three unknowns, ω_B, ω_C and β. It is necessary to introduce an additional equation such as that of the overall equilibrium of the structure.

The horizontal force causes a displacement in the horizontal direction producing a vertical bar rotation. The horizontal bar is considered as a diaphragm element, which undergoes a translation in its plane without rotation.

The horizontal equilibrium of the entire structure is written as

$$T_{AB}+T_{DC}=0$$

The horizontal reactions or shear stresses T_{AB} and T_{DC} of the vertical elements can be calculated due to the internal actions of each bar.

– *Column (AB)*

The equilibrium of bar (AB) allows us to write

$$\sum M_B = 0 \; : \; M_{BA} - T_{AB} L = 0$$

$$T_{AB} = \frac{M_{BA}}{L}$$

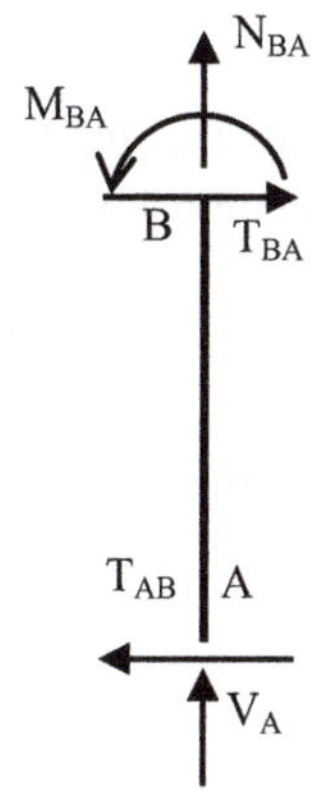

Figure 4.38. *Bar (AB)*

Substituting the value of M_{BA}, we find

$$T_{AB} = \frac{3EI}{L^2}\omega_B - \frac{3EI}{L^2}\beta$$

In the same way, we can deduce the horizontal reaction in bar (CD).

$$\sum M_C = 0 : \; M_{CD} + M_{DC} - T_{DC} L = 0$$

$$T_{DC} = \frac{M_{DC} + M_{CD}}{L}$$

$$T_{DC} = \frac{6EI}{L^2}\omega_C - \frac{12EI}{L^2}\beta$$

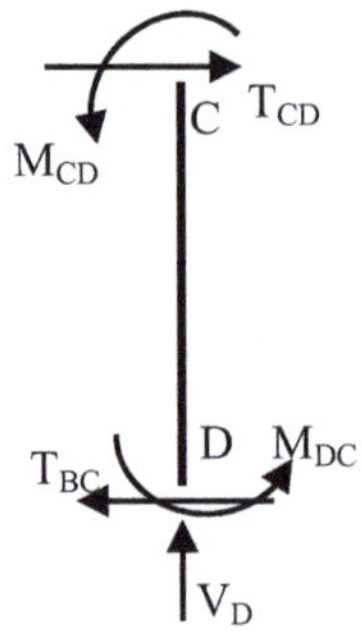

Figure 4.39. *Bar (DC)*

Substituting the values of T_{AB} and T_{DC} in the relationship of the horizontal equilibrium, we obtain the following relationship.

$$\omega_B + 2\omega_C - 5\beta = 0$$

Resolving the equations, which give the slopes, leads to calculating the unknowns of the problem, which are

$$\omega_B = -0.636\frac{qL^3}{EI}$$

$$\omega_C = -1.160\frac{qL^3}{EI}$$

$$\beta = -0.60\frac{qL^3}{EI}$$

– *Moments at the ends of bars*

After determining the rotations of the displaced joints, the expressions of moments at the ends of each bar are

$$M_{AB} = 0$$

$$M_{BA} = \frac{3EI}{L}\omega_B - \frac{3EI}{L}\beta = -0.00228qL^2$$

$$M_{BC} = \frac{1}{12}qL^2 + \frac{4EI}{L}\omega_B + \frac{2EI}{L}\omega_C = 0.00228qL^2$$

$$M_{CB} = -\frac{1}{12}qL^2 + \frac{2EI}{L}\omega_B + \frac{4EI}{L}\omega_C = -0.182qL^2$$

$$M_{CD} = \frac{4EI}{L}\omega_C - \frac{6EI}{L}\beta = -0.0183qL^2$$

$$M_{DC} = \frac{2EI}{L}\omega_C - \frac{6EI}{L}\beta = 0.0205qL^2$$

Finally, we find that the results obtained by the method of forces (Figures 3.31 and 3.32) are in line with those obtained by the slope-deflection method.

Support reactions can be deduced by the following:

Bar (AB)

$$\sum M_{/B} = 0 \ : \ H_A L - M_{BA} = 0$$

$$H_A = 2.28.10^{-3} qL$$

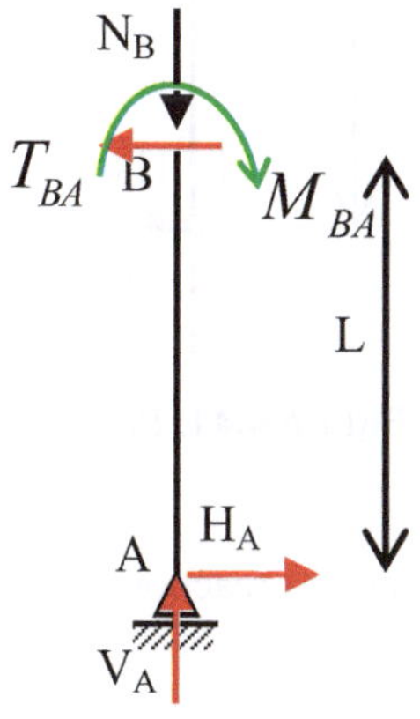

Figure 4.40. *Bar (AB)*

Bar (DC)

$$\sum M_{/C} = 0 \ : H_D L - M_{CD} + M_{DC} = 0$$

$$V_D = -2.28.10^{-3} qL$$

$$\sum F_{/yy} = 0 \ : \ V_A + V_D - P = 0.$$

$$\sum M_{/D} = 0 \ : \ V_A L - \frac{1}{2} qL^2 + P\frac{L}{5} - M_{DC} = 0.$$

$$V_A = 0.32qL$$

$$V_D = P - 0.32qL = 0.68qL$$

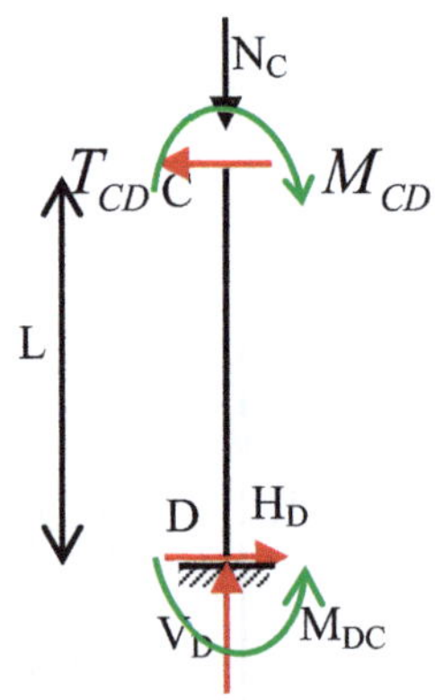

Figure 4.41. *Bar (DC)*

Finally, the deflection and support reactions acting on the frame are presented in Figure 4.42.

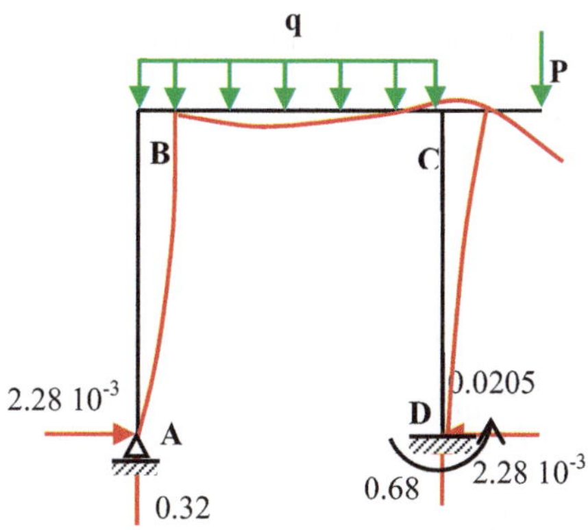

Figure 4.42. *Deflection of frame, fixing moment *(qL²) and support reactions*(qL)*

4.5.2.2. *Frames with columns of different lengths*

We re-analyze the frame shown in Figure 3.33 using the slope-deflection method. We assume that the flexural rigidity of the members is constant. The applied loading causes the frame to deform as shown in Figure 4.43.

The angular rotations of columns (AB) and (CD) are given as

$$\beta_{AB} = \beta_{BA} = \beta$$

$$\beta_{CD} = \beta_{DC} = \frac{8}{5}\beta$$

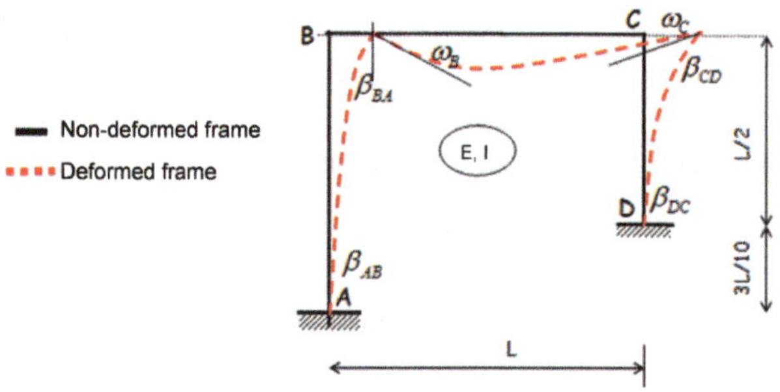

Figure 4.43. *Deformed frame*

The expressions of the moments at the joints are given by

$$M_{AB}=\frac{5EI}{2L}\omega_B-\frac{15EI}{2L}\beta$$

$$M_{BA}=\frac{5EI}{L}\omega_B-\frac{15EI}{2L}\beta$$

$$M_{BC}=\frac{1}{12}qL^2+\frac{4EI}{L}\omega_B+\frac{2EI}{L}\omega_C$$

$$M_{CB}=-\frac{1}{12}qL^2+\frac{2EI}{L}\omega_B+\frac{4EI}{L}\omega_C$$

$$M_{CD}=\frac{8EI}{L}\omega_C-\frac{96EI}{5L}\beta$$

$$M_{DC}=\frac{4EI}{L}\omega_C-\frac{96EI}{5L}\beta$$

The equilibrium of joints B and C allows us to write

$M_{BA}+M_{BC}=0$

$$9\omega_B+2\omega_C-\frac{15}{2}\beta=-\frac{qL^3}{12EI}$$

$M_{CB}+M_{CD}=0$

$$2\omega_B+12\omega_C-\frac{96}{5}\beta=\frac{1.qL^3}{12.EI}$$

The horizontal equilibrium of the structure is written as

$T_{AB}+T_{DC}=0$

The shear forces T_{AB} and T_{DC} of the vertical elements can be deduced by:

– Extracting element (AB).

The bar (AB) is in equilibrium; we can write

$$\sum M_B = 0$$

$$M_{AB} + M_{BA} - T_{AB}(\frac{4}{5}L) = 0$$

$$T_{AB} = \frac{M_{AB} + M_{BA}}{\frac{4}{5}L}$$

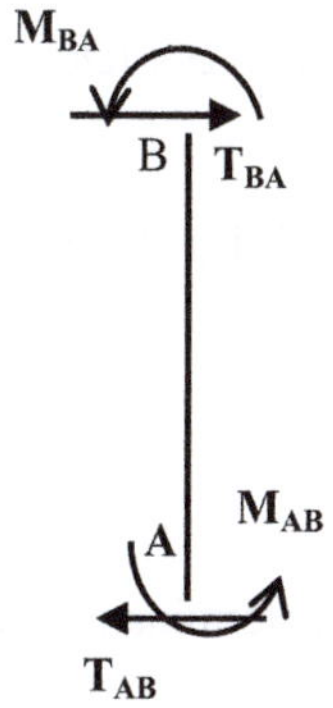

Figure 4.44

Substituting the values of M_{AB} and M_{BA}, the shear force in bar (AB) is written as

$$T_{AB} = \frac{75EI}{8L^2}\omega_B - \frac{75EI}{4L^2}\beta$$

– In the same way, we can deduce the horizontal reaction in bar (CD).

$$\sum M_C = 0$$

$$M_{CD} + M_{DC} - T_{DC}(\frac{L}{2}) = 0$$

$$T_{DC} = \frac{M_{DC} + M_{CD}}{\frac{L}{2}}$$

$$T_{DC} = \frac{24EI}{L^2}\omega_C - \frac{384EI}{5L^2}\beta$$

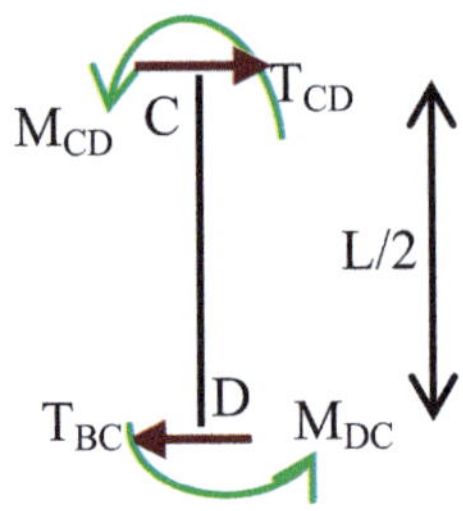

Figure 4.45. *Bar (DC)*

We substitute the values of T_{AB} and T_{DC} in the relationship of the horizontal equilibrium.

$$\frac{75}{8}\omega_B + 24\omega_C - \frac{1911}{20}\beta = 0$$

Resolving the equations leads to the determination of the slopes at joints B and C.

$$\omega_B = -0.0102\frac{qL^3}{EI}$$

$$\omega_C = 0.0118\frac{qL^3}{EI}$$

$$\beta = 0.002\frac{qL^3}{EI}$$

Substituting the slopes ω_B, ω_C and β with the bending moment expressions, we obtain

$$M_{AB} = \frac{5EI}{2L}\omega_B - \frac{15EI}{2L}\beta = -0.0403qL^2$$

$$M_{BA} = \frac{5EI}{L}\omega_B - \frac{15EI}{2L}\beta = -0.0659qL^2$$

$$M_{BC} = \frac{1}{12}qL^2 + \frac{4EI}{L}\omega_B + \frac{2EI}{L}\omega_C = 0.0659qL^2$$

$$M_{CB} = -\frac{1}{12}qL^2 + \frac{2EI}{L}\omega_B + \frac{4EI}{L}\omega_C = -0.0567qL^2$$

$$M_{CD} = \frac{8EI}{L}\omega_C - \frac{96EI}{5L}\beta = 0.0567qL^2$$

$$M_{DC} = \frac{4EI}{L}\omega_C - \frac{96EI}{5L}\beta = 0.0096qL^2$$

The diagrams of the bending moment and the shear force are shown, respectively, in Figures 3.38 and 3.39.

To calculate support reactions, consider bar (AB).

$$\sum M_{/B} = 0 \; : \; H_A(\frac{4L}{5}) - M_{BA} - M_{BA} = 0$$

$$H_A = 0.1327qL$$

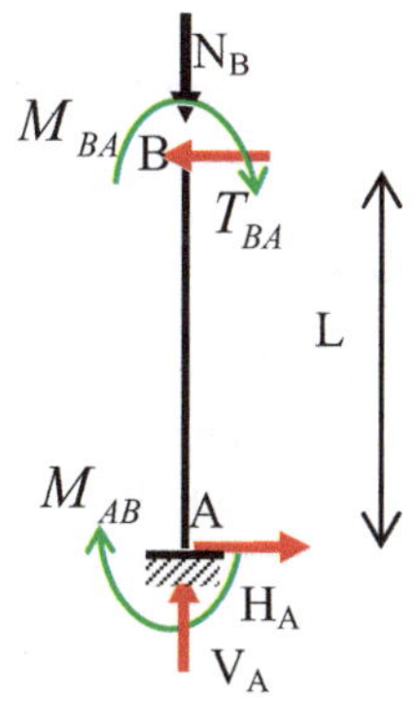

Figure 4.46. *Bar (AB)*

In the same way, we take bar (CD) into account.

$$\sum M_{/C} = 0 \; : H_D \frac{L}{2} - M_{CD} - M_{DC} = 0$$

$$H_D = 0.1326qL$$

$$\sum M_{/A} = 0 : V_D L - V_D \frac{3L}{10} + M_{DC} - M_{AB} - \frac{1}{2}qL^2 = 0$$

$$V_D = 0.49qL$$

$$V_A = 0.51qL$$

Figure 4.47 shows the deflection and the support reactions acting on the frame.

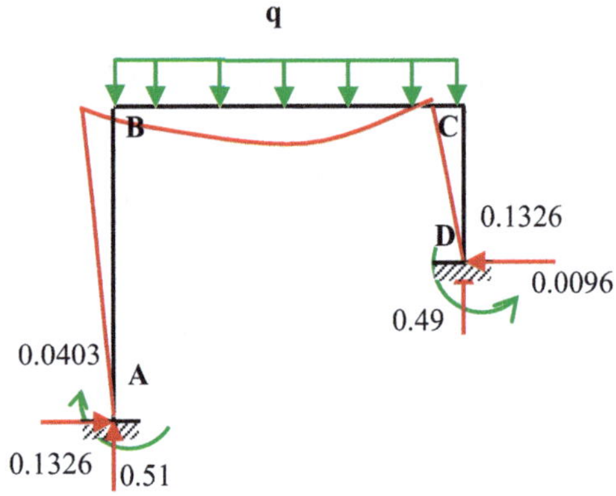

Figure 4.47. *Deflection of frame, fixed moments *(qL²) and support reactions *(qL)*

4.5.2.3. *Frames with oblique columns*

We analyze the frame with oblique bars shown in Figure 3.40 using the slope-deflection method knowing that it has already been studied using the method of forces. The given frame has inclined bars and has displaced joints. To analyze it using the slope-deflection method, it is necessary to calculate the slope due to the horizontal displacement of the frame.

The unknowns of the problem are the slopes ω_B, ω_C and the displacement Δ.

– *Fixed-end moment*

$$\gamma_{BC} = -\gamma_{CB} = \frac{1}{3}q.a^2$$

Due to the geometrical non-symmetry of the frame, its deflection is shown in Figure 4.48.

The slopes of the different elements due to displacements of the joints (Figure 4.48) are given as

$$\beta_{AB} = \frac{BB'}{L_{AB}} = \frac{\Delta_1}{L_{AB}} = \frac{\Delta}{L_{AB}\cos\alpha} = \frac{\Delta}{2a}$$

$$\beta_{DC} = \frac{\Delta_2}{L_{DC}} = \frac{\Delta}{L_{DC}\cos\psi} = \frac{\Delta}{a}$$

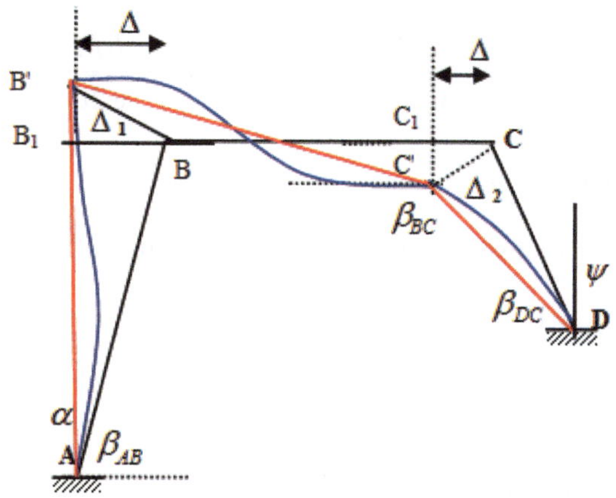

Figure 4.48. *Deformed frame*

$$\beta_{BC} = \frac{B'B_1 + C'C_1}{L_{BC}} = \frac{\Delta(tg\alpha + tg\psi)}{L_{BC}} = \frac{3\Delta}{4a}$$

The direct application of the slope-deflection method gives the moments at the ends of each bar.

$$M_{AB} = \frac{2EI}{\sqrt{5}a}\omega_B - \frac{6EI}{\sqrt{5}a}\beta_{AB} = \frac{2EI}{\sqrt{5}a}\omega_B - \frac{3EI}{\sqrt{5}a^2}\Delta$$

$$M_{BA} = \frac{4EI}{\sqrt{5}a}\omega_B - \frac{6EI}{\sqrt{5}a}\beta_{BA} = \frac{4EI}{\sqrt{5}a}\omega_B - \frac{3EI}{\sqrt{5}a^2}\Delta$$

$$M_{BC} = \frac{1}{3}qa^2 + \frac{4EI}{2a}\omega_B + \frac{2EI}{2a}\omega_C - \frac{6EI}{2a}\beta_{BC} =$$

$$\frac{1}{3}qa^2 + \frac{2EI}{a}\omega_B + \frac{EI}{a}\omega_C + \frac{9EI}{4a^2}\Delta$$

$$M_{CB} = -\frac{1}{3}qa^2 + \frac{2EI}{2a}\omega_B + \frac{4EI}{2a}\omega_C - \frac{6EI}{2a}\beta_{CB} =$$

$$-\frac{1}{3}qa^2 + \frac{EI}{a}\omega_B + \frac{2EI}{a}\omega_C + \frac{9EI}{4a^2}.\Delta$$

$$M_{CD} = \frac{4EI}{\sqrt{2}a}\omega_C - \frac{6EI}{\sqrt{2}a}\beta_{CD} = \frac{4EI}{\sqrt{2}a}\omega_C - \frac{6EI}{\sqrt{2}a^2}\Delta$$

$$M_{DC} = \frac{2EI}{\sqrt{2}a}\omega_C - \frac{6EI}{\sqrt{2}a}\beta_{DC} = \frac{2EI}{\sqrt{2}a}\omega_C - \frac{6EI}{\sqrt{2}a^2}\Delta$$

– *Equilibrium equations*

The equilibrium of displaced joints B and C gives the following relationships:

$$M_{BA} + M_{BC} = 0$$

$$M_{CB} + M_{CD} = 0$$

$$3.788\omega_B + \omega_C + 0.908\frac{\Delta}{a} = -\frac{1}{3}\frac{qa^3}{EI}$$

$$\omega_B + 4.828\omega_C - 1.992\frac{\Delta}{a} = \frac{1}{3}\frac{qa^3}{EI}$$

The third equilibrium equation can be obtained by considering the horizontal equilibrium of the frame.

$$T_{AB} + T_{DC} = 0$$

$$M_{AB} + M_{BA} - T_{AB}(2a) + V_A a = 0$$

or

$$T_{AB} = \frac{M_{AB} + M_{BA} - V_A.a}{2a}$$

$$T_{AB} = \frac{3EI}{\sqrt{5}a^2}\omega_B - \frac{3EI}{\sqrt{5}a^3}\Delta - \frac{V_A}{2}$$

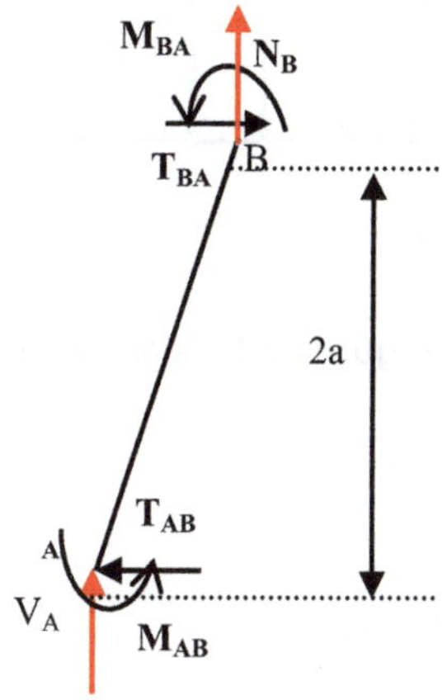

Figure 4.49. *Bar (AB)*

Similarly for bar CD, we can deduce that

$$M_{CD} + M_{DC} + V_D.a - T_{DC}\,a = 0$$

or

$$T_{DC} = \frac{M_{CD} + M_{DC} + V_D.a}{a}$$

$$T_{DC} = \frac{6EI}{\sqrt{2}a^2}\omega_C - \frac{12EI}{\sqrt{2}a^3}\Delta + V_D$$

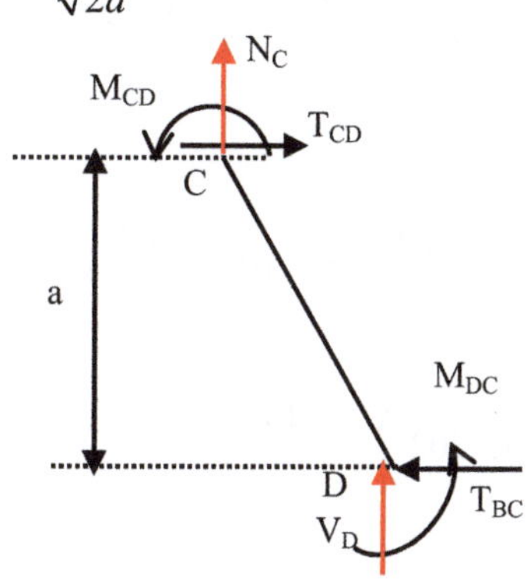

Figure 4.50. *Bar (DC)*

The third equation obtains the equilibrium of bar (BC).

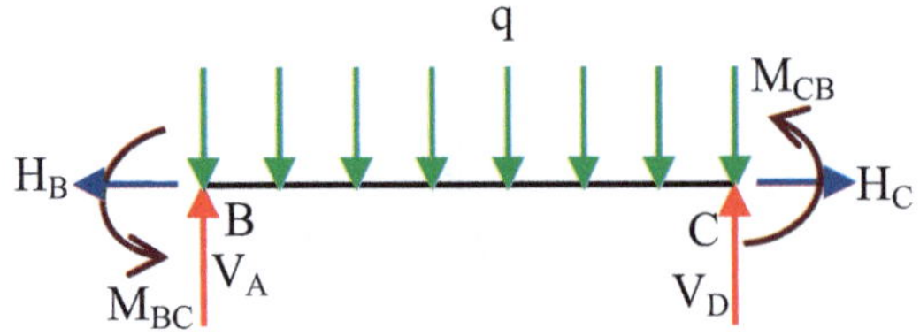

Figure 4.51. *Beam (BC)*

$$\sum M_C = 0$$

$M_{BC} + M_{CB} + 2qa^2 - 2\,V_A.a = 0$

$$V_A = \frac{M_{CB} + M_{BC} + 2qa^2}{2a} = qa + \frac{3EI}{2a^2}\omega_B + \frac{3EI}{2a^2}\omega_C + \frac{9EI}{4a^3}\Delta$$

$$\sum F_y = 0 : V_A + V_D = 2qa$$

$$V_D = qa - \frac{3EI}{2a^2}\omega_B - \frac{3EI}{2a^2}\omega_C - \frac{9EI}{4a^3}\Delta$$

By substituting the values of reactions V_A and V_D with the expressions giving the shear forces, T_{AB} and T_{DC}, we get

$$T_{AB} = 0.591\frac{EI}{a^2}\omega_B - 0.75\frac{EI}{a^2}\omega_C - 2.466\frac{EI}{a^3}\Delta - \frac{qa}{2}$$

Similarly,

$$T_{DC} = qa - 1.5\frac{EI}{a^2}\omega_B + 2.742\frac{EI}{a^2}\omega_C - 10.735\frac{EI}{a^3}\Delta$$

$$-0.909\omega_B + 1.993\omega_C - 13.202\frac{\Delta}{a} = -\frac{1}{2}\frac{qa^3}{EI}$$

The rotations of displaced joints and the horizontal displacement become

$$\omega_B = -0.137\frac{qa^3}{EI}$$

$$\omega_C = 0.125\frac{qa^3}{EI}$$

$$\Delta = 0.067\frac{qa^4}{EI}$$

The slopes of the bars can be deduced by

$$\beta_{AB} = \frac{\Delta}{2a} = 0.0335\frac{qa^3}{EI}$$

$$\beta_{DC} = \frac{\Delta}{a} = 0.067\frac{qa^3}{EI}$$

$$\beta_{BC} = \frac{3\Delta}{4a} = 0.050\frac{qa^3}{EI}$$

Substituting the values of slopes ω_B, ω_C and Δ with the bending moment expressions, we obtain the moments at the ends of bars.

$$M_{AB} = \frac{2EI}{\sqrt{5}a}\omega_B + \frac{3EI}{\sqrt{5}a^2}\Delta = -0.216\ qa^2$$

$$M_{BA} = \frac{4EI}{\sqrt{5}a}\omega_B - \frac{3EI}{\sqrt{5}a^2}\Delta = -0.335\,qa^2$$

$$M_{BC} = \frac{1}{3}qa^2 + \frac{2EI}{a}\omega_B + \frac{EI}{a}\omega_C + \frac{9EI}{4a^2}\Delta = 0.335\ qa^2$$

$$M_{CB} = -\frac{1}{3}qa^2 + \frac{EI}{a}\omega_B + \frac{2EI}{a}\omega_C + \frac{9EI}{4a^2}.\Delta = -0.073\ qa^2$$

$$M_{CD} = \frac{4EI}{\sqrt{2}a}\omega_C - \frac{6EI}{\sqrt{2}a^2}\Delta = 0.073\ qa^2$$

$$M_{DC} = \frac{2EI}{\sqrt{2}a}\omega_C - \frac{6EI}{\sqrt{2}a^2}\Delta = -0.105\ qa^2$$

The bending moment and shear force moment variation can be found in the corresponding section established by the method of forces.

Applying the procedure allows us to determine the support reactions, which can be used in the same way and lead to the results, as shown in Figure 4.52.

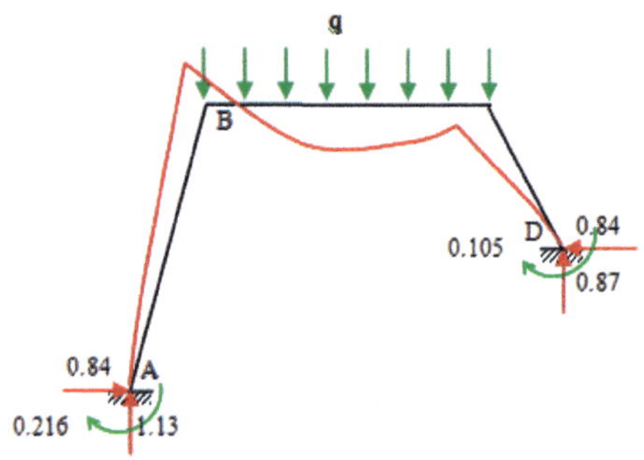

Figure 4.52. *Deflection of frame, fixed moments *(qa²) and support reactions *(qa)*

4.5.2.4. *Multistory frame*

Wed determine the moments at the joints and plot the diagrams of the bending moment and shear force of a multistory frame, as shown in Figure 4.53, using the slope-deflection method.

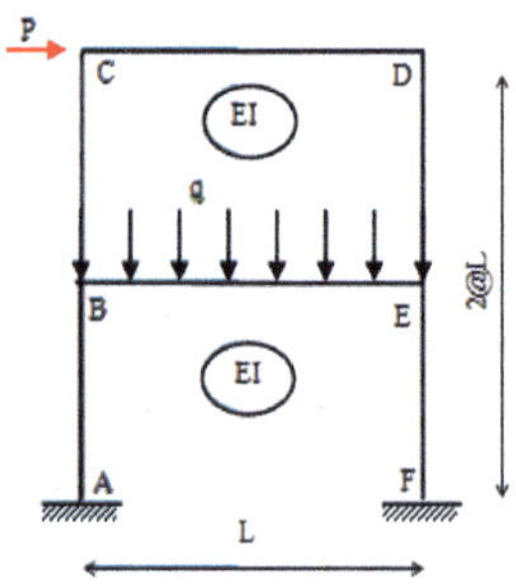

Figure 4.53. *Multistory frame*

Under the effect of the horizontal force P and load q, the frame is deformed as shown in Figure 4.54.

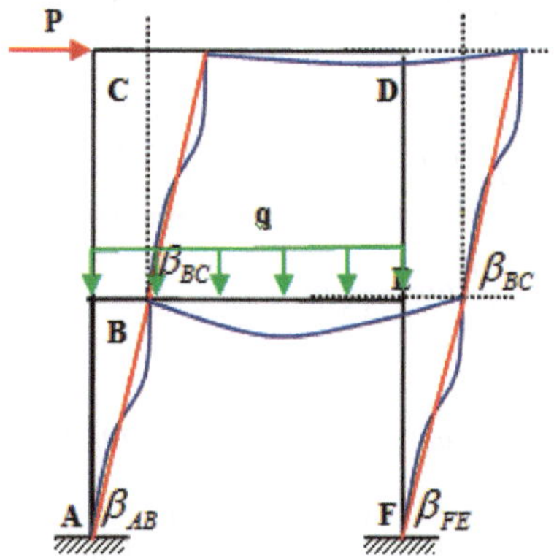

Figure 4.54. *Deformed frame*

The kinematic unknowns are the six unknowns of the problem posed, which are ω_B, ω_C, ω_D, ω_E, Δ_1 and Δ_2. The displaced joints can turn with rotations ω_B, ω_C, ω_D, ω_E and under the effect of horizontal loads F_1 and F_2, the frame can move in the horizontal direction of Δ_1 and Δ_2. The horizontal displacement Δ_2 is considered to be the relative displacement of the second level with respect to the first level characterized by displacement Δ_1.

– *Fixed-end moments*

$$\gamma_{BE} = -\gamma_{EB} = \frac{1}{12} q.L^2$$

– *Slopes of columns*

$$\beta_{AB} = \beta_{BA} = \beta_{FE} = \beta_{EF} = -\frac{\Delta_1}{L}$$

$$\beta_{BC} = \beta_{CB} = \beta_{ED} = \beta_{DE} = -\frac{\Delta_2 - \Delta_1}{L}$$

– *The moments at the ends of each beam*

$$M_{AB} = \frac{2EI}{L}\omega_B - \frac{6EI}{L}\beta_{AB} = \frac{2EI}{L}\omega_B - \frac{6EI}{L^2}\Delta_1$$

$$M_{BA} = \frac{4EI}{L}\omega_B - \frac{6EI}{L}\beta_{BA} = \frac{4EI}{L}\omega_B - \frac{6EI}{L^2}\Delta_1$$

$$M_{BE} = \frac{4EI}{L}\omega_B + \frac{2EI}{L}\omega_C + \frac{1}{12}qL^2$$

$$M_{BC} = \frac{4EI}{L}\omega_B + \frac{2EI}{L}\omega_C - \frac{6EI}{L}\beta_{BC} = \frac{4EI}{L}\omega_B + \frac{2EI}{L}\omega_C - \frac{6EI}{L^2}\Delta_2$$

$$M_{CB} = \frac{2EI}{L}\omega_B + \frac{4EI}{L}\omega_C - \frac{6EI}{L}\beta_{CB} = \frac{2EI}{L}\omega_B + \frac{4EI}{L}\omega_C - \frac{6EI}{L^2}\Delta_2$$

$$M_{CD} = \frac{4EI}{L}\omega_C + \frac{2EI}{L}\omega_D$$

$$M_{DC} = \frac{2EI}{L}\omega_C + \frac{4EI}{L}\omega_D$$

$$M_{DE} = \frac{4EI}{L}\omega_D + \frac{2EI}{L}\omega_{EC} - \frac{6EI}{L}\beta_{DE} = \frac{4EI}{L}\omega_D + \frac{2EI}{L}\omega_E - \frac{6EI}{L^2}\Delta_2$$

$$M_{ED} = \frac{4EI}{L}\omega_E + \frac{2EI}{L}\omega_D - \frac{6EI}{L}\beta_{ED} = \frac{4EI}{L}\omega_E + \frac{2EI}{L}\omega_D - \frac{6EI}{L^2}\Delta_2$$

$$M_{EB} = -\frac{1}{12}q.L^2 + \frac{4EI}{L}\omega_E + \frac{2EI}{L}\omega_B$$

$$M_{EF} = \frac{4EI}{L}\omega_E - \frac{6EI}{L}\beta_{EF} = \frac{4EI}{L}\omega_E - \frac{6EI}{L^2}\Delta_1$$

$$M_{FE} = \frac{2EI}{L}\omega_E - \frac{6EI}{L}\beta_{EF} = \frac{2EI}{L}\omega_E - \frac{6EI}{L^2}\Delta_1$$

– *Joint equilibrium*

Joint B: $M_{BA} + M_{BC} + M_{BE} = 0$

Joint C: $M_{CB} + M_{CD} = 0$

Joint D: $M_{DC} + M_{DE} = 0$

Joint E: $M_{EB} + M_{ED} + M_{EF} = 0$

The equations, which give the slopes, are written as

$$12\omega_B + 2\omega_C + 2\omega_E - 6\frac{\Delta_1}{L} - 6\frac{\Delta_2}{L} = -\frac{1}{12}\frac{qL^3}{EI}$$

$$2\omega_B + 8\omega_C + 2\omega_D - 6\frac{\Delta_2}{L} = 0.$$

$$2\omega_C + 8\omega_D + 2\omega_E - 6\frac{\Delta_2}{L} = 0.$$

$$2\omega_B + 2\omega_D + 12\omega_E - 6\frac{\Delta_1}{L} - 6\frac{\Delta_2}{L} = \frac{1}{12}\frac{qL^3}{EI}$$

The number of equations equals 4 with 6 unknowns; however, it is necessary to provide two additional equations to solve the given problem. To achieve this goal, we consider the horizontal equilibrium at the base of each level.

$$T_{BC} + T_{ED} = P$$

and

$$T_{AB} + T_{FE} = P$$

In the same way, we can deduce the shear forces at the base of each column by using

$$\sum M_B = 0$$

$$\sum M_E = 0$$

$$\sum M_C = 0$$

and $\sum M_D = 0$

$$T_{AB} = \frac{M_{AB} + M_{BA}}{L}$$

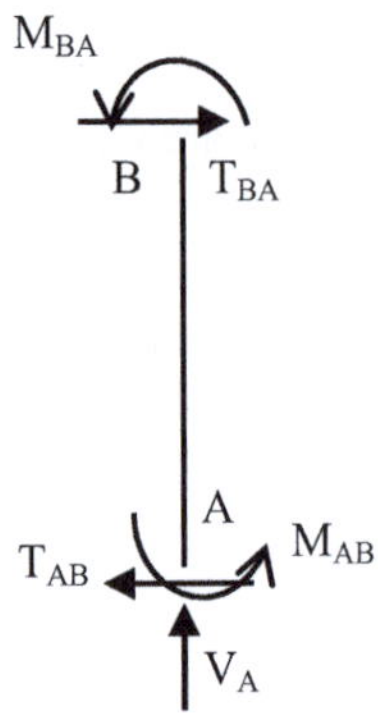

Figure 4.55. *Column (AB)*

Using the last equation makes it possible to deduce the shear force expressions at each level.

– *Level 1*

$$T_{AB} = \frac{6EI}{L^2}\omega_B - \frac{12EI}{L^3}\Delta_1$$

$$T_{FE} = \frac{6EI}{L^2}\omega_E - \frac{12EI}{L^3}\Delta_1$$

– *Level 2*

$$T_{BC} = \frac{6EI}{L^2}\omega_B + \frac{6EI}{L^2}\omega_c - \frac{12EI}{L^3}\Delta_2$$

$$T_{ED} = \frac{6EI}{L^2}\omega_E + \frac{6EI}{L^2}\omega_D - \frac{12EI}{L^3}\Delta_2$$

Substituting the shear force expressions with the horizontal equilibrium relationships, we obtain two additional equations.

$$\omega_B + \omega_C + \omega_D + \omega_E - 4\frac{\Delta_2}{L} = \frac{PL^2}{6EI}$$

$$\omega_B + \omega_E - 4\frac{\Delta_1}{L} = \frac{PL^2}{6EI}$$

Resolving the equations obtained makes it possible to evaluate the slopes at the displaced joints, namely:

$$\omega_B = -0.077\frac{qL^3}{EI}, \quad \omega_C = -0.043\frac{qL^3}{EI}, \quad \omega_D = -0.048\frac{qL^3}{EI}, \quad \omega_E = -0.06\frac{qL^3}{EI},$$

$$\Delta_1 = -0.076\frac{qL^4}{EI}, \qquad \Delta_2 = -0.1\frac{qL^4}{EI}$$

The moments at the ends of bars have the following expressions:

$$M_{AB} = \frac{2EI}{L}\omega_B - \frac{6EI}{L^2}\Delta_1 = 0.30qL^2$$

$$M_{BA} = \frac{4EI}{L}\omega_B - \frac{6EI}{L^2}\Delta_1 = 0.146qL^2$$

$$M_{BE} = \frac{1}{12}q.L^2 + \frac{4EI}{L}\omega_B + \frac{2EI}{L}\omega_E = -0.343qL^2$$

$$M_{BC} = \frac{4EI}{L}\omega_B + \frac{2EI}{L}\omega_C - \frac{6EI}{L^2}\Delta_2 = 0.197qL^2$$

$$M_{CB} = \frac{2EI}{L}\omega_B + \frac{4EI}{L}\omega_C - \frac{6EI}{L^2}\Delta_2 = 0.267qL^2$$

$$M_{CD} = \frac{4EI}{L}\omega_C + \frac{2EI}{L}\omega_D = -0.267qL^2$$

$$M_{DC} = \frac{2EI}{L}\omega_C + \frac{4EI}{L}\omega_D = -0.267qL^2$$

$$M_{DE} = \frac{4EI}{L}\omega_D + \frac{2EI}{L}\omega_E - \frac{6EI}{L^2}\Delta_2 = 0.267qL^2$$

$$M_{ED} = \frac{4EI}{L}\omega_E + \frac{2EI}{L}\omega_D - \frac{6EI}{L^2}\Delta_2 = 0.257qL^2$$

$$M_{EB} = -\frac{1}{12}q.L^2 + \frac{4EI}{L}\omega_E + \frac{2EI}{L}\omega_B = -0.475qL^2$$

$$M_{EF} = \frac{4EI}{L}\omega_E - \frac{6EI}{L^2}\Delta_1 = 0.218qL^2$$

$$M_{FE} = \frac{2EI}{L}\omega_E - \frac{6EI}{L^2}\Delta_1 = 0.336qL^2$$

The moments at the ends of each bar mean we can draw the bending moment and shear force diagrams using

$$M_{AB}(x) = \mu_{AB}(x) - M_{AB}(1 - \frac{x}{L_{AB}}) + M_{AB}(\frac{x}{L_{AB}})$$

$$T_{AB}(x) = \frac{d\mu_{AB}(x)}{dx} + \frac{M_{AB} + M_{BA}}{L_{AB}}$$

The bending moment diagram is shown in Figure 4.56.

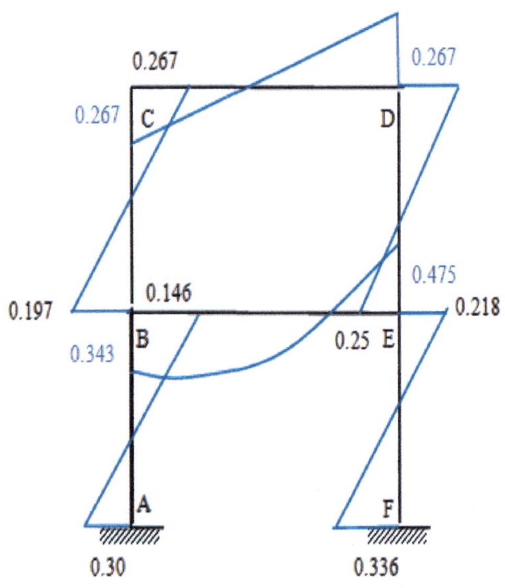

Figure 4.56. *Bending moment diagram *(qL²)*

And the shear force diagram is shown in Figure 4.57.

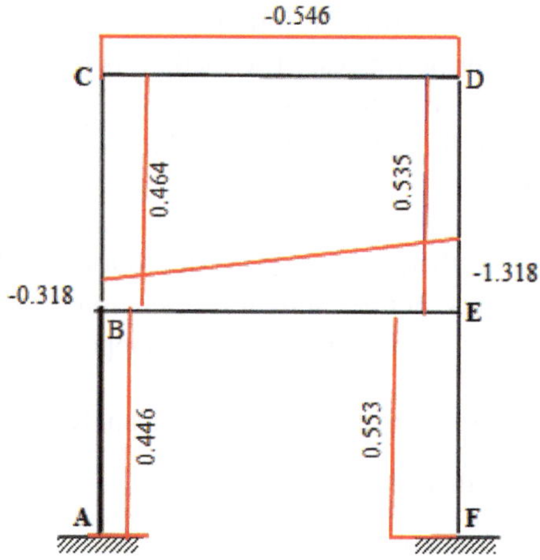

Figure 4.57. *Shear force diagram *(qL)*

The deflection of the frame and the support reactions are shown schematically in Figure 4.58.

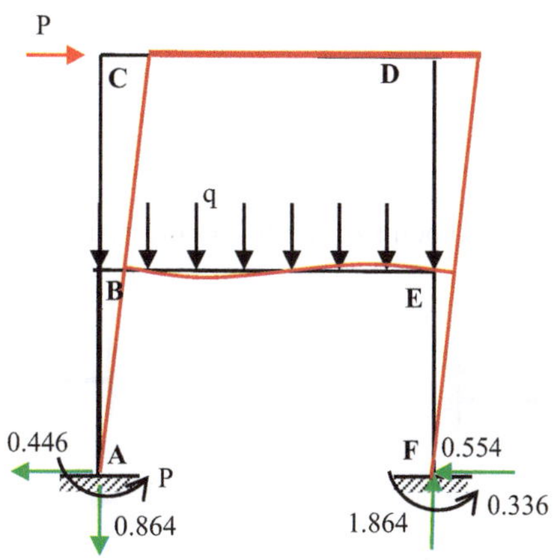

Figure 4.58. *Deflection and support reactions*

4.6. Conclusion

In this chapter, we have shown the formulation of the slope-deflection method for analyzing beams and plane frames. The method is based on the formulation of the expression of the moments at the ends of each bar. This expression is a function of the slopes of the ends of the bar and the slope due to the support settlement if it is taken into consideration.

The slope-deflection method is a very powerful tool for analyzing beams without and with support settlements. Furthermore, it is often used to analyze frames without sidesway and with sidesway, and with and without support settlements.

Finally, the slope-deflection method allows us to analyze:

– continuous beams with and without cantilever;

– beams with support settlements;

– frames without sidesway and frames with sidesway;

– frames with support settlements;

– frames with oblique columns and of different lengths;

– multistory frames.

4.7. Problems

Exercise 1

Calculate the fixed-end moments of the following beams:

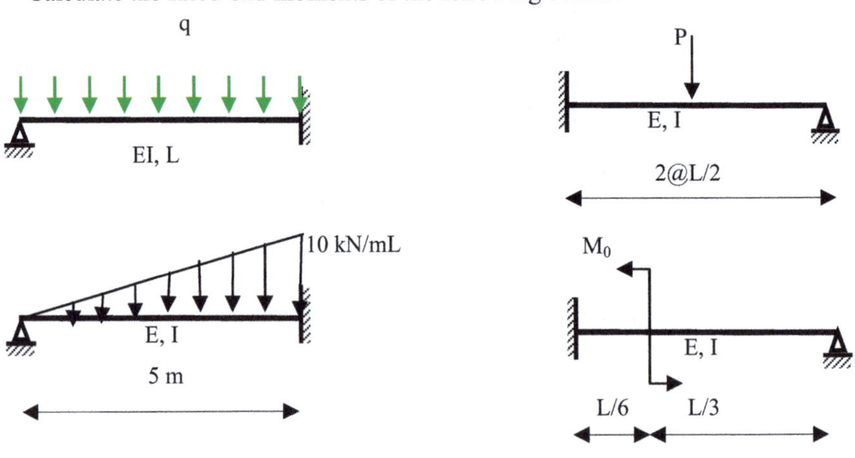

Exercise 2

Calculate the moments at the ends of each element of the following structures:

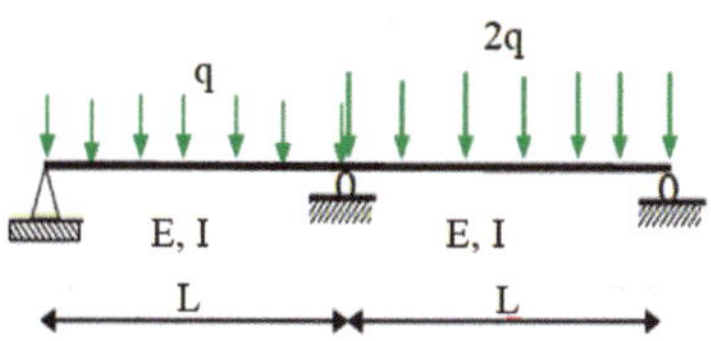

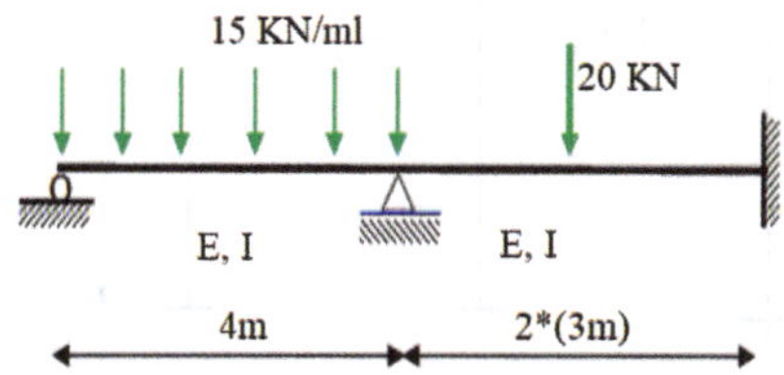

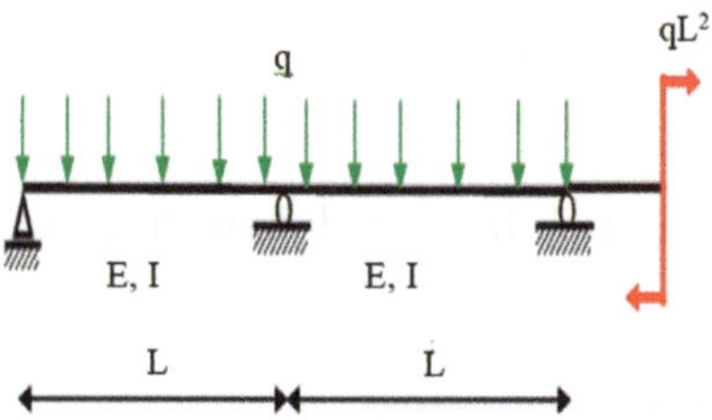

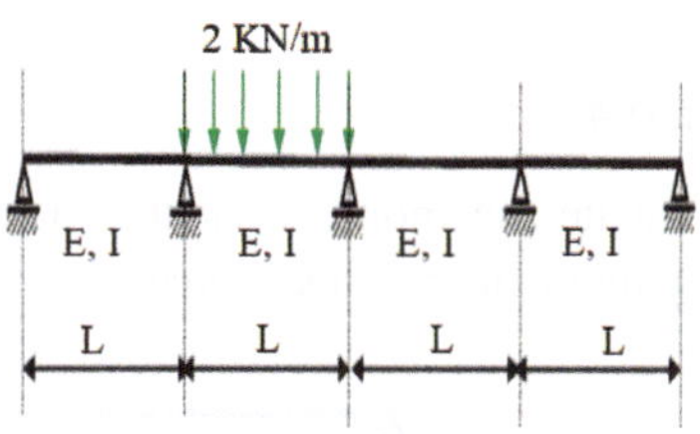

Exercise 3

Calculate the moments at the ends of each bar of the following frames:

Draw diagrams of the bending moment and the shear force for each frame.

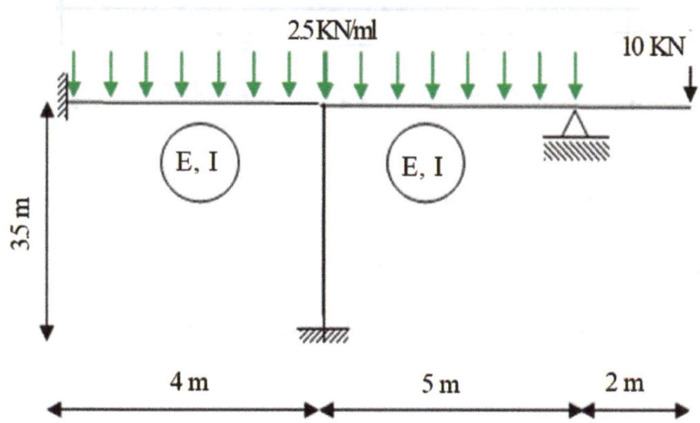

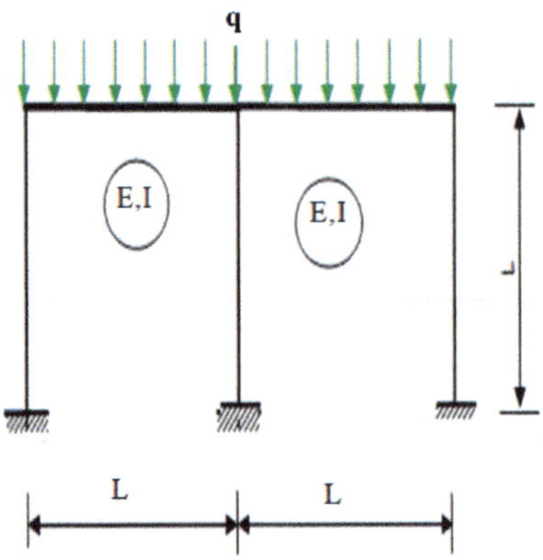

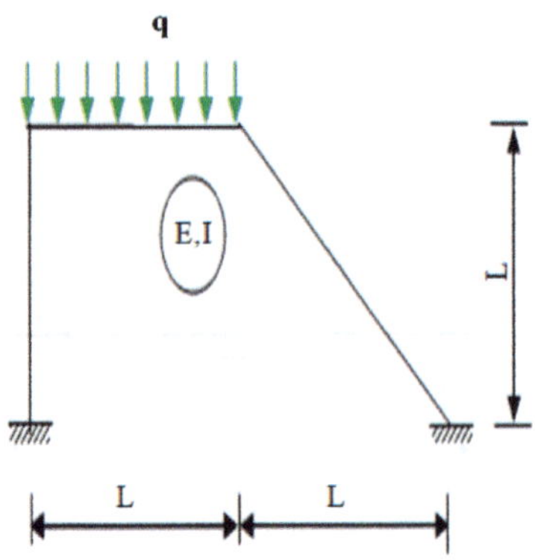

Exercise 4

Calculate the moments at the joints of each frame and draw diagrams of the bending moment and shear force.

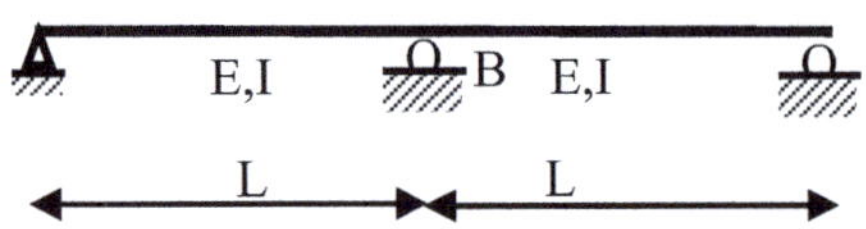

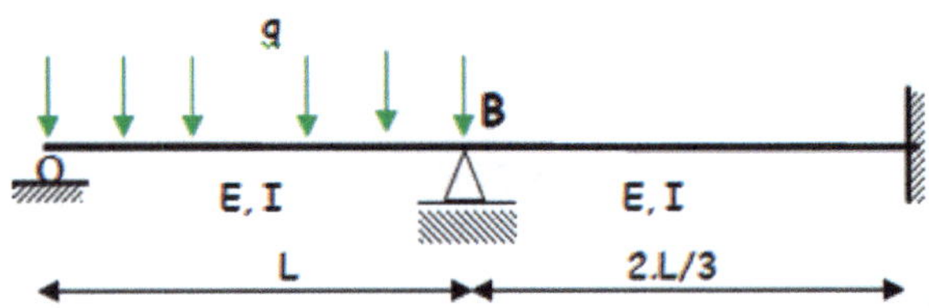

Settlement of support B of 75 mm.

Settlement of support B (Δ) and load q.

Settlement of support B of 80 mm and load q = 2.5 kN/mL.

Exercise 5

Calculate the moments at the joints of each frame.

Draw diagrams of the bending moment and the shear force.

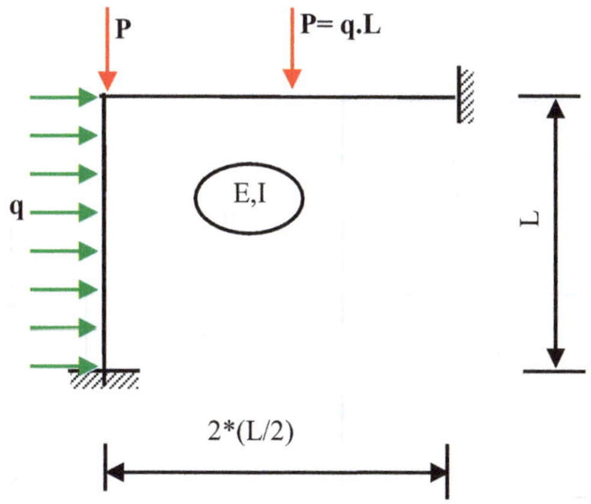
P
P= q.L
E,I
q
L
2*(L/2)

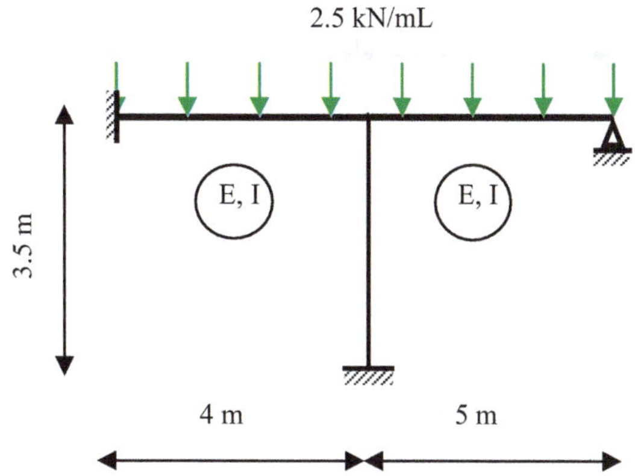
2.5 kN/mL
E, I
E, I
3.5 m
4 m
5 m

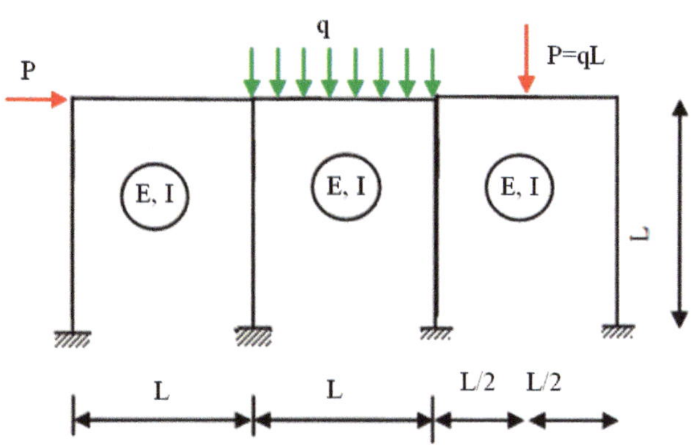
q
P
P=qL
E, I
E, I
E, I
L
L
L
L/2
L/2

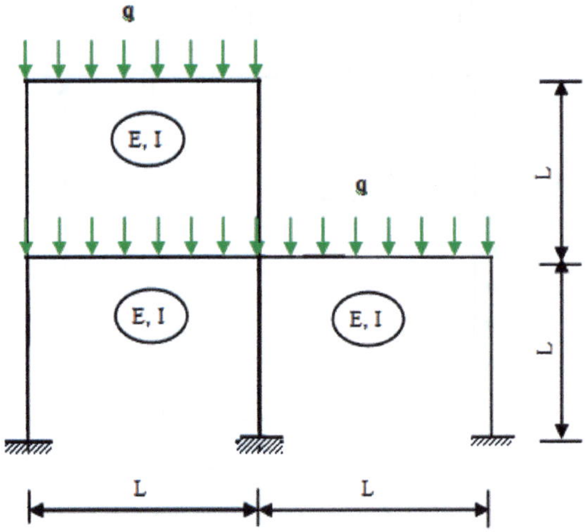
q
E, I
q
E, I
E, I
L
L
L
L

5

Moment-Distribution Method

The teaching objectives for this chapter are as follows:

– identifying the law of distribution of moments around a joint;

– applying the moment-distribution method to:

- the analysis of continuous beams with and without support settlements;

- the analysis of frames with and without sidesway;

- the analysis of simple and multistory frames with and without support settlements;

– using this method in practice.

First, we present the basic concepts of the moment-distribution method. Second, we discuss the application of this method in the analysis of continuous beams, and frames with and without displaced joints.

5.1. Hypotheses of the moment-distribution method

To establish the moment-distribution method, it is necessary to consider the following hypotheses:

1) only the bending deflections are taken into account, axial and shear deflections due to the normal and shearforces are neglected;

2) the flexural rigidity of each bar is assumed to be constant.

5.2. Presentation of the moment-distribution method

5.2.1. *Distribution of a moment around a rigid joint*

We consider moment M_A applied around a rigid joint A (Figure 5.1). Moment M_A is distributed among the bars passing through joint A.

Applying the slope-deflection method to the structure (Figure 5.1), we get:

$$M_{AB} = k_{AB}\omega_A \tag{5.1}$$

$$M_{AC} = k'_{AC}\omega_A \tag{5.2}$$

$$M_{AD} = k_{AD}\omega_A \tag{5.3}$$

$$M_{AE} = k'_{AE}\omega_A \tag{5.4}$$

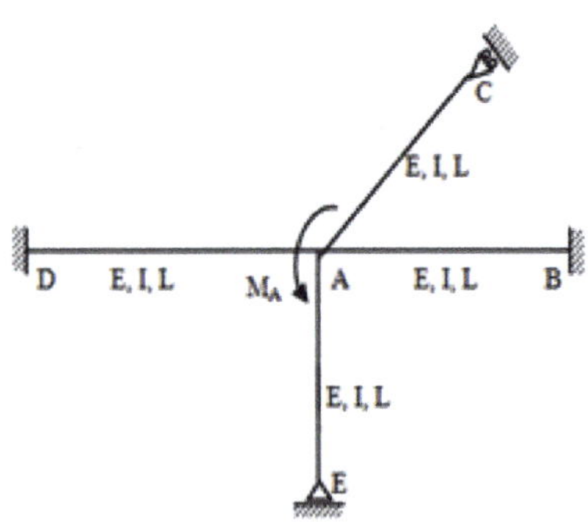

Figure 5.1. *Distribution of nodal moment M_A*

The equilibrium joint A allows us to write

$$M_A = M_{AB} + M_{AC} + M_{AD} + M_{AE} \tag{5.5}$$

Equation [5.5] is written in the general form

$$M_A = \sum_{i=1}^{n} M_{Ai} \tag{5.6}$$

where n is the number of bars passing through joint A.

We deduce the slope at joint A by

$$\omega_A = \frac{M_{AB}}{K_{AB}} = \frac{M_{AC}}{K'_{AC}} = \frac{M_{AD}}{K_{AD}} = \frac{M_{AE}}{K'_{AE}} \qquad [5.7]$$

The slope at joint A can also be written as

$$\omega_A = \frac{M_{AB} + M_{AC} + M_{AD} + M_{AE}}{K_{AB} + K'_{AC} + K_{AD} + K'_{AE}} = \frac{M_A}{\sum K} \qquad [5.8]$$

with $\sum K = K_{AB} + K'_{AC} + K_{AD} + K'_{AE}$

Taking equations [5.7] and [5.8] into account, we deduce the contribution of the distributed moment to each bar by the following expressions:

$$M_{AB} = \frac{K_{AB}}{\sum K} M_A \qquad [5.9]$$

In the same way, the moment returning to each bar is deduced by

$$M_{AC} = \frac{K'_{AB}}{\sum K} M_A \qquad [5.10]$$

$$M_{AD} = \frac{K_{AD}}{\sum K} M_A \qquad [5.11]$$

$$M'_{AE} = \frac{K'_{AE}}{\sum K} M_A \qquad [5.12]$$

We conclude that moment M_A is distributed between the elements in proportion to the rigidity factors of the bars.

The distribution factor of a bar is defined by the quotient of the rigidity factor of the bar and the sum of the rigidity factors of all the bars. It is written as

$$r_{ij} = \frac{K_{ij}}{\sum K} \qquad [5.13]$$

For example, the distribution coefficient of bar (AD) is

$$r_{AD} = \frac{K_{AD}}{\sum K}$$

We state that

$$r_{AB} + r_{Ac} + r_{AD} + r_{AE} = \frac{K_{AB}}{\sum K} + \frac{K'_{AC}}{\sum K} + \frac{K_{AD}}{\sum K} + \frac{K'_{AE}}{\sum K} \quad [5.14]$$

with $\sum K = K_{AB} + K'_{Ac} + K_{AD} + K'_{AE}$

Hence, the sum of the distribution coefficients is equal to unit, $\sum r = 1$.

This implies that the distribution of the moment around a joint is perfect.

Relationships [5.9–5.12] are written as follows:

$$M_{AB} = r_{AB} M_A \quad [5.15]$$

$$M_{AC} = r_{AC} M_A \quad [5.16]$$

$$M_{AD} = r_{AD} M_A \quad [5.17]$$

$$M'_{AE} = r_{AE} M_A \quad [5.18]$$

Relationships [5.15–5.18] describe the distribution of the nodal moment M_A in the bars passing through joint A.

5.2.2. *Distribution procedure*

The first stage of the distribution of moments is to consider that the structure is a set of bi-fixed bars where each member must be studied individually. The moment-distribution method is an iterative procedure, in which the joints have the freedom to rotate but this rotation movement is temporarily blocked by imaginary clamps.

After applying external actions or a support settlement, fixed-end moments occur at the ends of each bar. When these moments arise, it generates an imbalance in the structure at the joints, which causes their rotations.

In a general way, we consider a bar (ij) stressed by any load q(x) or subjected to a support settlement creating fixed-end moments γ_{ij} and γ_{ji} at ends i and j of the bar (Figure 5.2).

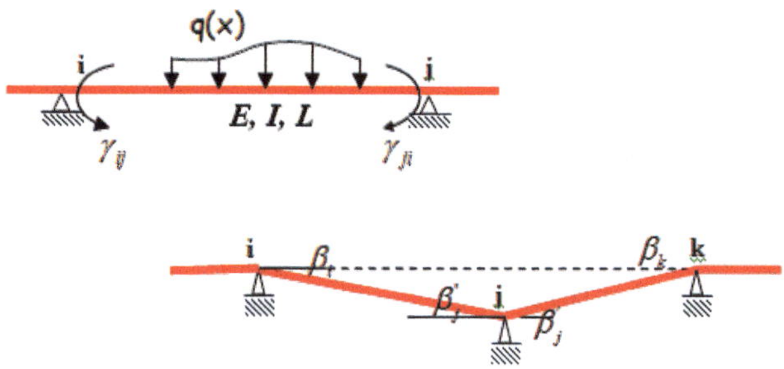

Figure 5.2. *Loading or support settlement*[1]

The clamps placed at joints i and j exert moments γ_{ij} and γ_{ji} on the bar. Conversely, the bar generates moments at the ends of the bar i and j.

$$M_i = -\gamma_{ij}$$

$$M_j = -\gamma_{ji}$$

The moment-distribution method is established as below.

– *Stage 1*

- *Distribution of moments*

We release the clamp at joint *i*, the moment $M_i = -\gamma_{ij}$ must be distributed in all the bars passing through joint *i*.

$$M_{ij}^{(1)} = -r_{ij} \cdot \gamma_{ij}$$

1 All of the figures in this chapter are available to view in full color at www.iste.co.uk/khalfallah/analysis2.zip.

- *Transmission*

The moment of joint i of expression $M_{ij}^{(1)} = -r_{ij}\gamma_{ij}$ transmits at joint j a moment equal to

$$M_{ji}^{(1)} = -\lambda_{ij} r_{ij} \gamma_{ij}$$

We state that joint i is in equilibrium.

– *Stage 2*

- *Distribution of moments*

After the transmission of moment $M_{ij}^{(1)}$, the resulting moment at joint j exerted by the clamp at joint j on bar (ji) is

$$M_{j}^{(2)} = \gamma_{ji} - \lambda_{ij} r_{ij} \gamma_{ij}$$

Conversely, the bar (ji) acts on clamp j by a moment equal to ($-M_{j}^{(2)}$).

Now, we block joint i with a clamp and we release joint j, the moment $-M_{j}^{(2)}$ must be distributed between the bars passing through joint j.

$$M_{ji}^{(2)} = r_{ji}(-M_{j}^{(2)}) = -r_{ji}(\gamma_{ji} - \lambda_{ij} r_{ij} \gamma_{ij})$$

- *Transmission*

The resulting moment at the end (*ji*) of the bar transmits to joint i a moment equal to

$$M_{ij}^{(2)} = \lambda_{ji} M_{ji}^{(2)} = -\lambda_{ji} r_{ji}(\gamma_{ji} - \lambda_{ij} r_{ij} \gamma_{ij})$$

In this case, joint j is in equilibrium and the moment exerted by clamp i on the bar (*ij*) becomes

$$M_{i}^{(3)} = M_{i}^{(1)} + M_{ij}^{(1)} + M_{ij}^{(2)}$$

or

$$M_i^{(3)} = \gamma_{ij} - r_{ij}\gamma_{ij} - \lambda_{ji}r_{ji}(\gamma_{ji} - \lambda_{ij}r_{ij}\gamma_{ij})$$

The above iterations of equilibrium and transmission can be repeated until the moments released after removing the clamps become rather negligible. Finally, the final moment at each end is calculated by the sum of all the moments obtained during the various iterations. Although the quality of the obtainable results depends on the number of iterations chosen, for this reason, it is called an iterative method.

5.3. Continuous beam analysis

Analyze the beam (Figure 2.21) using the moment-distribution method. The value of the flexural rigidity (EI) is assumed to be constant along the beam.

– *Rigidity factors*

$$K_{AB} = K_{BA} = \frac{4EI}{L}$$

$$K_{BC} = K_{CB} = \frac{4EI}{L}$$

– *Transmission coefficients*

$$\lambda_{AB} = \lambda_{BA} = \frac{1}{2}$$

$$\lambda_{BC} = \lambda_{CB} = \frac{1}{2}$$

– *Distribution factors*

$$r_{BA} = \frac{K_{BA}}{K_{BA} + K_{BC}} = \frac{4}{4+4} = \frac{1}{2}$$

$$r_{BC} = \frac{K'_{BC}}{K_{BA} + K'_{BC}} = \frac{4}{4+4} = \frac{1}{2}$$

– *Fixed-end moments*

The fixed-end moments of loaded bars (AB) and (BC) are

$$\gamma_{AB} = -\gamma_{BA} = \frac{1}{12} qL^2$$

$$\gamma_{BC} = \frac{5}{27} PL = \frac{5}{27} qL^2$$

The moment-distribution method is illustrated in Table 5.1.

No. of iteration	Joints	A	B		C
	Bars	AB	BA	BC	CB
	R		$\frac{1}{2}$	$\frac{1}{2}$	
	$\gamma(\frac{qL^2}{108})$	9	-9 (7)	16	-8
1	Equilibrium		–7/2	–7/2	
	Transmission	–7/4			–7/4
	M $(\frac{qL^2}{108})$	29/4	–25/2	25/2	–39/4

Table 5.1. *Equilibrium and transmission of moments. For a color version of this table, see www.iste.co.uk/khalfallah/analysis 2.zip*

The results have already been obtained using the slope-deflection method (section 4.4.2) by the method of forces (section 3.2) and by the method of three moments (section 2.3.1). The moment-distribution method converges in this case, since no moments remain to be redistributed after the first iteration.

5.3.1. *Beam with support settlement*

We analyze the continuous beam using the moment-distribution method (see Figure 2.24) due to the settlement at supports 2 and 3 of Δ_1 and $\Delta_2 (\Delta_1 = 2\Delta_2)$. The beam has a constant flexural rigidity (EI).

The support settlement generates fixed-end moments that are at the ends of bar (*ij*) which are given by

$$M_{ij} = \gamma_{ij} + k_{ij}\omega_i + k_{ij}.\lambda_{ij}.\omega_j - k_{ij}(1+\lambda_{ij})\beta_i$$

$$M_{ji} = \gamma_{ji} + k_{ji}\omega_j + k_{ji}.\lambda_{ji}.\omega_i - k_{ji}(1+\lambda_{ji})\beta_j$$

The fixed-end moments due to the settlement at supports 2 and 3 are

$$\gamma_{12} = \gamma_{21} = -K_{12}(1+\lambda_{12})\beta_{12} = -\frac{6EI}{L}(-\frac{\Delta_1}{L}) = \frac{6EI}{L^2}\Delta_1$$

$$\gamma_{23} = \gamma_{32} = K_{23}(1+\lambda_{23})\beta_{23} = -\frac{6EI}{L^2}(\Delta_1 - \Delta_2)$$

$$\gamma_{34} = \gamma_{43} = -K_{34}(1+\lambda_{34})\beta_{34} = -\frac{6EI}{L^2}\Delta_2$$

Knowing that it is possible to write Δ_1 according to Δ_2, the fixed-end moments become

$$\gamma_{12} = \gamma_{21} = -\frac{12EI}{L^2}\Delta_2$$

$$\gamma_{23} = \gamma_{32} = \frac{6EI}{L^2}\Delta_2$$

$$\gamma_{34} = \gamma_{43} = -\frac{6EI}{L^2}\Delta_2$$

Applying the moment-distribution method consists of calculating:

– *Rigidity factors*

$$K_{12} = K_{21} = K_{23} = K_{32} = \frac{4EI}{L}$$

– *Transmission coefficients*

$$\lambda_{21} = \lambda_{23} = \lambda_{32} = \frac{1}{2}$$

– *Distribution factors*

$$r_{21} = \frac{K_{21}}{K_{21} + K_{23}} = \frac{4}{4+4} = \frac{1}{2}$$

$$r_{23} = \frac{K_{23}}{K_{21} + K_{23}} = \frac{4}{4+4} = \frac{1}{2}$$

$$r_{32} = \frac{K_{32}}{K_{32} + K_{34}} = \frac{4}{4+3} = \frac{4}{7}$$

$$r_{34} = \frac{K_{34}}{K_{32} + K_{34}} = \frac{3}{4+3} = \frac{3}{7}$$

The implementation of the moment-distribution method is described in Table 5.2.

No. of Iteration	Joint	1	2		3		4
	Bars	1-2	2-1	2-3	3-2	3-4	4-3
	R		$\frac{1}{2}$	$\frac{1}{2}$	$\frac{4}{7}$	$\frac{3}{7}$	
	γ	2	2	–1	–1	–1/2	
1	Equilibrium		–1/2	–1/2	6/7	9/14	
	Transmission	–1/4		3/7	–1/4		0
2	Equilibrium		–3/14	–3/14	1/7	3/28	
	Transmission	–3/28		1/14	–3/28		0
	Equilibrium		–1/28	–1/28	3/49	9/196	
	M	23/14	35/28	–35/28	–29/98	29/98	0

Table 5.2. *Moments at joints* $*(\frac{6EI}{L^2}\Delta_2)$

Table 5.3 shows a comparison between the results obtained from using the moment-distribution method and the slope-deflection method.

Joint	1	2		3		4
Bars	1-2	2-1	2-3	3-2	3-4	4-3
Moment-distribution method	9.8571	7.500	–7.500	–1.7755	1.7755	0
Slope-deflection method	9.6923	7.3846	–7.3846	–1.8461	1.8461	0
Convergence rate	98.32%	98.46%	98.46%	96.17%	96.17%	100%

Table 5.3. *Comparison between the slope-deflection and moment-distribution methods* $(\frac{EI}{L^2}\Delta_2)$

The distribution process is continued by adding another iteration; the moments at the joints are presented in Table 5.4.

No. of iteration	Joint	1	2		3		4
	Bars	1-2	2-1	2-3	3-2	3-4	4-3
	R		$\frac{1}{2}$	$\frac{1}{2}$	$\frac{4}{7}$	$\frac{3}{7}$	
	γ	2	2	–1	–1	–1/2	
1	Equilibrium		–1/2	–1/2	6/7	9/14	
	Transmission	–1/4		3/7	–1/4		0
2	Equilibrium		–3/14	–3/14	1/7	3/28	
	Transmission	–3/28		1/14	–3/28		0
3	Equilibrium		–1/28	–1/28	3/49	9/196	
	Transmission	–1/56		3/98	–1/56		0
	Equilibrium		–3/196	–3/196	1/98	3/392	
	M	91/56	121/98	–121/98	–119/392	119/392	0

Table 5.4. *Moments at joints after three iterations* $*(\frac{6EI}{L^2}\Delta_2)$

Table 5.5 summarizes the comparison between the results obtained using the slope-deflection method and the moment-distribution method after three iterations.

No. of iteration	Joint	1	2		3		4
	Bars	1-2	2-1	2-3	3-2	3-4	4-3
After 2 iterations	Moment-distribution method	9.8571	7.500	–7.500	–1.7755	1.7755	0
After 3 iterations	Moment-distribution method	9.75	7.4081	–7.4081	–1.8214	1.8214	0
	Slope-deflection method	9.6923	7.3846	–7.3846	–1.8461	1.8461	0
Convergence rate	After two iterations	98.32%	98.46%	98.46%	96.17%	96.17%	100%
	After three iterations	99.40%	99.68%	99.68%	98.66%	98.66%	100%

Table 5.5. *Comparison between the slope-deflection and moment-distribution methods* $(\frac{EI}{L^2}\Delta_2)$. For a color version of this table, see www.iste.co.uk/khalfallah/analysis 2.zip*

We find that after the third iteration of the distribution of moments, the results are very satisfactory compared with those of the slope-deflection method.

5.3.2. *Beam with cantilever*

We use the moment-distribution method to analyze the beam (Figure 2.27). The cantilever beam was analyzed by the method of three moments (section 2.3.3) and the slope-deflection method (section 4.4.3). The flexural rigidity (EI) is assumed to be constant along the beam. The moment-distribution method is carried out by establishing the calculation steps.

– *Rigidity factors*

$$K_{12} = K_{21} = \frac{4EI}{L}$$

$$K_{23} = \frac{4EI}{L}$$

– *Transmission coefficients*

$$\lambda_{12} = \lambda_{21} = \frac{1}{2}$$

$$\lambda_{23} = 0$$

– *Distribution factors*

$$r_{21} = \frac{K_{21}}{K_{21} + K_{23}} = \frac{1}{2}$$

$$r_{23} = \frac{K_{23}}{K_{21} + K_{23}} = \frac{1}{2}$$

– *Fixed-end moments*

For simplification reasons, we assume that P = qL.

$$\gamma_{23} = -\gamma_{32} = \frac{1}{12} q.L^2$$

Concentrated force P (Figure 2.27) generates a moment at joint 3 equal to $\frac{PL}{5}$.

No. of iteration	Joint	1	2		3	
	Bars	1-2	2-1	2-3	3-2	3-4
	r		$\frac{1}{2}$	$\frac{1}{2}$	1	
	γ			5	-5	12
1	Equilibrium		–5/2	–5/2	–7	
	Transmission	–5/4		–7/2	–5/4	
2	Equilibrium		7/4	7/4	5/4	
	Transmission	7/8		5/8	7/8	
	Equilibrium		–5/16	–5/16	–7/8	
	M	–3/8	–17/16	17/16	–12	12

Table 5.6. *Moments at joints of a cantilever beam* $*(\frac{qL^2}{60})$

Table 5.7 shows the convergence rate of the moment-distribution method in relation to the slope-deflection method. The convergence rate varies between 80% and 100% for results taken after two calculation iterations.

Joint	1	2		3	
Bars	1-2	2-1	2-3	3-2	3-4
Moment-distribution method	–0.00625	–0.0177	0.0177	–0.20	0.20
Slope-deflection method	–0.0071	–0.0142	0.0142	–0.20	0.20
Convergence rate	88%	80.22%	80.22%	100%	100%

Table 5.7. *Convergence rate and moment-distribution method* (qL^2)

To achieve very satisfactory results, it is necessary to continue the iterative process until the convergence rate is respected.

5.3.3. *Beam subjected to a moment*

We reanalyze the beam shown in Figure 2.40. We assume that the flexural rigidity EI and the length of the spans are equal.

– *Rigidity factors*

$$K_{21} = \frac{3EI}{L}$$

$$K_{23} = K_{32} = K_{34} = \frac{4EI}{L}$$

– *Transmission coefficients*

$$\lambda_{21} = 0$$

$$\lambda_{23} = \lambda_{32} = \lambda_{34} = \frac{1}{2}$$

– *Distribution factors*

$$r_{21} = \frac{K_{21}}{K_{21} + K_{23}} = \frac{3}{7}$$

$$r_{23} = \frac{K_{23}}{K_{21} + K_{23}} = \frac{4}{7}$$

$$r_{32} = \frac{K_{21}}{K_{21} + K_{23}} = \frac{1}{2}$$

$$r_{34} = \frac{K_{23}}{K_{21} + K_{23}} = \frac{1}{2}$$

The purpose of this exercise is to present how moment M_0 = 15 kN·m is distributed along the beam (1-4) using the moment-distribution method.

No. of iteration	Joint	1	2		3		4
	Bars	1-2	2-1	2-3	3-2	3-4	4-3
	r		$\frac{3}{7}$	$\frac{4}{7}$	$\frac{1}{2}$	$\frac{1}{2}$	
1	Equilibrium						1
	Transmission					1/2	
2	Equilibrium				–1/4	–1/4	
	Transmission			–1/8			
3	Equilibrium		3/56	4/56			
	Transmission	0			2/56		
4	Equilibrium				–1/56	–1/56	
	Transmission			–1/112			
	Equilibrium		3/784	4/784			
	$M(M_0)$		45/784	–45/784	–13/56	13/56	1

Table 5.8. *Distribution of shore moment*

Table 5.8 shows the distribution of a shore moment at the joints of a continuous beam using the moment-distribution method. The comparison between the results obtained by this method and the slope-deflection method is presented in Table 5.9.

Joint	1	2		3		4
Bars	1-2	2-1	2-3	3-2	3-4	4-3
Moment-distribution method	0	0.861	–0.861	–3.4821	3.4821	15
Slope-deflection method	0	1	–1	–4	4	15
Convergence rate	100%	86.10%	86.10%	87.05%	87.05%	100%

Table 5.9. *Comparison between the slope-deflection and moment-distribution methods*

The convergence rate varies between 86.10% and 100%. It should be noted that the convergence rate depends on the number of iterations used.

5.4. Analysis of frames

5.4.1. *Frame without sidesway*

5.4.1.1. *Single span frame*

We reanalyze the simple frame (Figure 3.19) using the moment-distribution method. The flexural rigidities of bars (AB) and (BC) are, respectively, (EI) and (2EI).

– *Rigidity factors*

$$K_{AB} = K_{BA} = \frac{4EI}{L}$$

$$K_{BC} = \frac{3E(2I)}{L} = \frac{6EI}{L}$$

– *Distribution factor*

$$r_{BA} = \frac{K_{BA}}{K_{BA} + K_{BC}} = \frac{2}{5}$$

$$r_{BC} = \frac{K_{BC}}{K_{BA} + K_{BC}} = \frac{3}{5}$$

– *Transmission coefficients*

$$\lambda_{AB} = \lambda_{BA} = \frac{1}{2}$$

$$\lambda_{BC} = 0$$

– *Fixed-end moments*

$$\gamma_{BC} = \frac{1}{8} qL^2$$

No. of iteration	Joint	A	B		C
	Bars	AB	BA	BC	CB
	r		$\frac{2}{5}$	$\frac{3}{5}$	
	$\gamma(\frac{qL^2}{8})$			1	
1	Equilibrium		–2/5	–3/5	
	Transmission	–1/5			0
	M($(\frac{qL^2}{8})$	–1/5	–2/5	2/5	0
	M((qL^2)	–1/40	–1/20	1/20	0
	Slope-deflection method	–1/40	–1/20	1/20	0
	Convergence rate	100%			

Table 5.10. *Frame analysis using the moment-distribution method*

5.4.1.2. *Multiple span frame*

We reanalyze the symmetrical frame whose mechanical and geometrical characteristics are shown in Figure 4.24.

– *Rigidity factor*

$$K'_{BA} = K'_{CD} = K'_{CE} = K'_{BF} = \frac{3EI}{L}$$

$$K_{BC} = K_{CB} = \frac{8EI}{L}$$

– *Transmission coefficients*

$$\lambda'_{AB} = \lambda'_{BA} = \lambda'_{BC} = \lambda'_{CB} = \lambda'_{CD} = \lambda'_{DC} = 0$$

$$\lambda_{BC} = \lambda_{CB} = \frac{1}{2}$$

– *Distribution factors*

$$r_{BA} = r_{BF} = r_{CD} = r_{CE} = \frac{K'_{BA}}{K'_{BA} + K'_{BC} + K'_{BF}} = \frac{3}{3+3+8} = \frac{3}{14}$$

$$r_{BC} = r_{CB} = \frac{K_{BC}}{K'_{BA} + K'_{BF} + K_{BC}} = \frac{8}{14}$$

– *Fixed-end moments*

The fixed-end moments of loaded bars are

$$\gamma_{BC} = -\gamma_{CB} = \frac{1}{12} q.L^2$$

$$\gamma_{BF} = -\frac{3}{16} q.L^2$$

$$\gamma_{CE} = \frac{3}{16} q.L^2$$

The equilibrium and the transmission of the iterations are grouped in Table 5.11.

Joint	B			C		
Bars	BA	BF	BC	CB	CE	CD
r	$\frac{3}{14}$	$\frac{3}{14}$	$\frac{8}{14}$	$\frac{8}{14}$	$\frac{3}{14}$	$\frac{3}{14}$
$\gamma(\frac{q.L^2}{48})$		–9	4	–4	9	
Equilibrium	15/14	15/14	40/14	–40/14	–15/14	–15/14
Transmission			–20/14	20/14		
Equilibrium	60/196	60/196	160/196	–160/196	–60/196	–60/196
Transmission			–80/196	80/196		
Equilibrium	240/2744	240/2744	640/2744	–640/2744	–240/2744	–240/2744
M $(\frac{q.L^2}{131,712})$	4020	–20,676	16,656	–16,676	–4,020	20,676
M $(q.L^2)$	0.0305	–0.1569	0.1264	–0.1264	–0.0305	0.1569
M$(q.L^2)$ Slope-deflection method	0.0312	–0.1562	0.125	–0.125	–0.0312	0.1562
Convergence	97.75%	99.55%	98.89%	98.89%	97.75%	99.55%

Table 5.11. *Analysis of frame without sidesway. For a color version of this table, see www.iste.co.uk/khalfallah/analysis 2.zip*

Table 5.11 also shows the convergence rate of the moment-distribution method. This rate is fairly convergent, varying between 97.75% and almost 100%.

5.4.1.3. *Frame subjected to a support settlement*

We use the moment-distribution method to reanalyze the frame subjected to a support settlement (Figure 4.30). The flexural rigidity of the bars is constant and their lengths are equal. Support C has a value of Δ.

– *Distribution factors*

$$r_{BA} = \frac{K_{BA}}{K_{Ba} + K_{BC} + K'_{BD}} = \frac{4}{4+4+3} = \frac{4}{11}$$

$$r_{BC} = \frac{K_{BC}}{K_{Ba} + K_{BC} + K'_{BD}} = \frac{4}{4+4+3} = \frac{4}{11}$$

$$r_{BD} = \frac{K_{BD}}{K_{Ba} + K_{BC} + K'_{BD}} = \frac{3}{4+4+3} = \frac{3}{11}$$

– *Fixed-end moments*

$$\gamma_{AB} = \gamma_{BA} = -\frac{6EI}{L}\psi = \frac{6EI}{L^2}\Delta$$

$$\gamma_{BD} = -\frac{3EI}{L}\psi = -\frac{3EI}{L^2}\Delta$$

Implementing the moment-distribution method to analyze frames subjected to a support settlement is shown in Table 5.12.

Joint	A	B			C	D
Bars	AB	BA	BC	BD	CB	DB
R		4/11	4/11	3/11		
γ	6	6		–3		
Equilibrium		–12/11	–12/11	–9/11		
Transmission	–6/11				–6/11	
M	60/11	54/11	–12/11	–42/11	–6/11	0

Table 5.12. *Analysis of frame with support settlement *($\frac{EI}{L^2}\Delta$). For a color version of this table, see www.iste.co.uk/khalfallah/analysis 2.zip*

In the same context, a comparison of the results obtained is given in Table 5.13. In this case, a perfect convergence of the moment-distribution method is observed.

Joint	A	B			C	D
Bars	AB	BA	BC	BD	CB	DB
Moment-distribution method	60	54	–12	–42	–6	0
Slope-deflection method	60	54	–12	–42	–6	0
Convergence rate	100 %					

Table 5.13. *Comparison between the slope-deflection and moment-distribution methods* ($\frac{EI}{11L^2}\Delta$). For a color version of this table, see www.iste.co.uk/khalfallah/analysis 2.zip*

5.4.2. Frame with sidesway

5.4.2.1. Frame with cantilever

We reanalyze the frame with cantilever (Figure 3.26) using the moment-distribution method. We assume that the flexural rigidity (EI) is constant and that P = q.L.

– *Rigidity factors*

$$K_{BA} = \frac{3EI}{L}$$

$$K_{BC} = K_{CB} = \frac{4EI}{L}$$

$$K_{CD} = K_{DC} = \frac{4EI}{L}$$

– *Transmission coefficients*

$$\lambda_{BA} = 0.$$

$$\lambda_{BC} = \lambda_{CB} = \lambda_{CD} = \lambda_{DC} = \frac{1}{2}$$

– *Distribution factors*

$$r_{BA} = \frac{K_{BA}}{K_{BA} + K_{BC}} = \frac{3}{3+4} = \frac{3}{7}$$

$$r_{BC} = \frac{K_{BC}}{K_{BA} + K_{BC}} = \frac{4}{4+3} = \frac{4}{7}$$

$$r_{CB} = r_{CD} = \frac{1}{2}$$

To analyze frames with sidesway, we use the superposition principle. The structure shown in Figure 3.26 can be divided into two systems: (1) the system of fixed joints (Figure 5.3) and (2) the system of displaced joints (Figure 5.6). Both systems were analyzed separately. At the end of the analyses, the effects are superimposed to evaluate the final answer.

The fixed joint structure is obtained by adding an artificial (imaginary) support to joint C to block horizontal movement. In the same way, the displaced joint structure is obtained by releasing the structure or the frames from the displacement; the previously blocked support is suspended.

In the first system, the external loads are applied to the frame and the moments at the joints are calculated using the steps of the moment-distribution method in the usual way.

After determining the moments at the joints, the reaction to the imaginary support can be calculated by the frame's overall equilibrium. In the second system, the frame is stressed by the reaction of the artificial support in the opposite direction (Figure 5.6). Finally, the values of the moments at the ends of each bar add up to the values of the moments at the joints, which have already been calculated in the first phase.

In this context, it is also possible to use an indirect approach, assuming that a force is applied at the imaginary support in the opposite direction to the support reaction (Figure 5.7), causing an arbitrary displacement Δ'.

The fixed-end moments obtained because of the displacement Δ' are distributed using this method. After determining the moments, the reaction to the imaginary support can be calculated using the equation of the frame's overall equilibrium.

Finally, a correction will be made on the arbitrary value that is already associated with the displacement value Δ'. The correction coefficient is deduced by applying the horizontal equilibrium.

$$\sum T_{base}^{0} + k\sum T_{base}^{\Delta'} = \sum F_x \qquad [5.19]$$

where $\sum T_{base}^{0}$ and $\sum T_{base}^{\Delta'}$ are, respectively, the shear forces at the base of the frame without and with sidesway.

The relationship [5.19] allows us to determine the correction coefficient by

$$k = \frac{\sum F_x - \sum T_{base}^{0}}{\sum T_{base}^{\Delta'}} \qquad [5.20]$$

The final moment at joint i can be obtained using the sum of the moments.

$$M_i = M_i^0 + kM_i^{\Delta'} \qquad [5.21]$$

where $M_i^0, M_i^{\Delta'}$ are, respectively, the moment at joint i of the frame without and with sidesway.

– *Frame without sidesway (Figure 5.3)*

The fixed-end moments due to load q are

$$\gamma_{BC} = -\gamma_{CB} = \frac{1}{12}qL^2$$

The moments due to the equilibrium and distribution of moments are presented in Table 5.14.

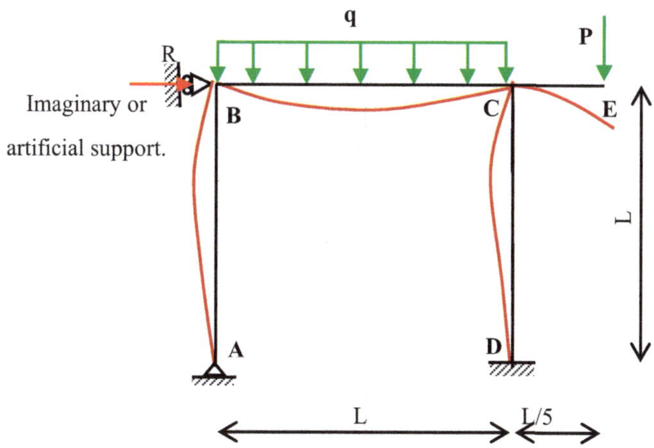

Figure 5.3. *Frame without sidesway*

Joint	A	B		C		D
Bars	AB	BA	BC	CB	CD	DC
r		$\frac{3}{7}$	$\frac{4}{7}$	$\frac{1}{2}$	$\frac{1}{2}$	
$\gamma(\frac{q.L^2}{12})$			1	2.4 –1		
Equilibrium		–3/7	–4/7	–7/10	–7/10	
Transmission	0		–7/20	–2/7		–7/20
Equilibrium		21/140	28/140	1/7	1/7	
Transmission	0		1/14	1/10		1/14
Equilibrium		–3/98	–4/98	–1/20	–1/20	
M (qL^2)	0	–0.0257	0.0257	–0.1493	–0.0506	–0.0232

Table 5.14. *Frame without sidesway. For a color version of this table, see www.iste.co.uk/khalfallah/analysis 2.zip*

We extract element (AB)

$$\sum M_B = 0$$

$$M_{AB} + M_{BA} - T_{AB} L = 0$$

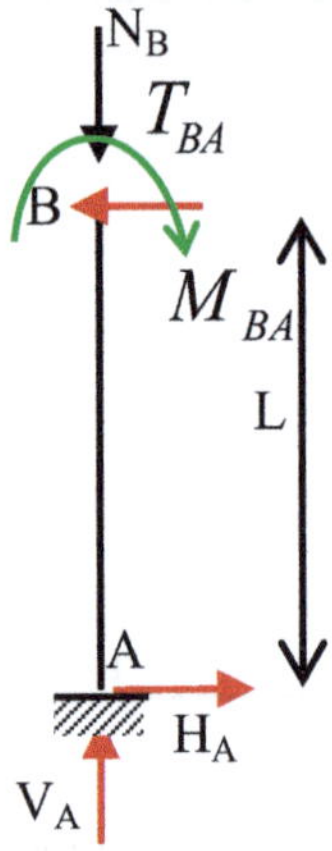

Figure 5.4. *Column (CB)*

$$T_{AB} = \frac{M_{BA}}{L}$$

Substituting the values of M_{AB} and M_{BA}, we find

$$T_{AB} = -0.0257qL$$

The horizontal reaction of bar (CD) is deduced using

$$\sum M_C = 0$$

$$M_{CD} + M_{DC} - T_{DC}.L = 0$$

$$T_{DC} = \frac{M_{DC} + M_{CD}}{L}$$

$$T_{DC} = -0.0738qL$$

$$(T_{AB} + T_{DC})^0 = -0.10qL$$

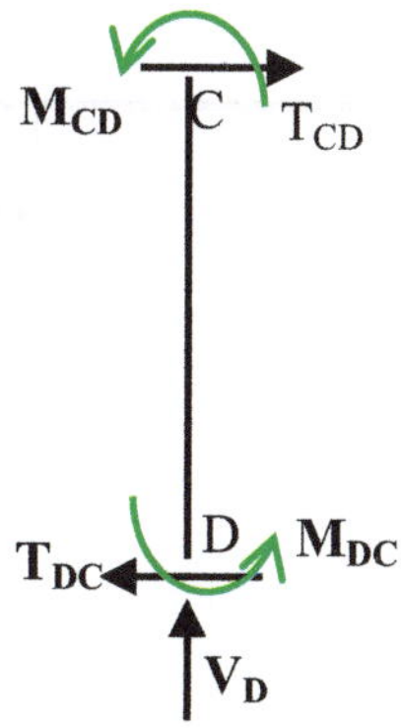

Figure 5.5. *Column (CD)*

The exponent (°) indicates that the calculated forces are from a frame without sidesway.

The horizontal equilibrium of the frame is

$$(T_{AB} + T_{DC})^{\circ} - R = 0$$

$$R = (T_{AB} + T_{DC})^{\circ} = = -0.10qL$$

– *Frame with sidesway*

In this area, there are two different methods.

- *Direct method*

The fixed-end moments due to the displacements of the columns (Figure 5.6) are

$$\gamma_{CD} = \gamma_{DC} = -\frac{6EI}{L}\beta$$

$$\gamma_{BA} = -\frac{3EI}{L}\beta$$

Applying the moment-distribution method leads us to evaluate the moments at the ends of each bar of the frame with sidesway (Table 5.15).

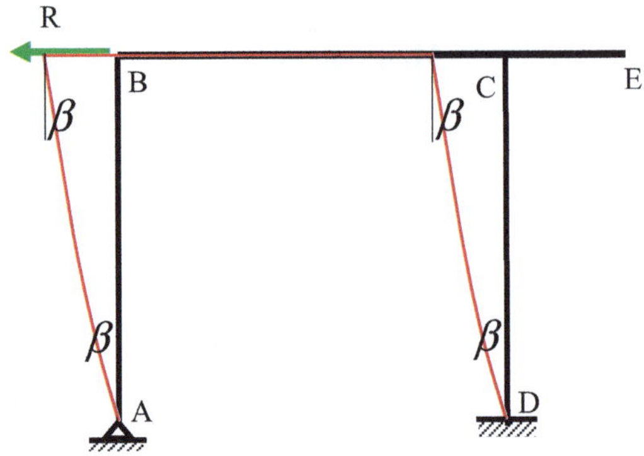

Figure 5.6. *Frame with sidesway – direct method*

Joint	A	B		C		D
Bars	AB	BA	BC	CB	CD	DC
r		$\frac{3}{7}$	$\frac{4}{7}$	$\frac{1}{2}$	$\frac{1}{2}$	
γ		–3			–6	–6
Equilibrium		9/7	12/7	3	3	
Transmission	0		3/2	6/7		3/2
Equilibrium		–9/14	–6/7	–3/7	–3/7	
Transmission	0		–3/14	–3/7		–3/14
Equilibrium		9/98	6/49	3/14	3/14	
M	0	–111/49	111/49	45/14	–45/14	–33/7

Table 5.15. *Frame with sidesway* $*(\frac{EI}{L}\beta)$

The shear forces at the base of the frame are calculated as follows.

The equilibrium equation in relation to point B:

$$M_{BA} - T_{AB}.L = 0$$

$$T_{AB} = \frac{M_{BA}}{L} = -\frac{111}{49}\frac{EI}{L^2}\beta$$

The same procedure is applied to bar (CD):

$$M_{CD} + M_{DC} - T_{DC}.L = 0$$

$$T_{DC} = \frac{M_{DC} + M_{CD}}{L} = -\frac{111}{14}\frac{EI}{L^2}\beta$$

The horizontal equilibrium of the frame (Figure 5.3) allows us to write:

$$T_{AB} + T_{DC} + R = 0.$$

$$\frac{111}{49}\frac{EI}{L^2}\beta + \frac{111}{14}\frac{EI}{L^2}\beta + 0.10\,qL = 0$$

It leads to the evaluation of the slope of the frame's columns.

$$\beta = -9.81.10^{-3}\frac{qL^3}{EI}$$

Table 5.16 groups the final moments at the ends of each bar of the given frame.

Joint	A	B		C		D
Bars	AB	BA	BC	CB	CD	DC
M (qL^2)	0	–0.0257	0.0257	–0.1493	–0.0506	–0.0232
$M^D(\frac{EI}{L}\beta)$	0	–111/49	111/49	45/14	–45/14	–33/7
	$\beta = -9.81.10^{-3}\frac{qL^3}{EI}$					
$M^D(qL^2)$	0	0.0222	–0.0222	–0.0315	0.0315	0.0462

$M^\circ (qL^2)$	0	–0.0257	0.0257	–0.1493	–0.0506	–0.0232
M. (final) $(10^{-3} * qL^2)$	0	–3.50	3.50	–117.76	–19.10	23.05

Table 5.16. *Moments at joints of frame*

- *Indirect method*

The fixed-end moments using an optional displacement Δ' (Figure 5.7) are

$$\gamma_{BA} = -\frac{3EI}{L}\beta$$

$$\gamma_{CD} = \gamma_{DC} = -\frac{6EI}{L}\beta$$

We deduce a relationship between the fixed-end moments due to the effect of the displacement Δ', a relationship which must be established between the fixed-end moments.

$$\gamma_{BA} : \gamma_{CD} : \gamma_{DC} = 1:2:2$$

A fixed value must be assigned to the fixed-end moment. For example: $\gamma_{BA} = 100$ unit of moment. Therefore, $\gamma_{CD} = \gamma_{DC} = 200$ unit of moment.

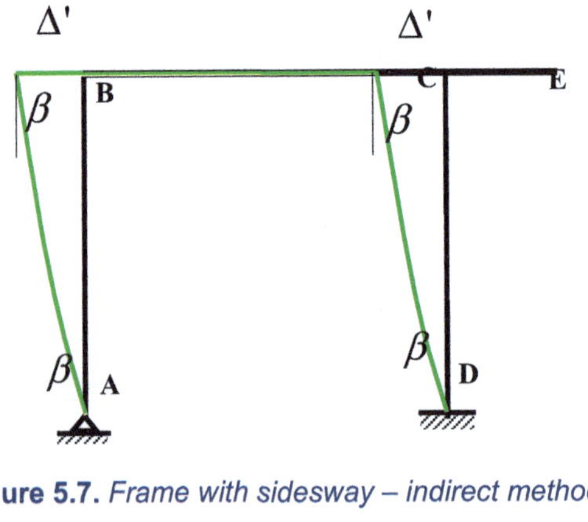

Figure 5.7. *Frame with sidesway – indirect method*

The moments at the ends of each bar are given in Table 5.17.

Joint	A	B		C		D
Bars	AB	BA	BC	CB	CD	DC
r		$\frac{3}{7}$	$\frac{4}{7}$	$\frac{1}{2}$	$\frac{1}{2}$	
γ		100			200	200
Equilibrium		–300/7	–400/7	–100	–100	
Transmission	0		–50	–200/7		–50
Equilibrium		150/7	200/7	100/7	100/7	
Transmission	0		50/7	100/7		50/7
Equilibrium		–150/49	–200/49	–50/7	–50/7	
M^D	0	75.51	–75.51	–107.142	107.142	157.142

Table 5.17. *Moments of a frame with sidesway*

The horizontal reactions of supports B and D can be deduced by

$$T_{AB} = \frac{M_{BA}}{L} = \frac{75.51}{L}$$

$$T_{DC} = \frac{M_{DC} + M_{CD}}{L} = \frac{264.284}{L}$$

$$(T_{AB} + T_{DC})^{\Delta'} = \frac{339.80}{L}$$

Hence

$$(T_{AB} + T_{DC})^0 + k(T_{AB} + T_{DC})^{\Delta'} = 0$$

$$0.10qL - k(\frac{339.80}{L}) = 0$$

The correction constant is given by

$$k = 2.94 \times 10^{-4}.qL^2$$

Table 5.18 groups the moments at the ends of each bar using the indirect method.

Joint	A	B		C		D
Bars	AB	BA	BC	CB	CD	DC
M (qL^2)	0	–0.0257	0.0257	–0.1493	–0.0506	–0.0232
M^D	0	75.51	–75.51	–107.142	107.142	157.142
K	$2.94 \times 10^{-4} qL^2$					
M^D $(10^{-4}.qL^2)$	0	222.0	–222.0	–315.0	315.0	462.0
M^0 (qL^2)	0	–0.0257	0.0257	–0.1493	–0.0506	–0.0232
$M(q.L^2)$	0	–0.0035	0.0035	–0.1808	–0.0191	0.023

Table 5.18. *Moments of a frame with sidesway*

The two methods can be indifferently applied to the analysis of frames with sidesway.

5.4.2.2. *Frame with columns of different lengths*

We reanalyze the frame (Figure 3.33) with columns of different lengths using the moment-distribution method. The mechanical and geometrical characteristics are given in the figure. The analysis of structures with displaced joints is carried out in two steps: (1) the structure with fixed joints and (2) the structure with displaced joints.

– *Frame without sidesway*

The joints of the frame are blocked vis-à-vis the horizontal displacement by means of a roller. The frame thus becomes a frame without sidesway.

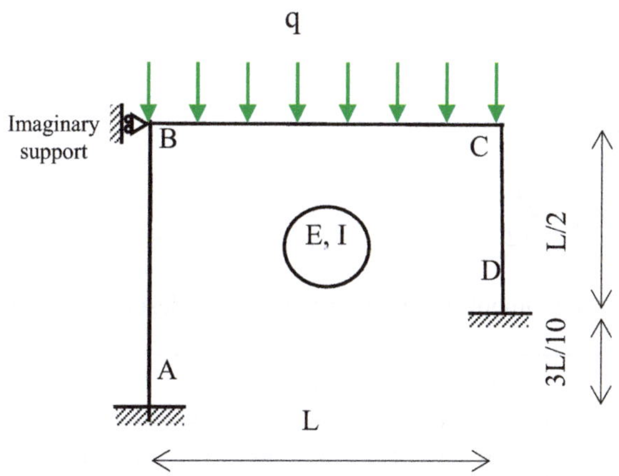

Figure 5.8. *Frame without sidesway*

– *Rigidity factor*

$$K_{AB} = K_{BA} = \frac{5EI}{L}$$

$$K_{BC} = K_{CB} = \frac{4EI}{L}$$

$$K_{CD} = K_{DC} = \frac{8EI}{L}$$

– *Distribution factors*

$$r_{BA} = \frac{K_{BA}}{K_{BA} + K_{BC}} = \frac{5}{5+4} = \frac{5}{9}$$

$$r_{BC} = \frac{K_{BC}}{K_{BA} + K_{BC}} = \frac{4}{4+5} = \frac{4}{9}$$

$$r_{CB} = \frac{1}{3}$$

$$r_{CD} = \frac{2}{3}$$

– *Fixed-end moments*

$$\gamma_{AB} = -\gamma_{BA} = \frac{qL^2}{12}$$

The steps of the moment-distribution method are described in Table 5.19.

Joint	A	B		C		D
Bars	AB	BA	BC	CB	CD	DC
R		$\frac{5}{9}$	$\frac{4}{9}$	$\frac{1}{3}$	$\frac{2}{3}$	
$\gamma(\frac{q.L^2}{12})$			1	–1		
Equilibrium		–5/9	–4/9	1/3	2/3	
Transmission	–5/18		1/6	–2/9		1/3
Equilibrium		–5/54	–4/54	2/27	4/27	
Transmission	–5/108		1/27	–2/54		2/27
Equilibrium		–5/243	–4/243	2/162	4/162	
M(qL^2)	–0.027	–0.0557	0.0557	–0.070	0.070	0.034

Table 5.19. *Moments of a frame without sidesway*

The horizontal reactions at supports A and D of a frame without sidesway are

$$T_{AB} = \frac{5(M_{AB} + M_{BA})}{4L} = -0.1034qL$$

$$T_{DC} = \frac{2(M_{DC} + M_{CD})}{L} = 0.208qL$$

$$(T_{AB} + T_{DC})^0 = 0.1046qL$$

– *Frame with sidesway*

We release the frame (Figure 5.8) to horizontal displacement. The vertical columns (AB) and (DC) are horizontally displaced Δ (Figure 5.9). The fixed-end moments are

$$\gamma_{AB}^d = \gamma_{BA}^d = -\frac{15.EI}{2L}\beta$$

We have

$$\beta = -\frac{5\Delta}{4L}$$

$$\beta' = -\frac{2\Delta}{L}$$

Hence

$$\beta' = \frac{8}{5}\beta$$

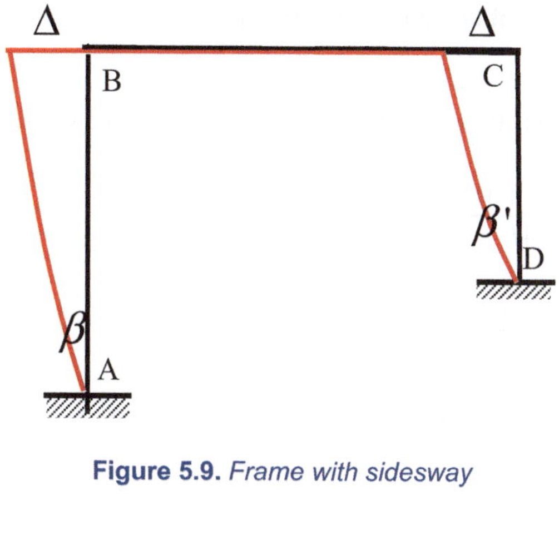

Figure 5.9. *Frame with sidesway*

$$\gamma^d_{CD} = \gamma^d_{DC} = \frac{12.EI}{L}\beta' = -\frac{96.EI}{5L}\beta$$

A relationship between fixed-end moments can be deduced as follows:

$$\gamma^d_{AB} : \gamma^d_{BA} : \gamma^d_{CD} : \gamma^d_{DC} = 1:1:2.56:2.56$$

Applying the moment-distribution method is presented in Table 5.20.

Joint	A	B		C		D
Bars	AB	BA	BC	CB	CD	DC
R		$\frac{5}{9}$	$\frac{4}{9}$	$\frac{1}{3}$	$\frac{2}{3}$	
γ^d	1	1			2.56	2.56
Equilibrium		–5/9	–4/9	–2.56/3	–5.12/3	
Transmission	–5/18		–1.28/3	–2/9		–2.56/3
Equilibrium		6.4/27	5.12/27	2/27	4/27	
Transmission	3.2/27		1/27	2.56/27		2/27
Equilibrium		–5/243	–4/243	–2.56/81	–5.12/81	
M	0.8407	0.6610	–0.6610	–0.9381	0.9381	1.781

Table 5.20. *Moments of a frame with sidesway*

The horizontal reactions at supports A and D of a frame with sidesway are

$$T_{AB} = \frac{5(M_{AB} + M_{BA})}{4L} = \frac{1.8771}{L}$$

$$T_{DC} = \frac{2(M_{DC} + M_{CD})}{L} = \frac{5.438}{L}$$

$$(T_{AB} + T_{DC})^d = \frac{7.315}{L}$$

Therefore

$$(T_{AB}+T_{DC})^0 - k(T_{AB}+T_{DC})^d = 0$$

$$k = -\frac{(T_{AB}+T_{DC})^0}{(T_{AB}+T_{DC})^d} = -0.0143\, qL^2$$

Table 5.16 groups the moments of the joints of the given frame.

Joint	A	B		C		D
Bars	AB	BA	BC	CB	CD	DC
$M^0(qL^2)$	–0.0231	–0.0557	0.0557	–0.070	0.070	0.034
K	–0.0143 qL^2					
M^D	0.8407	0.6610	–0.6610	–0.9381	0.9381	1.781
$M(q.L^2)$	–0.035	–0.0651	0.0651	–0.0566	0.0566	0.0085

Table 5.21. *Moments at joints of given frame*

5.4.2.3. *Frame with oblique columns*

Using the moment distribution method, calculate the moments at the ends of the bars of the frame (Figure 3.40), which have already been analyzed by the method of forces (section 3.6.4) and the slope-deflection method (section 4.5.2.3).

– *Frame without sidesway (Figure 5.10)*

- *Rigidity factor*

$$K_{AB} = K_{BA} = \frac{4EI}{\sqrt{5}a}$$

$$K_{BC} = K_{CB} = \frac{2EI}{a}$$

$$K_{CD} = K_{DC} = \frac{4EI}{\sqrt{2}a}$$

- *Transmission coefficient*

$$\lambda_{BA} = \lambda_{BC} = \lambda_{CB} = \lambda_{CD} = \frac{1}{2}$$

– *Distribution factors*

$$r_{BA} = \frac{K_{BA}}{K_{BA} + K_{BC}} = 0.472$$

$$r_{BC} = \frac{K_{BC}}{K_{BA} + K_{BC}} = 0.528$$

$$r_{CB} = 0.414$$

$$r_{CD} = 0.586$$

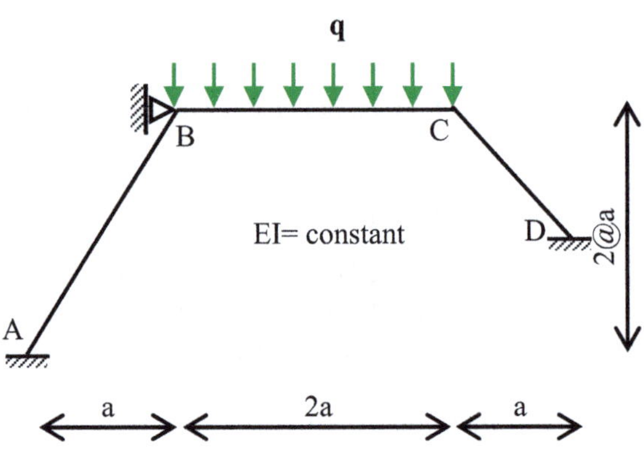

Figure 5.10. *Frame without sidesway*

The fixed-end moments are given as

$$\gamma_{BC} = -\gamma_{CB} = \frac{qa^2}{3}$$

The moments at the ends of bars of the frame without sidesway are presented in Table 5.22.

Joint	A	B		C		D
Bars	AB	BA	BC	CB	CD	DC
r		0.472	0.528	0.414	0.586	
$\gamma(\frac{qa^2}{3})$			1	–1		
Equilibrium		–0.472	–0.528	0.414	0.586	
Transmission	–0.236		0.207	–0.264		0.293
Equilibrium		–0.0977	–0.1093	0.1093	0.1547	
Transmission	–0.0488		0.0546	–0.0546		0.0773
Equilibrium		–0.0258	–0.0288	0.0226	0.0320	
M	–0.2848	–0.5955	0.5955	–0.7727	0.7727	0.3703

Table 5.22. *Moments of a frame without sidesway* $*(\frac{qa^2}{3})$

The horizontal equilibrium of the frame without sidesway allows us to write:

$$M_{AB} + M_{BA} - T_{AB}\,(2a) + V_A.a = 0$$

or

$$T_{AB} = \frac{M_{AB} + M_{BA} - V_A.a}{2a}$$

$$T_{AB} = 0.1467qa + \frac{V_A}{2}$$

In the same way, we consider bar CD:

$$M_{CD} + M_{DC} + V_D.a - T_{DC}\,a = 0$$

$$T_{DC} = \frac{M_{CD} + M_{DC} + V_D.a}{a}$$

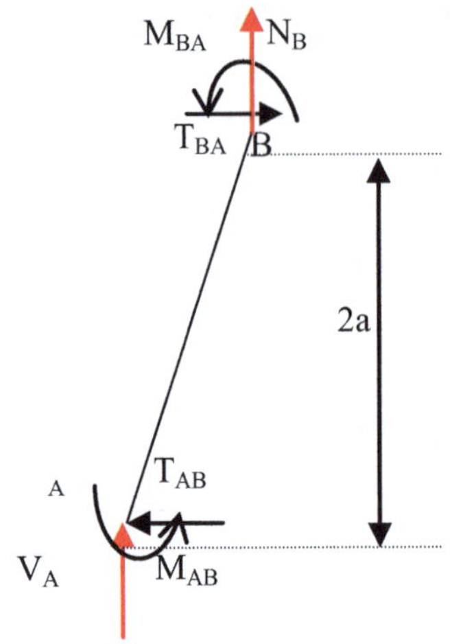

Figure 5.11. *Column (AB)*

$$T_{DC} = \frac{M_{CD} + M_{DC} + V_D.a}{a}$$

$$T_{DC} = 0.381qa + V_D$$

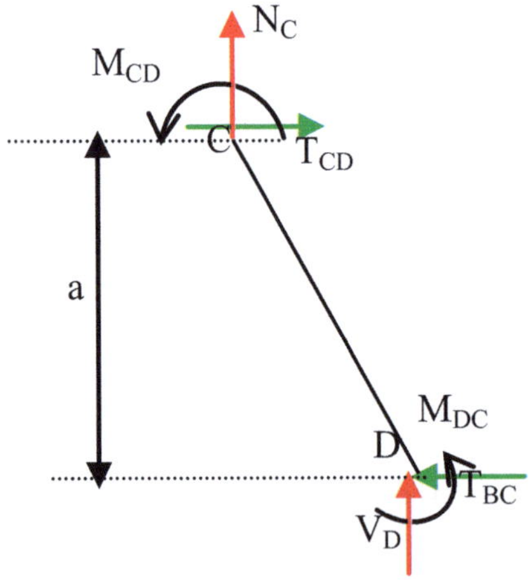

Figure 5.12. *Column (DC)*

Consider Bar (BC):

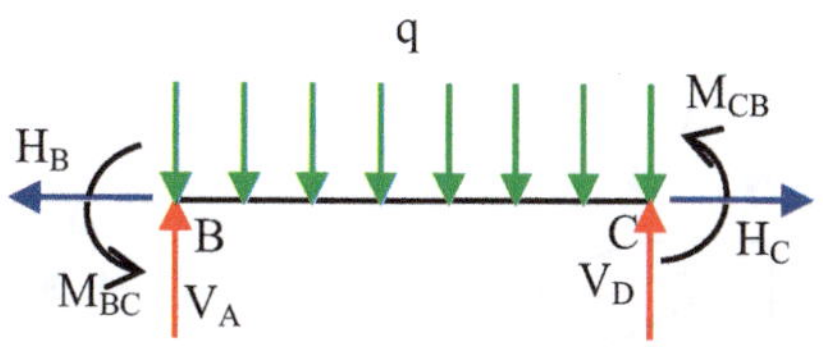

Figure 5.13. *Horizontal bar (BC)*

We write the sum of moments relating to point C.

$$\sum M_C = 0 : -M_{BC} + M_{CB} - 2qa^2 + 2\, T_{BC}.a = 0$$

$$T_{BC} = \frac{M_{CB} - M_{BC} + 2qa^2}{2a} = 0.915qa$$

$$\sum F_y = 0 : T_{BC} + T_{CB} = 2qa$$

$$T_{CB} = 2qa - 0.912qa = 1.085qa$$

$$V_A = T_{BC} = 0.915qa$$

$$V_D = T_{CB} = 1.085qa$$

$$(T_{AB} + T_{DC})^0 = 0.1467qa + \frac{V_A}{2} + 0.381qa + V_D = -0.863qa$$

– *Frame with sidesway (Figure 5.14)*

The deflection (Figure 5.14) allows us to calculate the slopes at bars (AB) and (CD).

$$\beta_{AB} = \frac{BB'}{L_{AB}} = \frac{\Delta_1}{L_{AB}} = \frac{\Delta}{L_{AB}\cos\alpha} = \frac{\Delta}{2a}$$

$$\beta_{DC} = \frac{\Delta_2}{L_{DC}} = \frac{\Delta}{L_{DC}\cos\psi} = \frac{\Delta}{a}$$

$$\beta_{BC} = \frac{B'B_1 + C'C_1}{L_{BC}} = \frac{\Delta(tg\alpha + tg\psi)}{L_{BC}} = \frac{3\Delta}{4a}$$

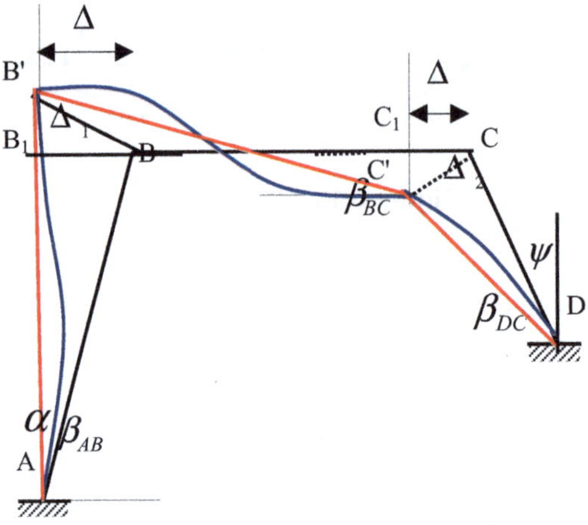

Figure 5.14. *Frame with sidesway*

The fixed-end moments due to the frame's horizontal displacement are

$$\gamma_{AB}^{d} = \gamma_{BA}^{d} = -K_{AB}(1+\lambda_{BA})\beta_{AB} = -\frac{3\sqrt{5}.EI}{5}\frac{\Delta}{a^2}$$

$$\gamma_{BC}^{d} = \gamma_{CB}^{d} = -K_{BC}(1+\lambda_{CB})\beta_{BC} = \frac{9.EI}{4}\frac{\Delta}{a^2}$$

$$\gamma_{CD}^{d} = \gamma_{DC}^{d} = -K_{DC}(1+\lambda_{CD})\beta_{DC} = -3\sqrt{2}.EI\frac{\Delta}{a^2}$$

The proportionality between the fixed-end moments relative to the displacement of the frame is

$$\gamma_{AB}^{d} : \gamma_{BA}^{d} : \gamma_{BC}^{d} : \gamma_{CB}^{d} : \gamma_{CD}^{d} : \gamma_{DC}^{d} = -1 : -1 : \frac{3\sqrt{5}}{4} : \frac{3\sqrt{5}}{4} : -\sqrt{10} : -\sqrt{10}$$

The moment-distribution method of joints allows us to build the table of the moment-distribution method (Table 5.23).

Joint	A	B		C		D
Bars	AB	BA	BC	CB	CD	DC
r		0.472	0.528	0.414	0.586	
γ	–1	–1	1.677	1.677	–3.162	–3.162
Equilibrium		–0.320	–0.360	0.615	0.870	
Transmission	–0.160		0.3075	–0.18		0.435
Equilibrium		-0.145	–0.162	0.075	0.106	
Transmission	–0.073		0.0375	–0.081		0.053
Equilibrium		0.017	0.020	–0.033	–0.048	
M	–1.233	–1.482	1.482	2.138	–2.138	–2.674

Table 5.23. *Moments of a frame with sidesway*

We apply the same method of a frame with sidesway to calculate $(T_{AB} + T_{DC})^d$.

$M_{AB} + M_{BA} - T_{AB}(2a) + V_A.a = 0$

$$T_{AB}^d = \frac{M_{AB} + M_{BA} - V_A a}{2a}$$

$$T_{AB}^d = -\frac{1.358}{a} - \frac{V_A}{2}$$

In the same way, we consider bar CD.

$M_{CD} + M_{DC} + V_D.a - T_{DC}\, a = 0$

$$T_{DC}^{d} = \frac{M_{CD} + M_{DC} + V_D a}{a}$$

$$T_{DC}^{d} = -\frac{4.812}{a} + V_D$$

Figure 5.15. *Shear forces of the bar (BC)*

We consider bar (BC).

$$\sum M_C = 0 : \ -M_{BC} - M_{CB} + T_{BC}^{d}(2a) = 0$$

$$T_{BC}^{d} = \frac{M_{CB} + M_{BC}}{2a}$$

$$T_{BC}^{d} = \frac{1.81}{a}$$

$$\sum F_y = 0 : T_{BC}^{d} + T_{CB}^{d} = 0$$

$$T_{CB}^{d} = -\frac{1.81}{a}$$

We can deduce that

$$(V_A + V_D)^d = 0$$

The shear force at the base of bar (AB) is given as

$$T_{AB}^{d} = \frac{1.358}{a} - \frac{V_A}{2} = \frac{3.168}{a}$$

$$T_{DC}^{d} = \frac{4.812}{a} + V_D = \frac{6.622}{a}$$

We know that

$$(T_{AB} + T_{DC})^0 + K(T_{AB} + T_{DC})^d = 0$$

$$K = -\frac{(T_{AB} + T_{DC})^0}{(T_{AB} + T_{DC})^d} = \frac{0.863qa}{(9.788/a)} = 0.0882qa^2$$

Table 5.24 shows the moments of the joints of the displaced frame.

Joint	A	B		C		D
Bars	AB	BA	BC	CB	CD	DC
$M(\frac{qa^2}{3})$	–0.2848	–0.5955	0.5955	–0.7727	0.7727	0.3703
M^D	–1.233	–1.482	1.482	2.138	–2.138	–2.674
K	0.0882 qa^2					
$M(q.a^2)$	–0.203	–0.330	0.330	–0.069	0.069	–0.112

Table 5.24. *Moments of given frame*

5.4.2.4. *Multistory frame*

In Chapter 4, we presented the analysis of multistory frames using the slope-deflection method. In the same way, the moment distribution approach can be applied to the analysis of multistory frames with several horizontal displacements. In this case, the moment-distribution method must go through several phases, namely (1) the analysis of the frame without sidesway and (2) the analysis of the frame with sidesway, which depends on the number of possible horizontal displacements.

In the first part, the displacements of the levels are assumed to be blocked by imaginary supports (Figure 5.16). The moments at the ends of each bar, M^0, are calculated by the moment-distribution method due to applying external loads and support reactions, R_1, R_2 and R_3, which can be calculated using the equilibrium of the frame.

In the second part of this analysis, we remove the support of the first level, the frame displaces Δ_1 due to the effect of reaction $-R_1$. In this case, we calculate the moments at the ends of the bars due to the displacement Δ_1 .

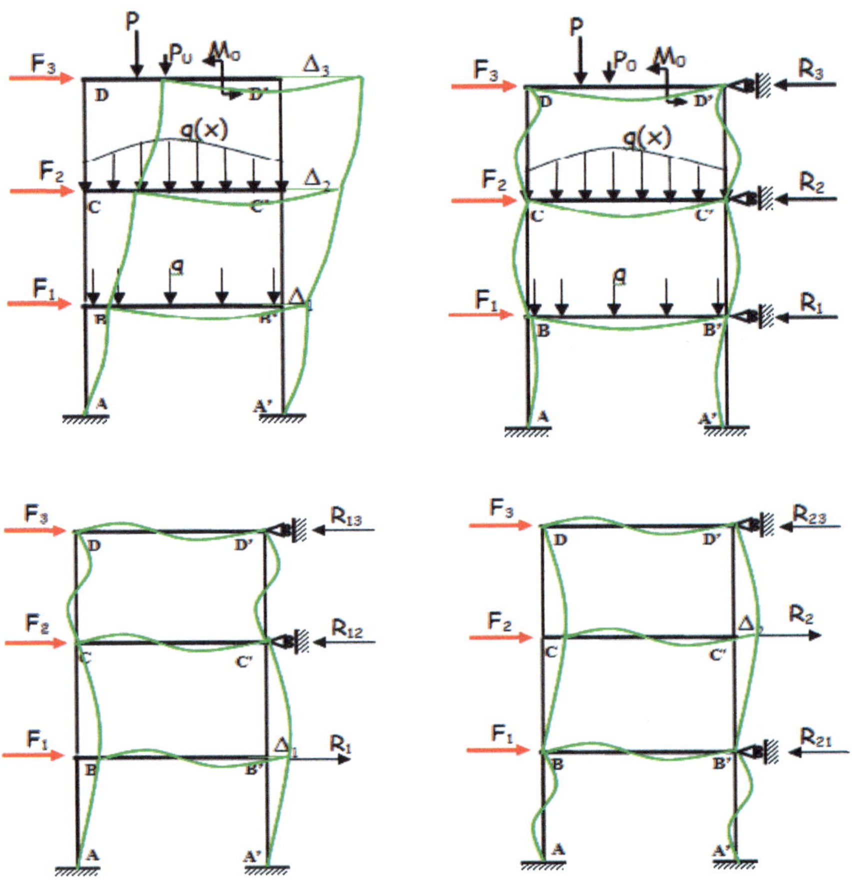

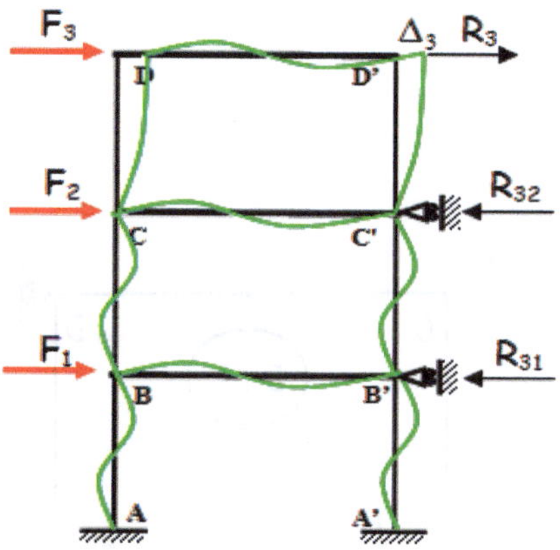

Figure 5.16. *Multistory frame*

This procedure is repeated for the displacements Δ_2 and Δ_3, which are due to reactions R_2 and R_3. The moments at the ends of the bars are calculated by a superposition of the moments calculated for four cases studied. The expression that gives the bending moment at each end is calculated using the principle of superposition of already calculated moments.

$$M_{ij} = M_{ij}^{0} + \sum_{k=1}^{3} M_{ij}^{k} \qquad [5.22]$$

– *Frame with two levels*

We reanalyze the frame given in Figure 4.53, which has already been studied by the slope-deflection method (section 4.5.2.3).

– *Frame without sidesway (Figure 5.17)*

Joint D is blocked by a roller preventing the horizontal displacements of the different levels.

- *Rigidity factors*

$$K_{ij} == \frac{4EI}{L}$$

- *Transmission coefficients*

$$\lambda_{ij} = \frac{1}{2}$$

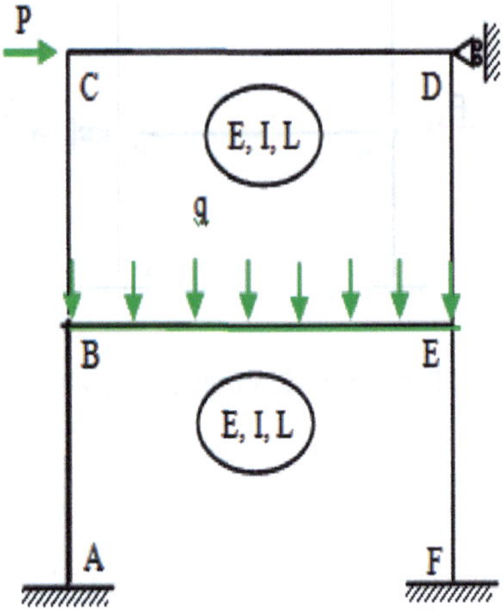

Figure 5.17. *Frame without sidesway*

- *Distribution factors*

$$r_{BA} = r_{BC} = r_{BE} = r_{EB} = r_{EF} = r_{ED} = \frac{1}{3}$$

$$r_{CB} = r_{CD} = r_{DC} = r_{DE} = \frac{1}{2}$$

- *Fixed-end moments*

$$\gamma_{BE} = -\gamma_{EB} = \frac{1}{12}qL^2 = \frac{1}{12}PL$$

- *Moment distribution method (Table 5.25)*

Joint	A		B		C		D		E		F	
Bars	AB	BA	BE	BC	CB	CD	DC	DE	ED	EB	EF	FE
R		$\frac{1}{3}$	$\frac{1}{3}$	$\frac{1}{3}$	$\frac{1}{2}$	$\frac{1}{2}$	$\frac{1}{2}$	$\frac{1}{2}$	$\frac{1}{3}$	$\frac{1}{3}$	$\frac{1}{3}$	
$\gamma(\frac{PL}{12})$			1							–1		
Equilibrium		–1/ 3	–1/3	–1/ 3					1/ 3	1/3	1/ 3	
Transmission	–1/6		1/6		–1/6			1/6		–1/6		1/6
Equilibrium		–1/18	–1/18	–1/18	1/12	1/12	–1/12	–1/12	1/18	1/18	1/18	
Transmission	–1/36		1/36	1/24	-1/36	-1/24	1/24	1/36	-1/24	-1/36		1/36
Equilibrium		–5/216	–5/216	–5/216	5/144	5/144	–5/144	–5/144	5/216	5/216	5/216	
$M(\frac{P.L}{12})$	–7/36	–89/216	169/216	–80/216	–11/144	11/144	–11/144	11/144	80/216	–169/216	89/216	7/36

Table 5.25. *Moments of fixed frame. For a color version of this table, see www.iste.co.uk/khalfallah/analysis 2.zip*

The results obtained in the table above show that

$(T_{AB} + T_{FE})^{N.F} = 0$

– *Frame with sidesway*

We release support D and calculate the moments at the ends of the bars due to the displacement effect Δ. The multistory frame is therefore subjected to force (–R) applied at point C (Figure 5.18).

The fixed-end moments due to displacement Δ are given as

$$\gamma_{AB} = \gamma_{BA} = \gamma_{EF} = \gamma_{FE} = -\frac{4EI}{L}(1+\frac{1}{2})\beta = -\frac{6EI}{L^2}\beta$$

The moments at the ends of each bar can be calculated using the moment-distribution method.

The optional moments are assigned to bars corresponding to the expression of the slope of each bar, as fixed-end moments, which are:

$$\gamma_{AB} : \gamma_{BA} : \gamma_{BC} : \gamma_{CB} : \gamma_{EF} : \gamma_{FE} : \gamma_{ED} : \gamma_{DE} = 1:1:1:1:1:1:1:1$$

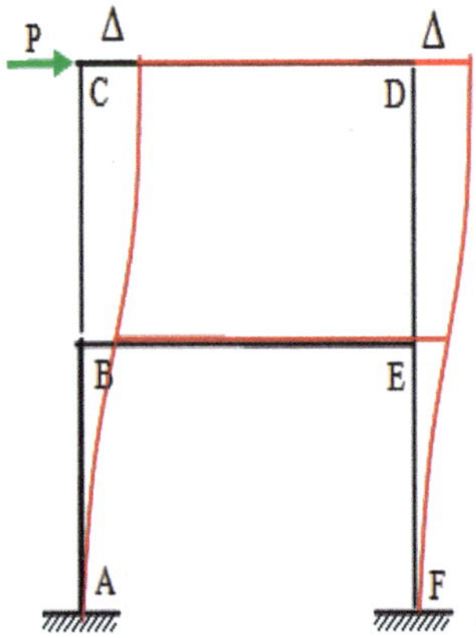

Figure 5.18. *Frame with sidesway due to displacement* Δ

The moments at the ends of bars of the frame with sidesway are presented in Table 5.26.

Joint	A		B		C		D		E		F	
Bars	AB	BA	BE	BC	CB	CD	DC	DE	ED	EB	EF	FE
r		1/3	1/3	1/3	1/2	1/2	1/2	1/2	1/3	1/3	1/3	
γ	1	1		1	1			1	1		1	1
Equilibrium		–2/3	–2/3	–2/3	–1/2	–1/2	–1/2	–1/2	–1/3	–1/3	–1/3	
Transmission	–1/3		–1/3	–1/4	–1/3	–1/4	–1/4	–1/3	–1/4	–1/3		–1/3
Equilibrium		7/36	7/36	7/36	7/24	7/24	7/24	7/24	7/36	7/36	7/36	
Transmission	7/72		7/72	7/48	7/72	7/48	7/48	7/72	7/48	7/72		7/72
Equilibrium		–35/432	–35/432	–35/432	–35/288	–35/288	–35/288	–35/288	–35/432	–35/432	–35/432	
M Δ	55/72	193/432	–341/432	148/432	125/288	–125/288	–125/288	125/288	148/432	–341/432	193/432	55/72

Table 5.26. *Moments of a frame with sidesway. For a color version of this table, see www.iste.co.uk/khalfallah/analysis 2.zip*

The horizontal equilibrium of the frame allows us to write

$$(T_{AB} + T_{FE})\,\Delta = 2(\frac{55}{72} + \frac{193}{432})\frac{1}{L} = \frac{523}{216L}$$

By writing the equilibrium equation in relation to a section passing through points A and F.

$$(T_{AB} + T_{FE})^{N.F} + k\,(T_{AB} + T_{FE})\,\Delta = P$$

The correction parameter is

$$k = 0.413\ PL$$

Now, the moments at the ends of the bars can be deduced using the following equation:

$$M_{ij} = M_{ij}^{0} + kM_{ij}^{\Delta}$$

For example:

$$M_{AB} = M_{AB}^{0} + kM_{AB}^{\Delta} = \frac{PL}{12}(-\frac{7}{36}) + 0.413PL.\frac{55}{72} = 0.299PL$$

$$M_{CB} = M_{CB}^{0} + kM_{CB}^{\Delta} = \frac{PL}{12}(-\frac{11}{144}) + 0.413PL.\frac{125}{288} = 0.1729PL$$

The diagram of the bending moment is shown in Figure 4.56 and the diagram of the shear force is shown in Figure 4.57.

5.5. Conclusion

In this chapter, we have shown the steps of the moment-distribution method for analyzing beams and plane frames.

The procedure for analyzing beams and plane frames without sidesway consists of distributing the fixed-end moments due to the applied loads or to a support settlement accompanied by a transmission of moments. In the case of beams and frames with sidesway, analyzing structures using the slope-deflection method is carried out in two phases. The first phase consists of analyzing the fixed joint structure by blocking the horizontal displacements of the structure by adding imaginary supports. Then, we continue the steps of the method until the values of moments at the ends of each bar are low enough.

In the second phase of analyzing the displaced joint structure using the moment-distribution method, the moments at the ends of each bar are calculated by applying the reaction of the imaginary support in the opposite direction. The structures move in this direction. The horizontal displacement of the structure induces fixed-end moments at the ends of each structural element. The fixed-end moments are distributed and they are transmitted in the same way as for frames without sidesway.

Lastly, the final moments at the ends of each bar are superimposed between the moments of the frame without sidesway and the moments of the frame with sidesway.

5.6. Problems

Exercise 1

Calculate the support reactions and draw diagrams of the bending moment and the shear force for the following beams:

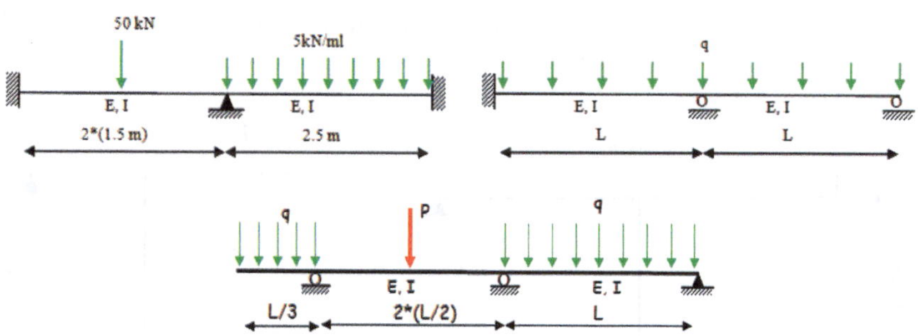

Exercise 2

Calculate the moments at the ends of each bar and deduce the support reactions of the following structures:

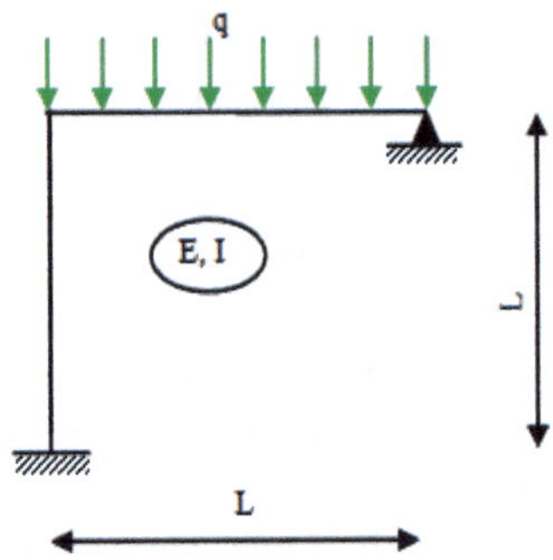

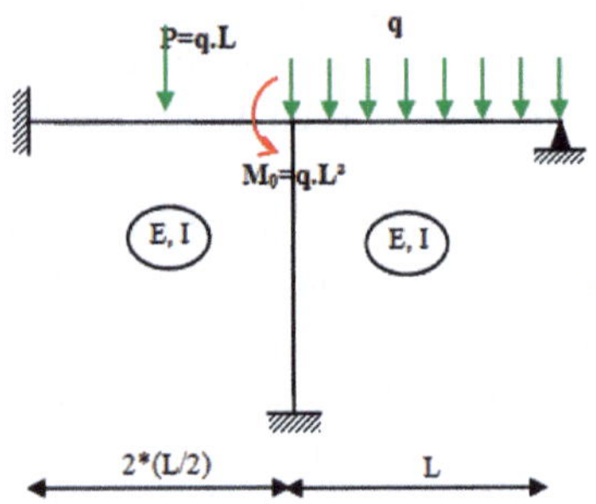

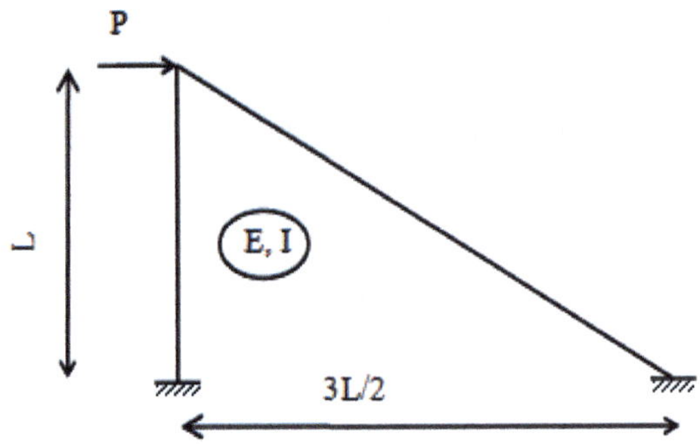

Exercise 3

Calculate the moments at the joints of each frame. Draw diagrams of the bending moment and the shear force for each frame.

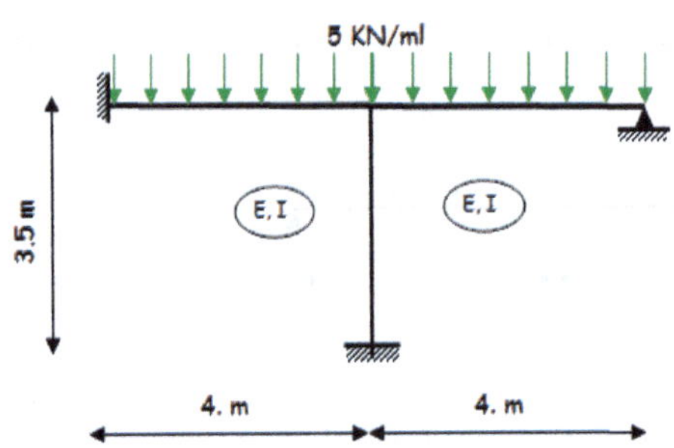

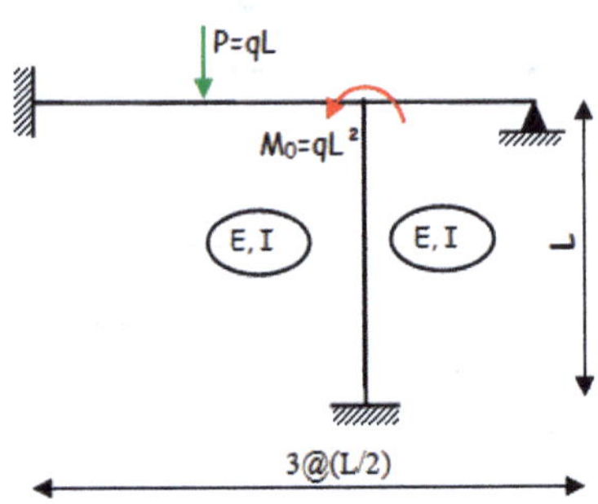

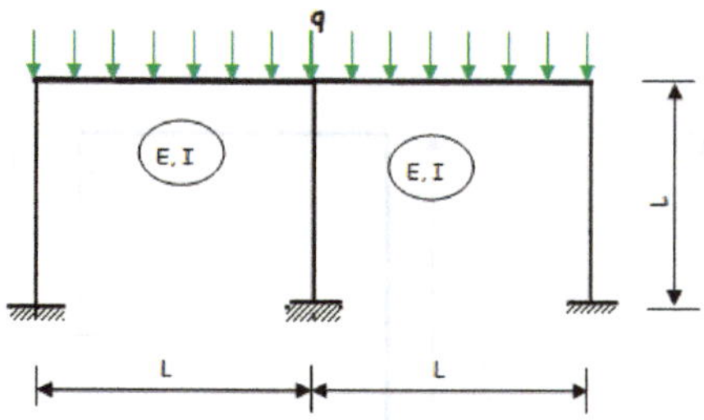

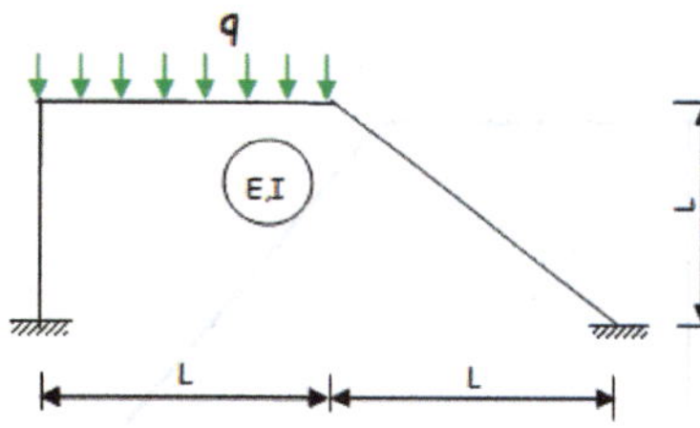

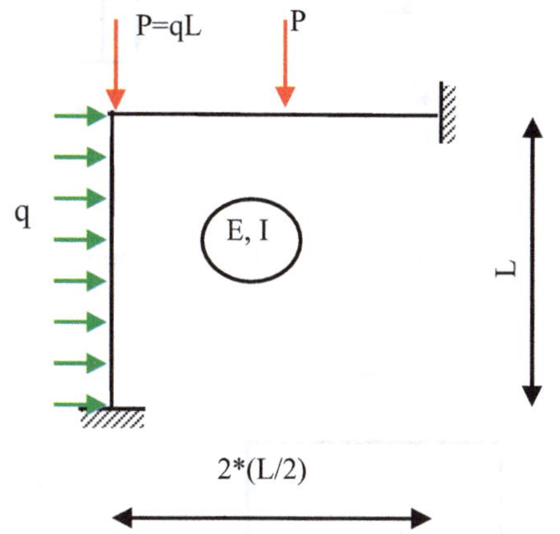

Exercise 4

Calculate the moments at the joints of each frame. Draw diagrams of the bending moment and the shear force for each frame.

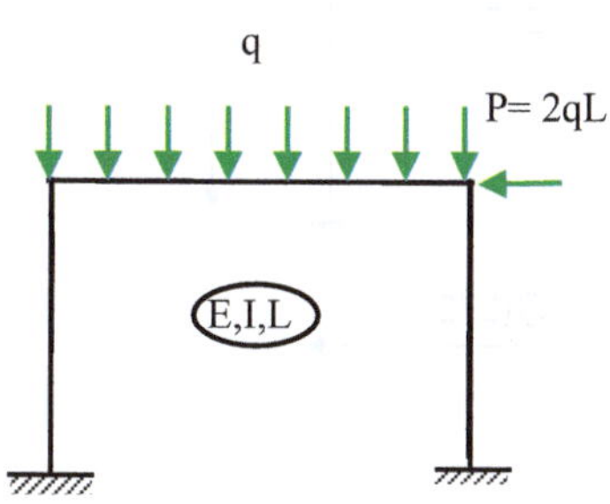

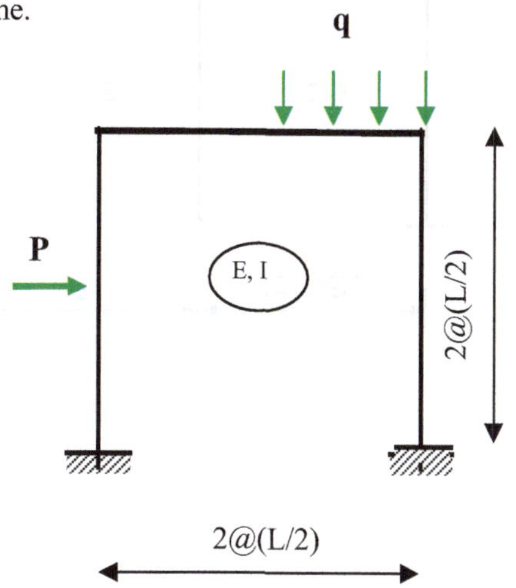

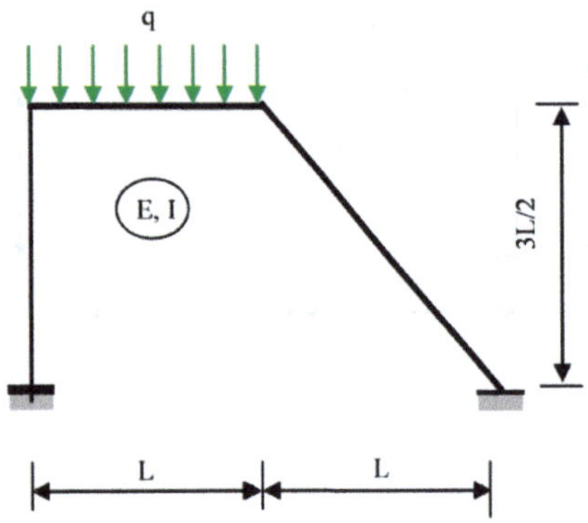

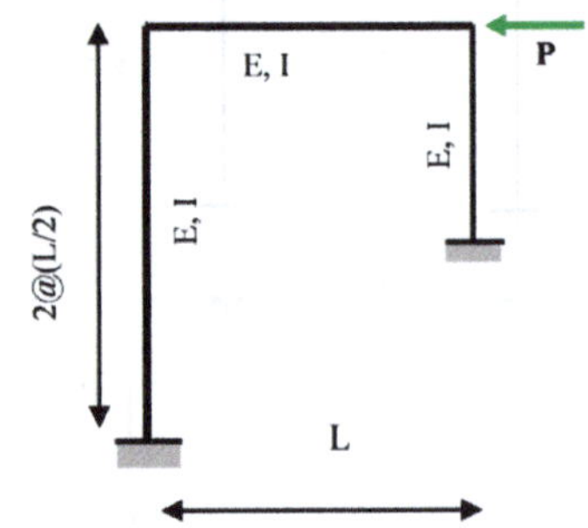

Exercise 5

Calculate the moments at the joints of each frame. Draw diagrams of the bending moment and the shear force.

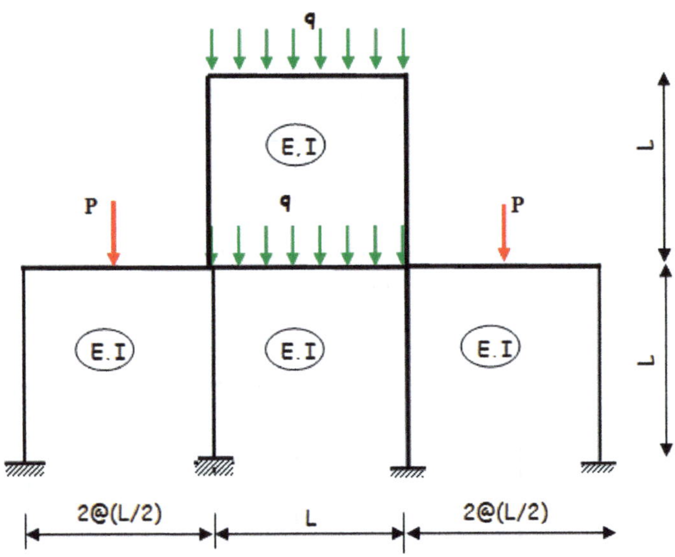

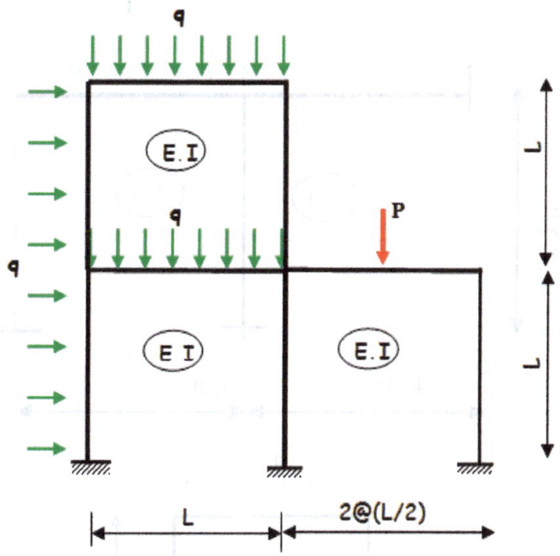

Exercise 6

Calculate the moments at the joints of each frame due to the support settlement of value Δ. Draw diagrams of the bending moment and the shear force.

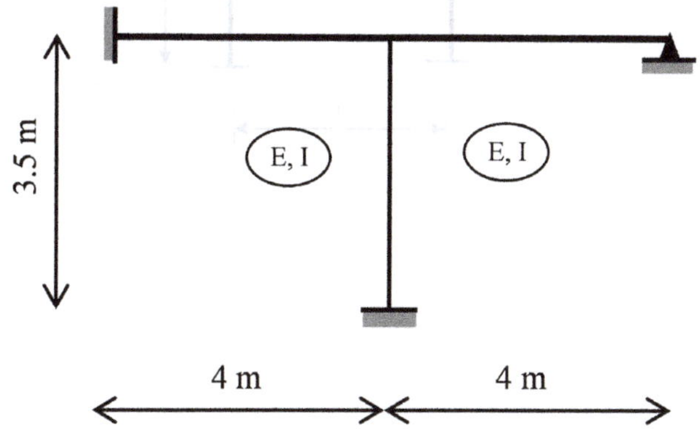

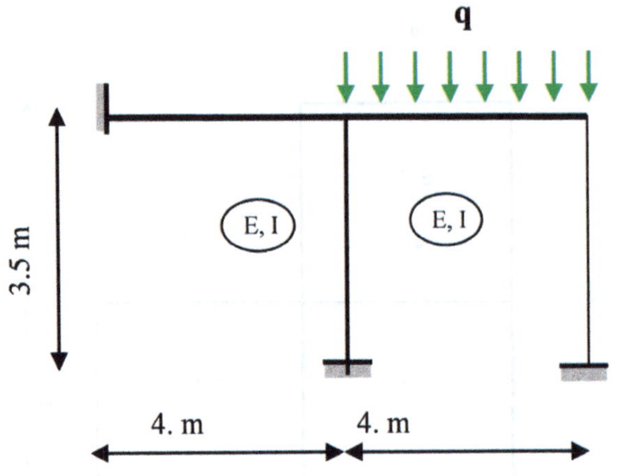
q
E, I
E, I
3.5 m
4. m
4. m

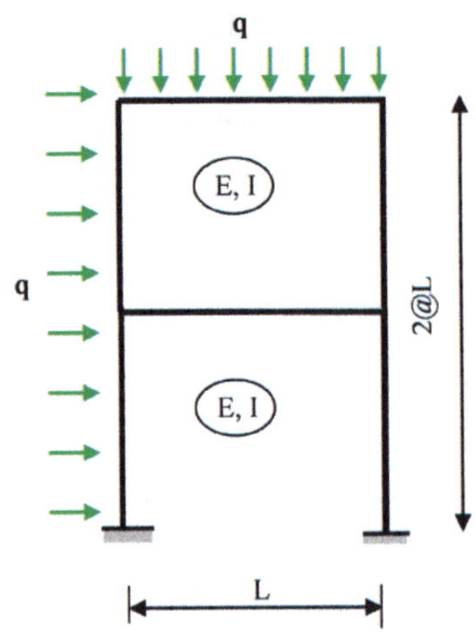
q
E, I
q
E, I
2@L
L

6

Influence Lines of Statically Indeterminate Structures

The teaching objectives for this chapter are as follows:

– the distinction between influence lines for statically determinate and statically indeterminate structures;

– the variation of support reactions and internal actions of a statically indeterminate structure subjected to a moving force;

– the establishment of influence lines;

– the knowledge of the positions and extreme stresses which can lead to dimensioning of the elements.

In this chapter, we present the basics of establishing an influence line based on a unit force moving along the statically indeterminate structure. With regard to this subject, we evaluate the influence lines of statically indeterminate beams, frames and trusses.

6.1. Introduction

The influence line of a structure is a curve, which shows the variation of an action (a support reaction, a shear force, a bending moment, etc.) due to the effect of a load moving along this structure. In general, the intensity of this moving force is a unit force. The study of influence lines of statically indeterminate structures is identical to that of statically determinate structures. The procedure for establishing influence lines involves calculating the value of the structural response for different positions of the applied unit load. In this case, they are generally curves, whereas in statically determinate structures they are straight lines.

To construct the influence line of a support reaction or an internal action of a statically indeterminate structure, we must use one of the statically indeterminate structural analysis methods (Chapters 2–5). In this chapter, we will use different statically indeterminate structural analysis methods to study the influence lines of beams, frames and truss structures.

6.2. Influence lines of beams

For simplification reasons, we study the influence lines of the internal actions and support reactions of a beam, which is once statically indeterminate and then twice statically indeterminate. Using the same concept, the analysis presented can easily be extended to beams with several degrees of static indeterminacy.

6.2.1. *Beam with one degree of static indeterminacy*

Consider the continuous beam (Figure 6.1) and draw the influence lines of the support reaction R_B and bending moment M_D.

The beam is stressed by a unit force with a variable position defined by the distance x measured from support A (Figure 6.1).

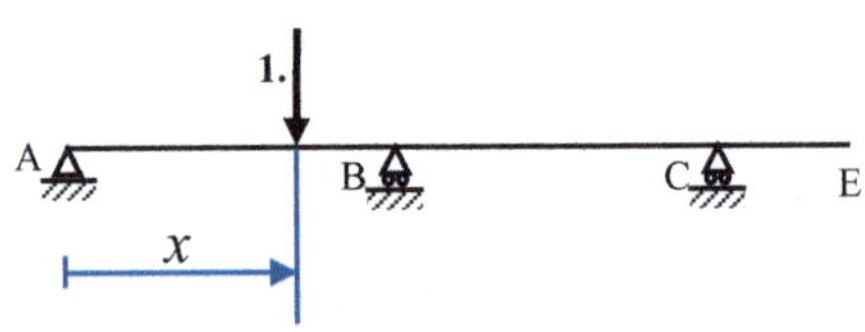

Figure 6.1. *Continuous beam with one degree of static indeterminacy*[1]

To determine the influence lines of support reaction R_B, it is necessary to formulate the expression of R_B according to position x of the unit force. In this case, the method of forces steps must be rigorously applied to achieve this objective. We therefore apply the method of forces steps.

– *Equivalent structures*

We substitute support B with reaction R_B (Figure 6.2).

1 All of the figures in this chapter are available to view in full color at www.iste.co.uk/khalfallah/analysis2.zip.

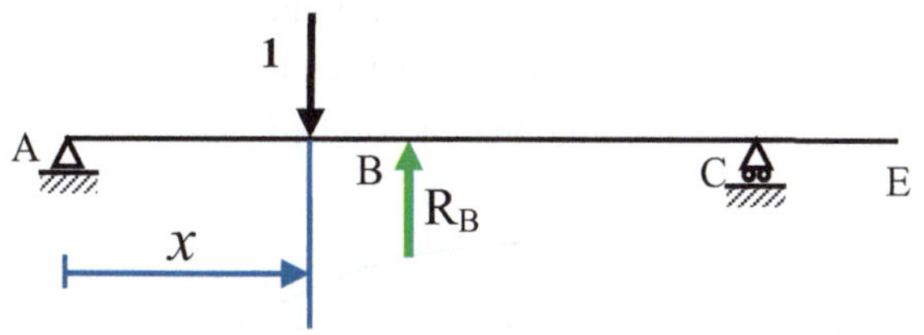

Figure 6.2. *Equivalent beam*

– *Fundamental structure*

We ignore the redundant force R_B to construct the fundamental structure. The structure obtained is a statically determinate structure (Figure 6.3).

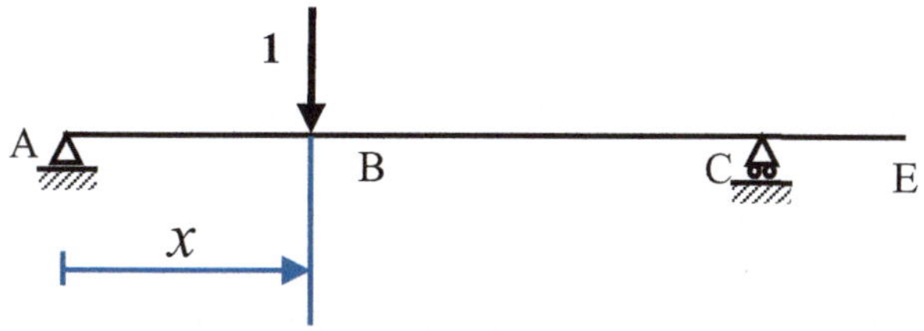

Figure 6.3. *Fundamental beam*

The unit force generates a deflection of the fundamental beam (Figure 6.4), which makes it possible to calculate the displacement of point B particularly, written as $\delta_B^0(x)$ (the exponent (^{0}) is introduced to show that the studied structure is zero times statically indeterminate or fundamental).

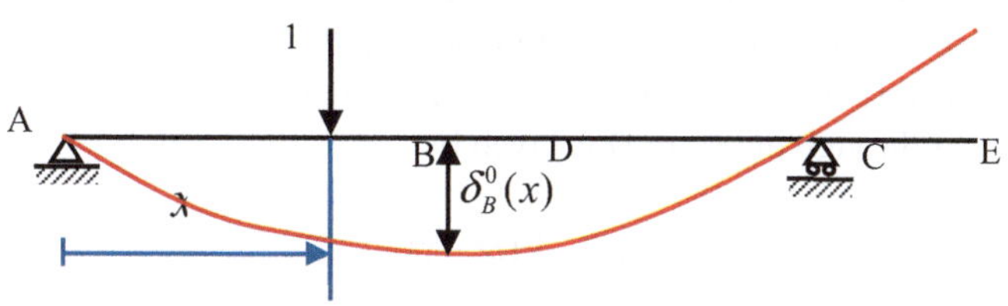

Figure 6.4. *Deflection of the fundamental beam*

We apply a unit force in the place of the redundant connection. This action generates a deflection of beam (AE) (Figure 6.5) and in particular δ_B^1 at point B (the exponent (^{1}) is introduced to highlight that the studied system is unitary).

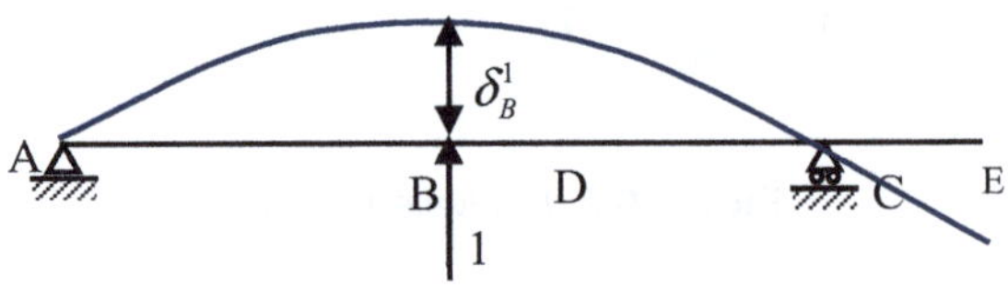

Figure 6.5. *Deflected unit system*

Initially, the vertical displacement of support B is zero.

$$\delta_B = 0$$

Applying the superposition principle allows us to write

$$\delta_B = \delta_B^0(x) + R_B.\delta_B^1 = 0 \qquad [6.1]$$

Using relationship [6.1], it is possible to calculate the redundant reaction.

$$R_B = -\frac{\delta_B^0(x)}{\delta_B^1} \qquad [6.2]$$

The relationship [6.2] describes the support reaction at joint B according to different positions of the unit force. In addition, it allows us to construct the influence line of this reaction by scanning the different positions of the unit force.

In the same way, the influence lines of the bending moment at point D can be obtained. In this case, it is necessary to calculate the reaction V_A or V_C using the static equations. From Figure 6.6, the expression of the bending moment at the section passing through the point D is given by

$$M(\alpha, x) = R_A.\alpha + R_B.(\alpha - x_B) - 1.(\alpha - x) \qquad [6.3]$$

where α is the abscissa of point D and x_B is the abscissa of point B.

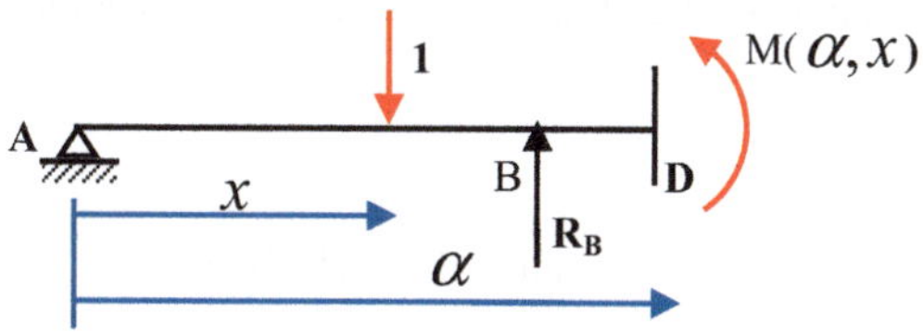

Figure 6.6. *Bending moment at point D*

The influence line is established by calculating $M(\alpha, x)$ in the different regions of beam (AE).

EXAMPLE 6.1.–

Construct the influence lines of support reaction R_B, shear force T_D and bending moment M_D of the beam (Figure 6.7) with constant EI.

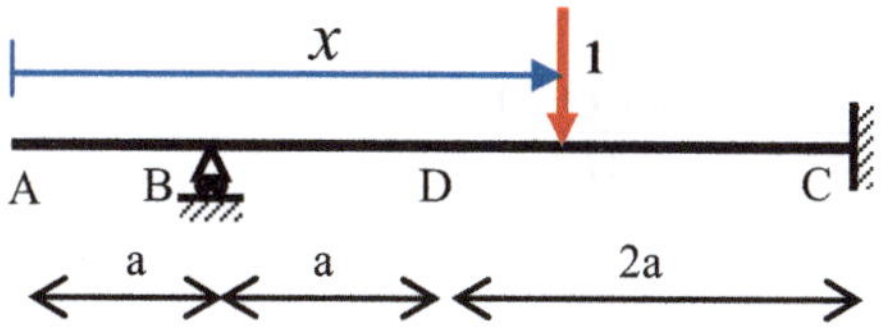

Figure 6.7. *Beam with cantilever*

To vary the statically indeterminate structural analysis methods, we use the slope-deflection method in this exercise.

We consider first the moving forces in different regions of application.

$0 \leq x \leq a$ (Figure 6.8)

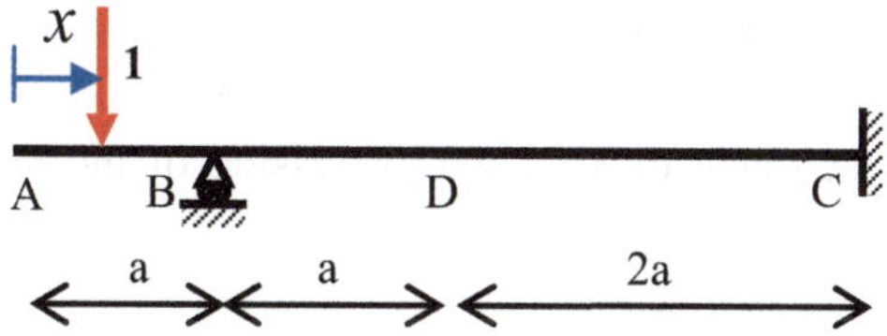

Figure 6.8. *Unit force on cantilever (AB)*

The unit force generates a bending moment at joint B, which is equal to

$$M_{BA} = -(a - x)$$

Similarly, the moment in the sections of bar (BC) is

$$M_{BC} = (a - x)$$

We apply the slope-deflection method.

$$M_{BC} = K_{BC}\omega_B + K_{BC}\lambda_{BC}\omega_C$$

Knowing that $\omega_C = 0$

$$M_{BC} = K_{BC}\omega_B$$

The slope of joint B can be deduced by

$$\omega_B = \frac{M_{BC}}{K_{BC}} = \frac{(a-x)}{\frac{4EI}{3a}} = \frac{3a(a-x)}{4EI}$$

The fixed-end moment can be deduced using the slope-deflection method.

$$M_{CB} = K_{CB}\lambda_{CB}\omega_B$$

or

$$M_{CB} = \frac{4EI}{3a}\frac{1}{2}\frac{3a(a-x)}{4EI} = \frac{1}{2}(a-x)$$

We take the equilibrium of forces relating to point C.

$$R_B(x)3a = M_{CB} + (4a - x)$$

By substituting the value of M_{CB}, the expression of the reaction at support B becomes

$$R_B(x) = \frac{3}{2} - \frac{x}{2a}$$

The shear force $T_B(x)$ and bending moment $M_C(x)$ expressions are

$$T_D(x) = R_B(x) - 1 = \frac{1}{2} - \frac{x}{2a}$$

$$M_C(x) = R_B(x)(3a) - (4a - x) = \frac{a}{2} - \frac{x}{2}$$

$a \leq x \leq 4a$ (Figure 6.9)

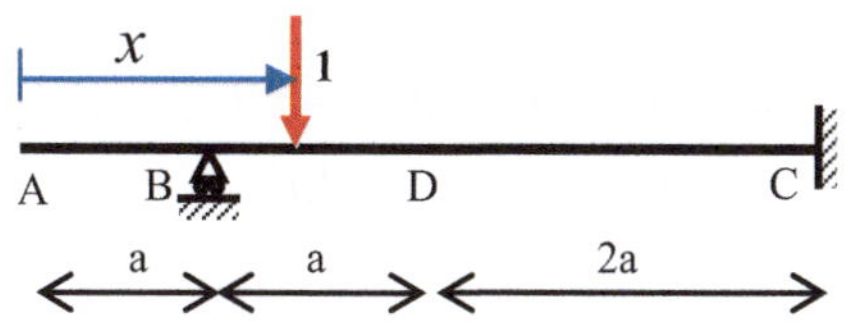

Figure 6.9. *Unit force on span (BC)*

We apply the Clapeyron method to determine the moment at fixed-end C.

$$\omega_C = \omega_C^{"} - bM_{BC} + cM_{CB} = 0$$

The moment at fixing point C is

$$M_{CB} = -\frac{\omega_C^{"}}{C}$$

Knowing that $\omega_C^{"} = \dfrac{(x-a)(4a-x)(2a+x)}{6EI(3a)}$ and C= $= \dfrac{a}{EI}$

$$M_{CB} = -\frac{(x-a)(4a-x)(2a+x)}{18a^2}$$

The reaction at support B, the shear force T_D and the bending moment M_C can be deduced using the equilibrium equations.

$$R_B(x)3a = -M_{CB} + (4a - x)$$

By substituting the expression of M_{CB}, the expression of the reaction at support B is

$$R_B(x) = \frac{4}{3} - \frac{x}{3a} - \frac{(x-a)(4a-x)(2a+x)}{54a^3}$$

To evaluate the shear force, we distinguish two cases according to the position of the applied force.

$a \leq x \leq 2a$

$$T_D(x) = R_B(x) - 1 = \frac{1}{3} - \frac{x}{3a} - \frac{(x-a)(4a-x)(2a+x)}{54a^3}$$

$2a \leq x \leq 4a$

$$T_D(x) = R_B(x) = \frac{4}{3} - \frac{x}{3a} - \frac{(x-a)(4a-x)(2a+x)}{54a^3}$$

and

$$M_{CB} = -\frac{(x-a)(4a-x)(2a+x)}{18a^2}$$

The diagrams of the influence lines of reaction R_B of the bending moment of the section passing through point C and the shear force of the section passing through point D are shown in Figures 6.10–6.12.

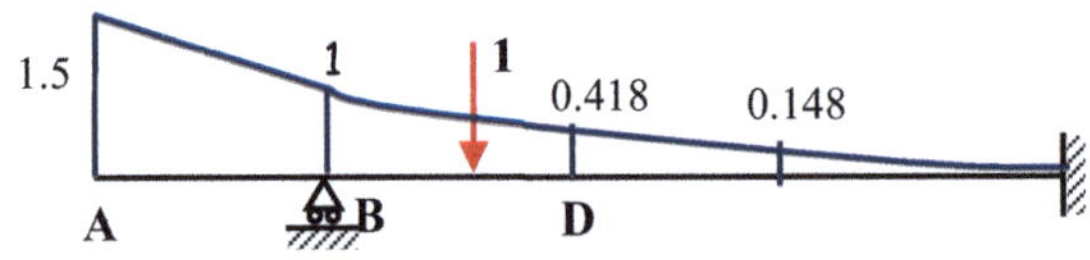

Figure 6.10. *Influence line of reaction RB*

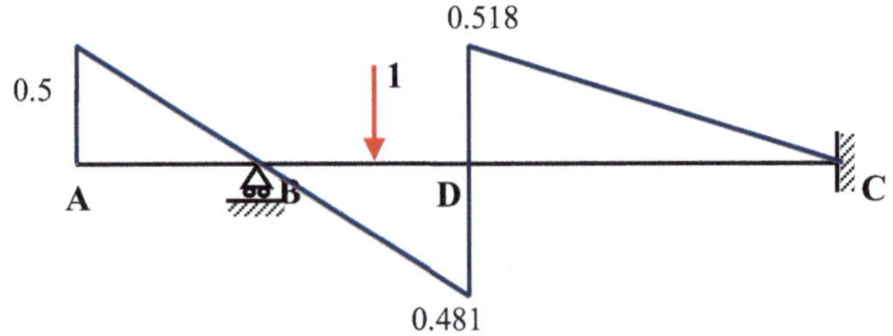

Figure 6.11. *Influence line of shear force TD*

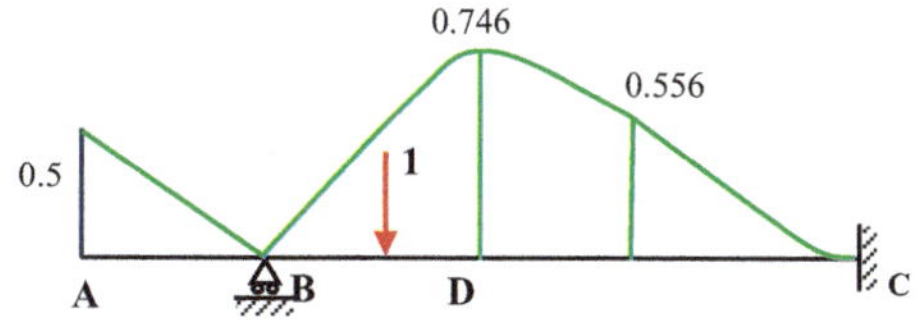

Figure 6.12. *Influence line of bending moment MC*(a)*

6.2.2. Beam with two degrees of static indeterminacy

Draw the influence lines of support reactions R_B and R_C and moment M_A of beam (AC) assuming that the flexural rigidity (EI) is constant. The beam is stressed by a moving unit load (Figure 6.13).

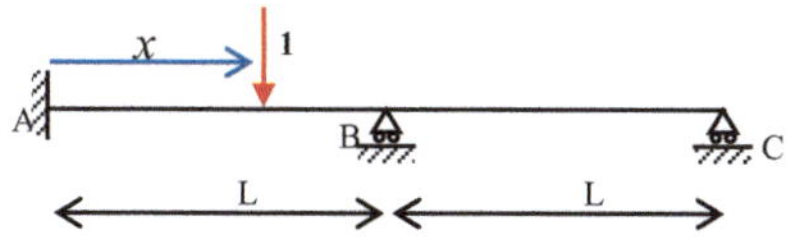

Figure 6.13. *Beam with two degrees of static indeterminacy*

To quantify reactions R_B and R_C and fixed-end moment M_A of the beam (Figure 6.13), we follow the procedure established for analyzing beams with a single degree of static indeterminacy using the method of forces.

The degree of static indeterminacy of the system is given by

$$f = (3 \cdot b + r) - (3 \cdot n + k) \quad [6.4]$$

$$f = (3 \cdot 1 + 5) - (3 \cdot 2 + 0) = 2$$

– *Equivalent system*

Supports B and C are substituted by equivalent support reactions V_B and V_C (Figure 6.14).

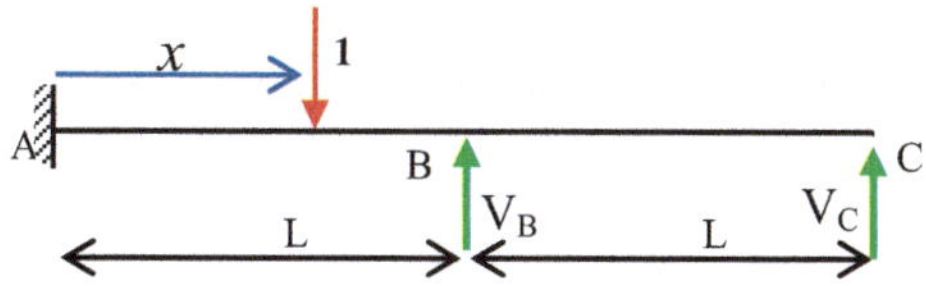

Figure 6.14. *Equivalent beam*

– *Fundamental system*

We ignore the effect of support reactions V_B and V_C, and the system becomes statically determinate (Figure 6.15).

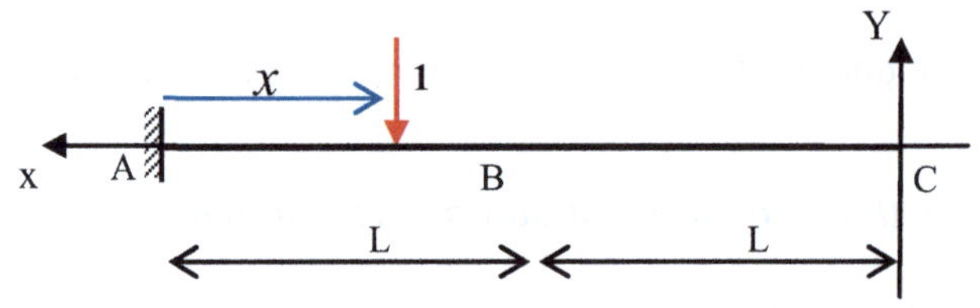

Figure 6.15. *Fundamental beam*

$0 \leq \alpha \leq x$

$$\mu(\alpha, x) = -(x - \alpha) \qquad [6.5a]$$

$x \leq \alpha \leq 2L$

$$\mu(\alpha, x) = 0 \qquad [6.5b]$$

The deformation of the elastic line method in regions ($0 \leq \alpha \leq x$) and ($x \leq \alpha \leq 2L$) allows us to formulate displacement expressions at points B and C.

The general expression of the vertical displacement of the cantilever at any section defined by α is

$0 \leq \alpha \leq x$

$$\delta^0(\alpha, x) = \frac{1}{6EI}(\alpha^3 - 3x\alpha^2) \qquad [6.6a]$$

$x \leq \alpha \leq 2L$

$$\delta^0(\alpha, x) = \frac{1.x^2}{6EI}(x - 3\alpha) \qquad [6.6b]$$

In particular, we calculate the displacements at point B ($\alpha = L$).

$0 \le x \le L$

$$\delta_B^0(x) = \frac{x^2}{6EI}(x - 3L) \qquad [6.7a]$$

$L \le x \le 2L$

$$\delta_B^0(x) = \frac{1}{6EI}(L^3 - 3L^2 x) \qquad [6.7b]$$

Similarly, we calculate the displacement at point C ($\alpha=2L$).

$0 \le x \le L$

$$\delta_C^0(x) = \frac{x^2}{6EI}(x - 6L) \qquad [6.8a]$$

$L \le x \le 2L$

$$\delta_C^0(x) = \frac{x^2}{6EI}(x - 3L) \qquad [6.8b]$$

– *Unit systems (Figures 6.16 and 6.18)*

The unit systems and the bending moment diagrams are shown, respectively, in Figures 6.16–6.19.

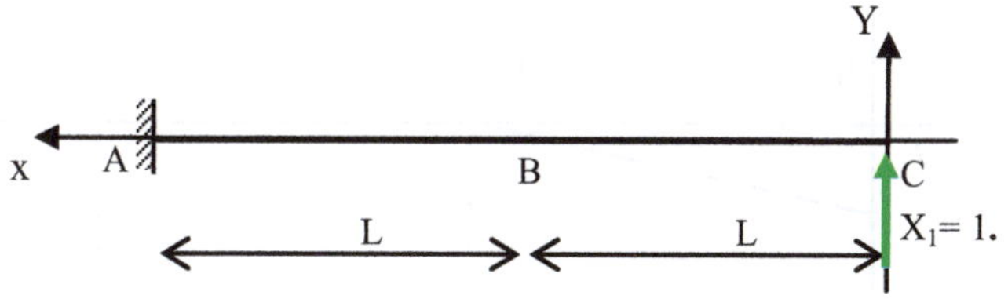

Figure 6.16. *First unit system*

The bending moment of a statically determinate beam (Figure 6.16) is given by

$$m_1(x) = x \qquad 0 \le x \le 2L \qquad [6.9]$$

The bending moment diagram is shown in Figure 6.17.

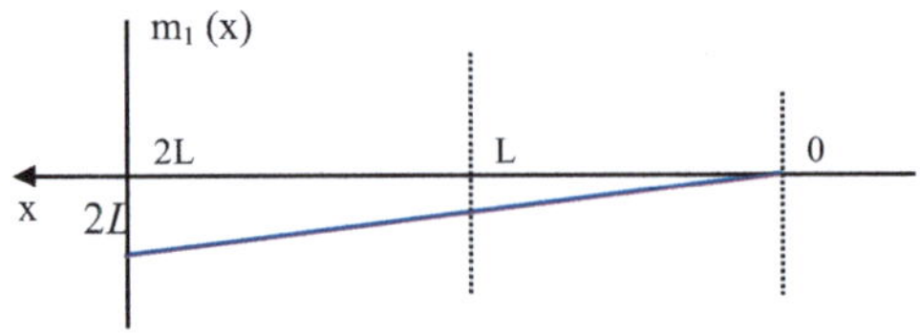

Figure 6.17. *Diagram of moment m1(x)*

We apply a unit moment X_2 =1 at point B (Figure 6.18).

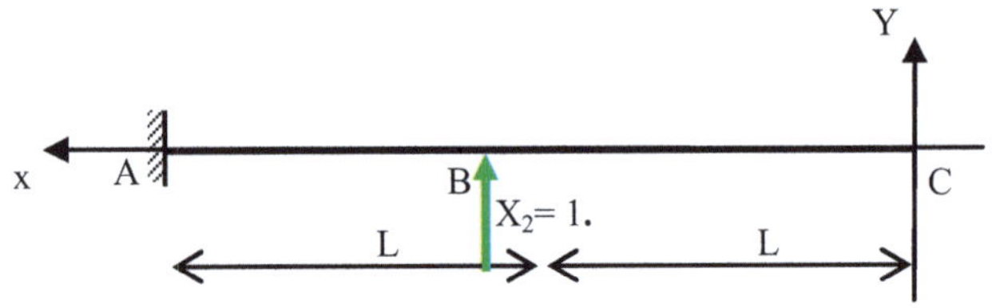

Figure 6.18. *Second unit system*

The variation of the bending moment and the corresponding diagram is shown in Figure 6.19.

$$m_2(x) = 1(x - L) \qquad L \le x \le 2.L \qquad [6.10]$$

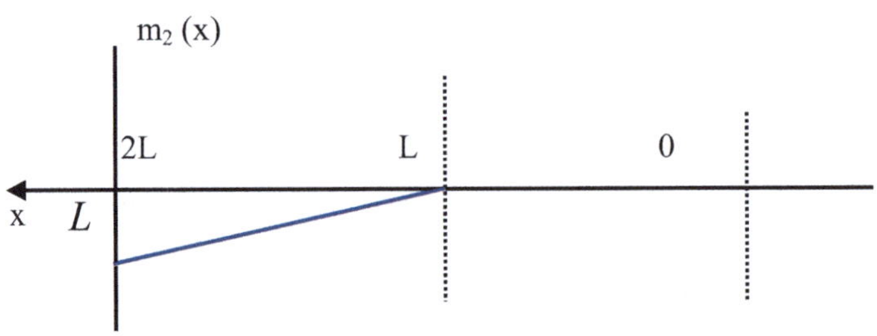

Figure 6.19. *Diagram of moment m₂(x)*

The displacements at points B and C of the unit systems in Figures 6.16 and 6.18 are calculated using the graphical method.

The displacements at points B and C are

$$\delta_{CC} = [\frac{1}{2}2L.2L(\frac{4}{3}L)] = \frac{8}{3}\frac{L^3}{EI}$$

$$\delta_{BC} = \frac{1}{EI}\left[\frac{1}{2}.L.L.(\frac{5L}{3})\right] = \frac{5L^3}{6EI}$$

$$\delta_{BB} = \frac{1}{EI}\left[\frac{1}{2}.L.(L).(\frac{2L}{3})\right] = \frac{L^3}{3EI}$$

$$\delta_{CB} = \delta_{BC} = \frac{5L}{6EI}$$

The compatibility conditions of the displacements at points B and C (Figure 6.20) are applied, and we obtain

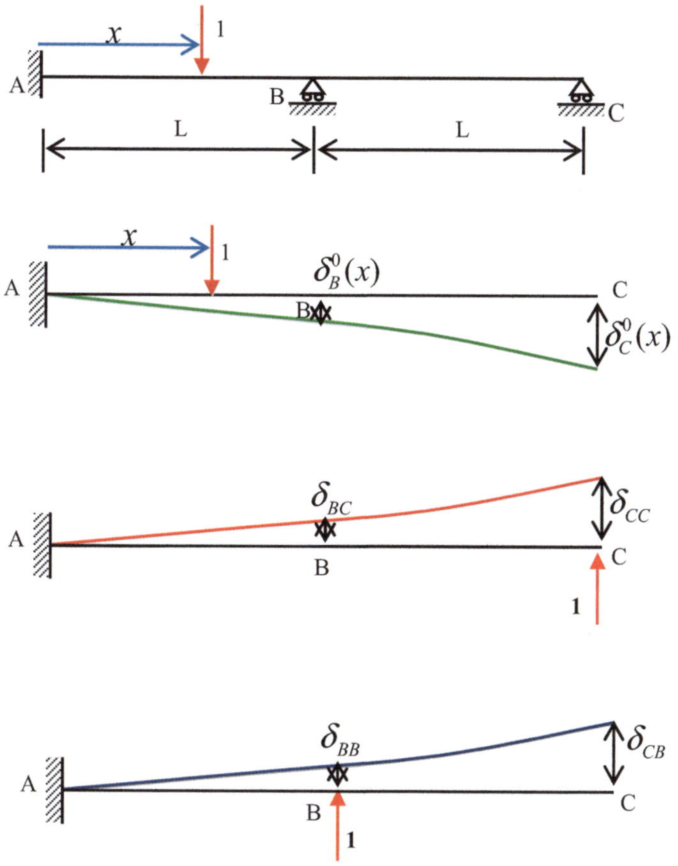

Figure 6.20. *Displacements at points B and C*

The displacements of the system given at points B and C are written as

$$\delta_B^0(x) + V_B.\delta_{BB} + V_C.\delta_{BC} = 0 \quad [6.11a]$$

$$\delta_C^0(x) + V_B.\delta_{CB} + V_C.\delta_{CC} = 0 \quad [6.11b]$$

The support reactions V_B and V_C can be deduced after resolving the system of equations ([6.11a] and [6.11b]).

$0 \le x \le L$

$$\frac{x^2}{6EI}(x-3L) + \frac{L^3}{3EI}V_B + \frac{5L^3}{6EI}V_C = 0$$

$$\frac{x^2}{6EI}(x-6L) + \frac{5L^3}{6EI}V_B + \frac{8L^3}{3EI}V_C = 0$$

The support reactions are

$$V_B = \frac{-1}{7L^3}\left[11x^3 - 18Lx^2\right]$$

$$V_C = \frac{3}{7L^3}\left[x^3 - Lx^2\right]$$

$L \le x \le 2L$

$$\frac{1}{6EI}(3L^2x - L^3) + \frac{L^3}{3EI}V_B + \frac{5L^3}{6EI}V_C = 0$$

$$\frac{x^2}{6EI}(x-6L) + \frac{5L^3}{6EI}V_B + \frac{8L^3}{3EI}V_C = 0$$

The support reactions become

$$V_B = \frac{1}{7L^3}(5x^3 - 30Lx^2 + 48L^2x - 16L^3)$$

$$V_C = -\frac{1}{7L^3}(2x^3 - 12Lx^2 + 15L^2x - 5L^3)$$

The influence lines of support reactions V_B and V_C are shown in Figure 6.21.

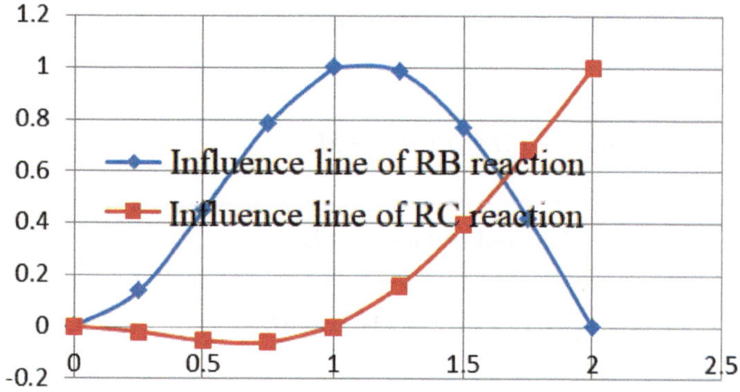

Figure 6.21. *Influence lines of reactions R_B and R_C*

The influence line of the bending moment at fixing A is

$$0 \le x \le L \quad M_A(x) = 1.x - R_B.L - R_C.(2L)$$

with

$$V_B = \frac{-1}{7L^3}\left[11x^3 - 18Lx^2\right]$$

$$V_C = \frac{3}{7L^3}\left[x^3 - Lx^2\right]$$

By substituting the support reactions values, the fixed-end moment expression becomes

$$M_A(x) = \frac{1}{7L^2}\left[5x^3 - 12Lx^2 + 7L^2x\right]$$

$$L \le x \le 2L$$

$$M_A(x) = 1.x - R_B.L - R_C.(2L)$$

with

$$V_B = \frac{1}{7L^3}(5x^3 - 30Lx^2 + 48L^2x - 16L^3)$$

$$V_C = -\frac{1}{7L^3}(2x^3 - 12Lx^2 + 15L^2x - 5L^3)$$

The expression of the fixed-end moment is

$$M_A(x) = -\frac{1}{7L^2}\left[x^3 - 6Lx^2 + 11L^2x - 6L^3\right]$$

The influence line of the fixed-end moment is shown in Figure 6.22.

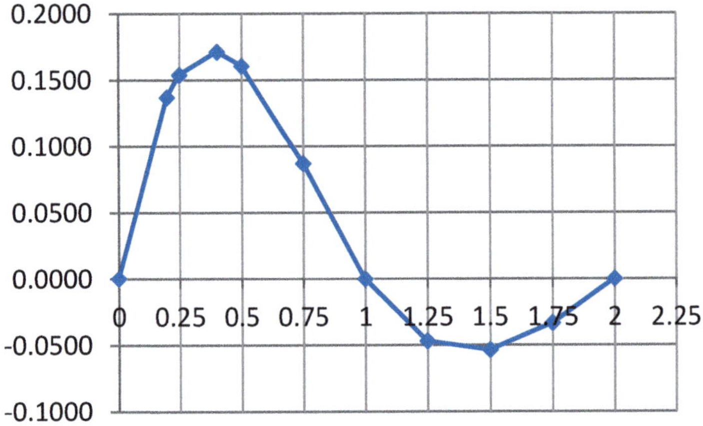

Figure 6.22. *Influence line of moment M_A*

6.3. Influence lines of frames

Draw the influence lines of fixed-end moment M_C and shear force T_C of the frame (Figure 6.23(a)) using the moment-distribution method. The given frame has displaced joints, so it is necessary to add a support at joint B to make it fixed joint (Figure 6.23(b)). We assume that the flexural rigidity (EI) is constant.

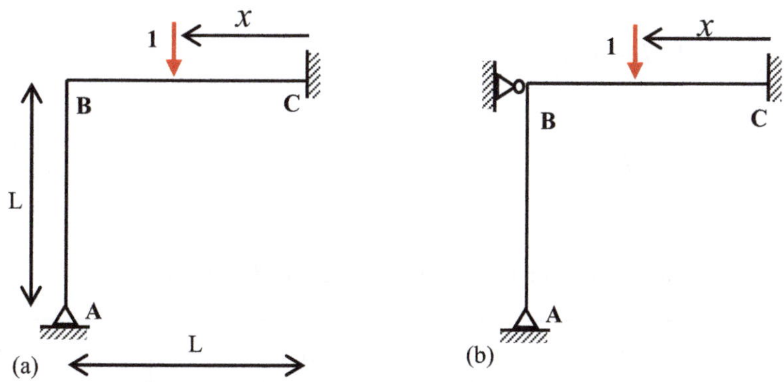

Figure 6.23. *a) Given frame and b) frame without sidesway*

– *Rigidity factors*

$$K_{BC} = K_{CB} = \frac{4EI}{L}$$

$$K_{BA}^{'} = \frac{3EI}{L}$$

– *Transmission coefficients*

$$\lambda_{BC} = \lambda_{CB} = \frac{1}{2}$$

$$\lambda_{BA}^{'} = 0$$

– *Distribution factors*

$$r_{BC} = \frac{K_{BA}}{K_{BA} + K_{BC}} = \frac{4}{4+3} = \frac{4}{7}$$

$$r_{BA} = \frac{K_{BC}^{'}}{K_{BA} + K_{BC}^{'}} = \frac{3}{3+4} = \frac{3}{7}$$

– *Fixed-end moments*

$$\gamma_{BC} = \frac{1.(L-x)x^2}{L^2}$$

$$\gamma_{CB} = -\frac{1.(L-x)^2 x}{L^2}$$

The method of distribution of moments is presented in Table 6.1.

Joints	A	B		C
Bars	AB	BA	BC	CB
r		$\frac{3}{7}$	$\frac{4}{7}$	
$\gamma(\frac{(L-x)x}{L^2})$			x	-(L-x)
Equilibrium		$-\frac{3x}{7}$	$-\frac{4x}{7}$	
Transmission	0			$-\frac{2x}{7}$
M $(\frac{x(L-x)}{L^2})$	0	$-\frac{3x}{7}$	$\frac{3x}{7}$	$\frac{5x-7L}{7}$

Table 6.1. *Moments at the ends of bars*

The influence line of moment M_C is shown in Figure 6.24.

The bending moment expression of bar (CB) is given as

$$M_{CB}(\alpha,x) = \mu(\alpha,x) - M_{BC}(0,x)(1-\frac{\alpha}{L}) + M_{BC}(L,x)(\frac{\alpha}{L})$$

where $\mu(\alpha,x)$ is the bending moment of bar (CB), which is considered statically determinate. We can distinguish

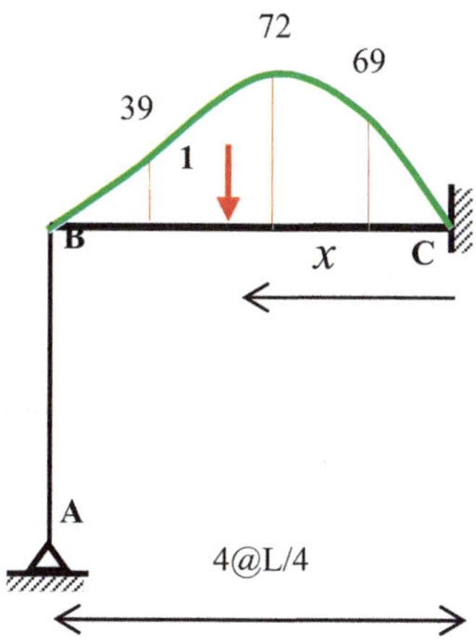

Figure 6.24. *Influence line of bending moment* M_C *$*(\frac{L}{448})$

$0 \leq \alpha \leq x$

$$\mu(\alpha,x) = \frac{L-x}{L}\alpha$$

$$\mathrm{M}_{CB}(\alpha,x) = \frac{L-x}{L}\alpha - \frac{5x-7L}{7}\frac{L-x}{L^2}x(1-\frac{\alpha}{L}) + \frac{3x}{7}\frac{L-x}{L^2}x(\frac{\alpha}{L})$$

$x \leq \alpha \leq L$

$$\mu(\alpha,x) = \frac{L-x}{L}\alpha - 1(\alpha - x) = x - \frac{x}{L}\alpha$$

$$\mathrm{M}_{CB}(\alpha,x) = x - \frac{x}{L}\alpha - \frac{5x-7L}{7}\frac{L-x}{L^2}x(1-\frac{\alpha}{L}) + \frac{3x}{7}\frac{L-x}{L^2}x(\frac{\alpha}{L})$$

The shear force expressions are deduced from the bending moment expressions.

$0 \leq \alpha \leq x$

$$T_{CB}(\alpha,x) = \frac{dM(\alpha,x)}{d\alpha} = \frac{L-x}{L} - \frac{5x-7L}{7}\frac{L-x}{L^2}x(-\frac{1}{L}) + \frac{3x}{7}\frac{L-x}{L^2}x(\frac{1}{L}) =$$
$$1 - \frac{x}{L} + \frac{(L-x)}{7L^3}x(8x-7L)$$

$x \le \alpha \le L$

$$T_{CB}(\alpha,x)=\frac{dM(\alpha,x)}{d\alpha}=-\frac{x}{L}+\frac{5x-7L}{7}\frac{L-x}{L^7}x+\frac{3x}{7}\frac{L-x}{L^3}x$$

$$=-\frac{x}{L}+\frac{L-x}{7L^3}x(8x-7L)$$

The influence line of shear stress T_C is shown in Figure 6.25.

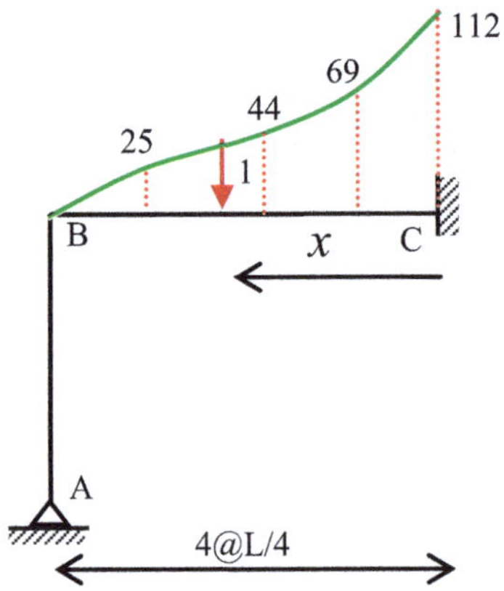

Figure 6.25. *Influence line of shear force TC*($\frac{1}{112}$)*

6.4. Influence lines of trusses

For truss systems, we distinguish three categories of structures that depend on the degree of static indeterminacy. They are (1) internally statically indeterminate truss systems, (2) externally statically indeterminate systems and (3) internally and externally statically indeterminate systems (Figure 6.26).

The degree of static indeterminacy of a truss system is given by the following formula:

$$f = b - (2 \cdot n - 3)$$

Table 6.2 groups the calculation of the degree of static indeterminacy of each truss system (Figure 6.26) and Table 6.3 shows the nature of the static indeterminacy of truss systems.

System	(a)	(b)	(C)
b	6	5	6
2n – 3	5	5	5
r	3	4	3
F	1	0	1

Table 6.2. *Degrees of static indeterminacy*

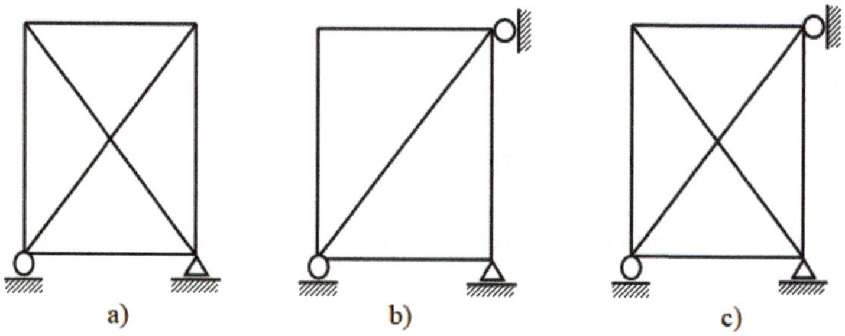

Figure 6.26. *Statically indeterminate truss structures. a) Internally statically indeterminate; b) externally statically indeterminate; c) externally and internally statically indeterminate*

System	Internal analysis	External analysis
(a)	Once statically indeterminate	Statically determinate
(b)	Statically determinate	Once statically indeterminate
(c)	Once statically indeterminate	Once statically indeterminate

Table 6.3. *Different static indeterminacy*

6.4.1. *Internally statically indeterminate truss*

Draw the influence lines of bars (BC) and (BD) of the truss system (Figure 6.27) knowing that all the bars have constant membrane rigidity ($E\Omega$). The load is moving on joints B and C.

– *External analysis*

Number of unknowns: 3

Number of static equations: 3

The system is externally statically determinate.

– *Internal analysis*

Number of bars: b = 6.

Number of equations: 2n – 3 = 5

Therefore, the system is once internally statically indeterminate (Figure 6.27).

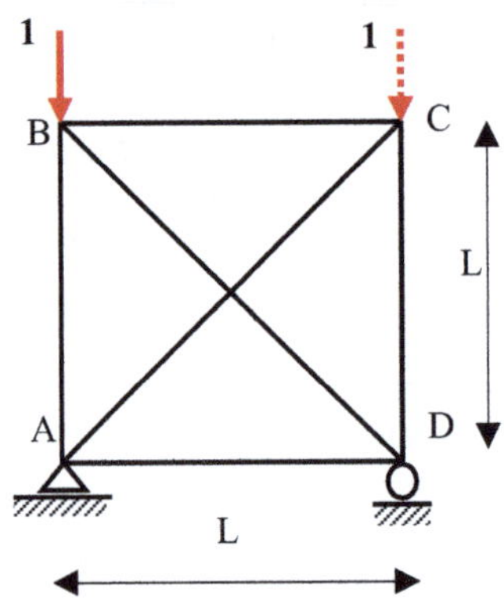

Figure 6.27. *Given system*

Truss system analysis of Figure 6.27 requires the application of the method of forces (Chapter 3).

– *Equivalent system*

In the equivalent truss system, it is necessary to replace bar (AC) with an equivalent force (Figure 6.28).

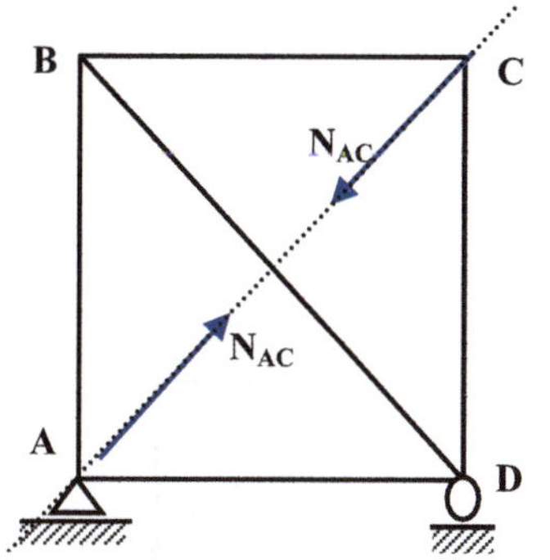

Figure 6.28. *Equivalent system*

– *Fundamental system*

We ignore the existence of bar (AC) to obtain the fundamental system (Figure 6.29). Then we apply a unit action to joint B and joint C alternatively.

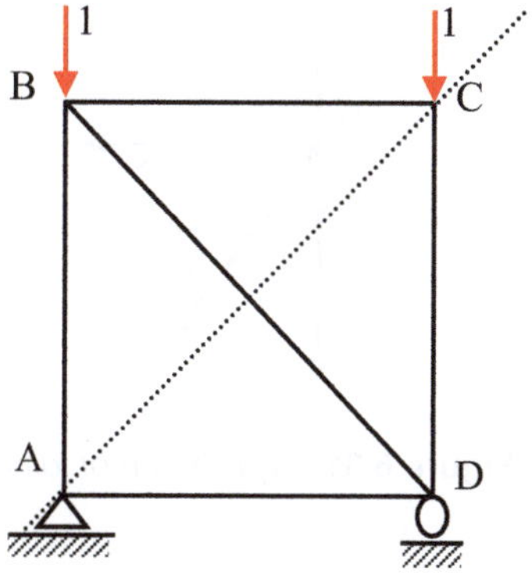

Figure 6.29. *Basic system*

The method of joint equilibrium is used to calculate the forces on the bars of the fundamental structure (Figure 6.29).

– *Case 1: Unit moment is at joint B*

- *Joint C*

$$\sum \vec{F} = \vec{0}$$

$F_2 + N_{CB} + N_{CD} = 0$

$N_{CB} = 0$

$N_{CD} = 0$

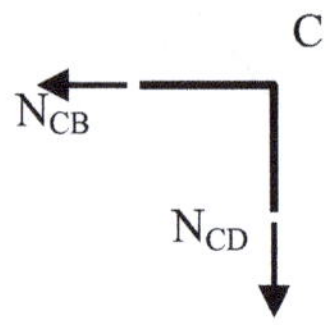

Figure 6.30. *Equilibrium of node C*

- *Joint B*

The equilibrium of joint B allows us to write

$$\sum F = 0$$

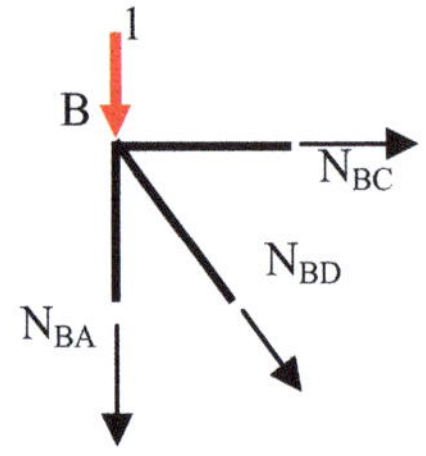

Figure 6.31. *Equilibrium of joint B*

The projection of internal forces along the axes gives

$$N_{BC} + N_{BD}\frac{\sqrt{2}}{2} = 0$$

$$N_{BA} + N_{BD}\frac{\sqrt{2}}{2} + 1 = 0$$

This leads to

$$N_{BC} = 0$$

$$N_{BA} = -1$$

- *Joint A*

The support reactions can be deduced using equilibrium equations of the fundamental system.

$$H_A = 0$$

$$V_A = +1$$

The equilibrium of joint A leads to

$$\sum \vec{F} = \vec{0}$$

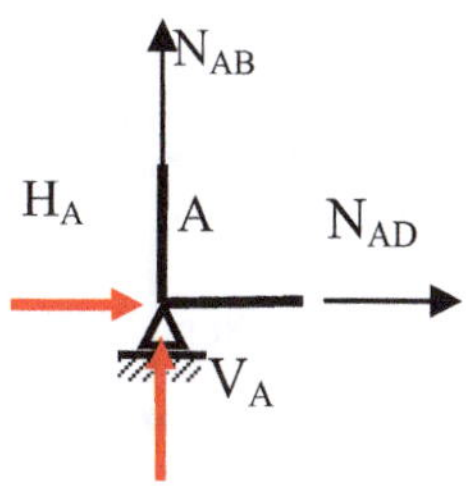

Figure 6.32. *Equilibrium of joint A*

$$H_A + V_A + N_{AD} + N_{AB} = 0$$

The projection of forces along the horizontal axis gives

$$N_{AD} = 0$$

The forces on the bars of the truss structure are grouped in Table 6.4.

Bar	N_B^0	Nature of force
AB	–1	Compression
AD	0	Neutral
BC	0	Neutral
BD	0	Neutral
CD	0	Neutral

Table 6.4. *Internal forces on the bars of the fundamental system*

– *Case 2: Unit moment is at joint C*

- *Joint C*

$N_{CB} = 0$

$N_{CD} = -1.$

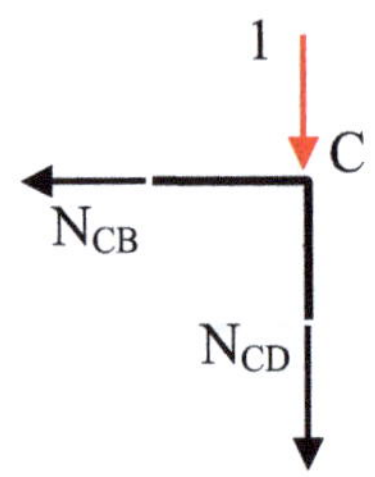

Figure 6.33. *Equilibrium of joint C*

- *Joint B*

The equilibrium of joint B leads to

$$\sum F = 0$$

$N_{BD} = N_{BA} = 0$

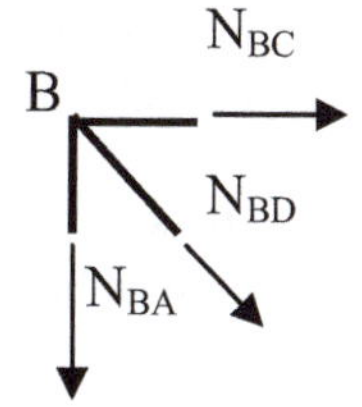

Figure 6.34. *Equilibrium of joint B*

- *Joint A*

Similarly, the force on bar (AD) is also zero.

$$N_{AD} = 0$$

The forces on the bars of truss structures when a unit force is applied at joint C are presented in Table 6.5.

Bar	N_C^0	Nature of force
AB	0	Neutral
AD	0	Neutral
BC	0	Neutral
BD	0	Neutral
CD	–1	Compression

Table 6.5. *Internal forces on the bars of the fundamental system*

- *Unit system analysis*

Instead of the ignored force (Figure 6.29), we apply a unit force in the direction of the removed bar (Figure 6.35).

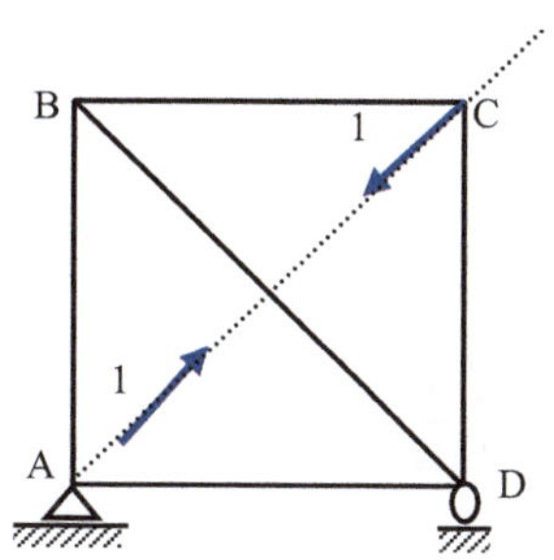

Figure 6.35. *Unit system*

The forces on the bars of truss structures when a unit force is applied at joints A and C alternatively are presented, respectively, in Tables 6.5 and 6.6.

Similarly, the internal force on the bars of the system (Figure 6.35) due to a unit action applied in the direction of the removed bar is presented in Table 6.6.

Bar	n	Nature of force
AB	$-\sqrt{2}/2$	Compression
AD	$-\sqrt{2}/2$	Compression
BC	$-\sqrt{2}/2$	Compression
BD	1	Traction
CD	$-\sqrt{2}/2$	Compression
AC	1	Traction

Table 6.6. *Internal forces on the bars of the unit system due to a unit action applied along the axis (AC)*

The displacement of bar (AC) is calculated when a unit force is applied to joint B. Applying the compatibility condition of the displacement along the diagonal (AC) makes it possible to write:

$$\delta^0_{AC,B} + \delta^1_{AC}.N^B_{AC} = 0$$

The force in bar (AC) is

$$N^B_{AC} = -\frac{\delta^0_{AC,C}}{\delta^1_{AC}}$$

with

$$\delta^0_{AC,C} = \frac{(-1)}{E\Omega}(-\frac{\sqrt{2}}{2})L$$

$$\delta^0_{AC,B} = \frac{(-1)}{E\Omega}(-\frac{\sqrt{2}}{2})L = \frac{\sqrt{2}L}{2E\Omega}$$

$$\delta^1_{AC} = \frac{1}{E\Omega}\left[4.(-\frac{\sqrt{2}}{2})^2 L + 2.\frac{\sqrt{2}}{2}L\right] = \frac{\sqrt{2}+2}{E\Omega}L$$

$$N_{AC}^{B} = -\frac{\sqrt{2}}{2(\sqrt{2}+2)}$$

We apply the same method to calculate the displacement at bar (AC) when a unit force is applied at joint C.

$$\delta_{AC,C}^{0} + \delta_{AC}^{1}.N_{AC}^{C} = 0$$

The force on bar (AC) under an applied unit force is

$$N_{AC}^{C} = -\frac{\delta_{AC,C}^{0}}{\delta_{AC}^{1}}$$

$$\delta_{AC,C}^{0} = \frac{(-1)}{E\Omega}(-\frac{\sqrt{2}}{2})L = \frac{\sqrt{2}\,L}{2E\Omega}$$

$$\delta_{AC}^{1} = \frac{\sqrt{2}+2}{E\Omega}L$$

$$N_{AC}^{C} = -\frac{\sqrt{2}}{2(\sqrt{2}+2)}$$

The influence line of the internal force on bar (AC) (Figure 6.36) when a unit force is successively applied between joint B and C is:

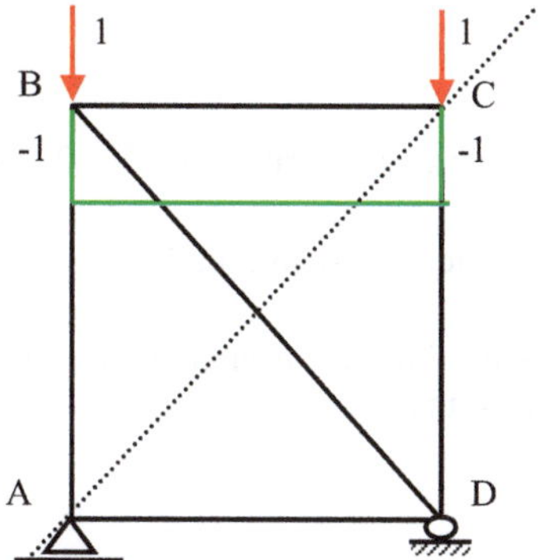

Figure 6.36. *Influence line and normal force on bar (AC)*(* $\frac{\sqrt{2}}{2(\sqrt{2}+2)}$ *)*

6.4.2. Externally statically indeterminate truss

Draw the influence lines of support reaction R_L and normal force N_{DE} of the truss structure (Figure 6.37). $E\Omega$ is constant and the unit force is assumed to be moving on the joints of the lower membrane.

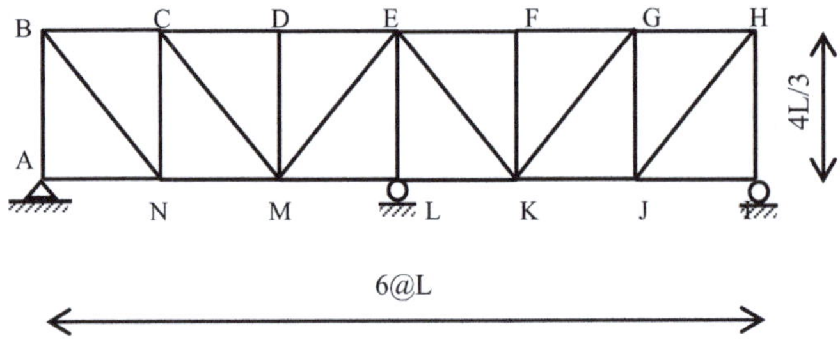

Figure 6.37. *Given truss*

– *External analysis*

Number of unknowns (support reactions) = 4

Number of static equations = 3

The system is once externally statically indeterminate.

– *Internal analysis*:

Number of bars: b = 25

Number of possible equations to write: $2n - 3 = 2\times 14 - 3 = 25$

The structure is statically determinate internally.

We choose the method of forces to analyze and evaluate the influence lines of the truss structure (Figure 6.37).

– *Equivalent system*

The given truss system is once statically indeterminate externally. We must replace the support reaction with an equivalent force, for example support reaction R_L (Figure 6.38).

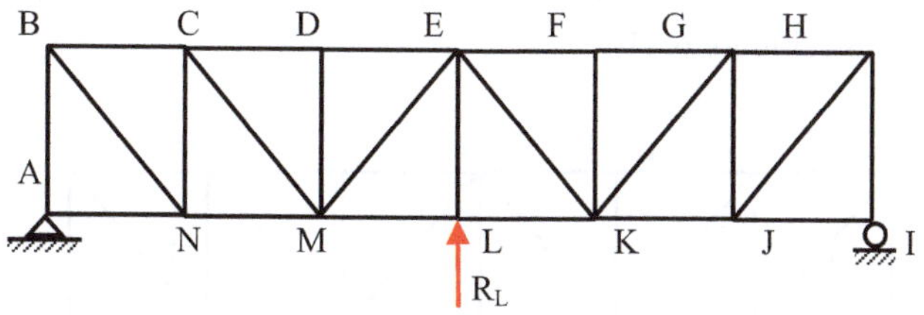

Figure 6.38. *Equivalent structure*

– *Fundamental system*

We ignore the support reaction replaced in the equivalent structure to obtain the fundamental system (Figure 6.39). Then we apply a unit action to joint N and joint J successively.

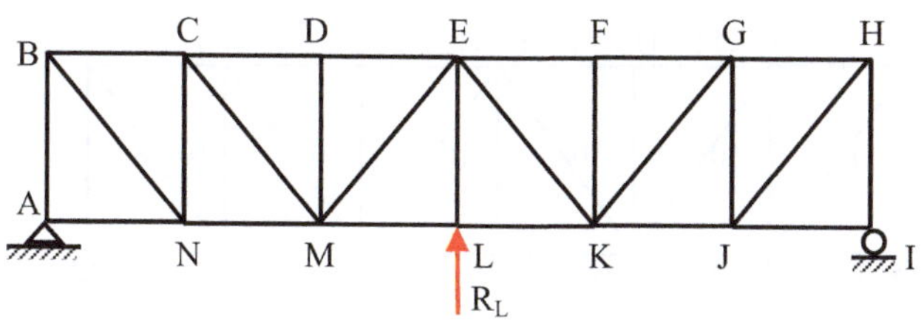

Figure 6.39. *Fundamental structure*

– *Case 1*: A unit force is applied to joint N (Figure 6.40).

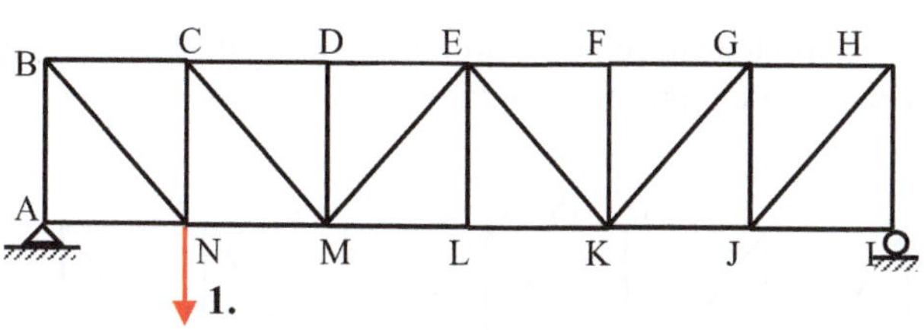

Figure 6.40. *Unit force at joint N*

The forces on the bars of the structure (Figure 6.40) are obtained by the method of joint equilibrium (Figure 6.41).

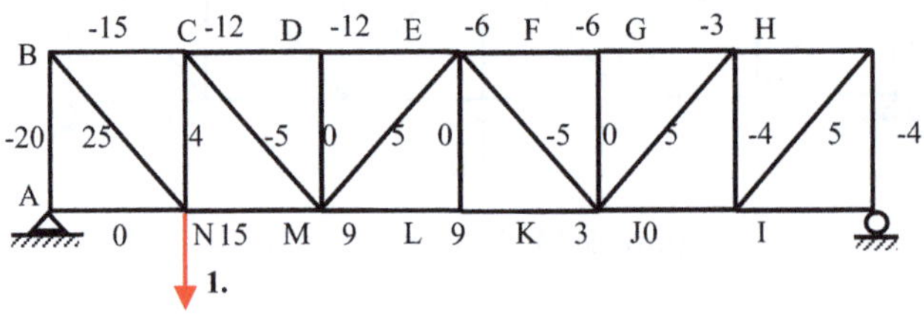

Figure 6.41. *Force on the bars due to a unit force applied at joint N*(1/24)*

We apply a unit force at point M. The internal forces on the bars are shown in Figure 6.41.

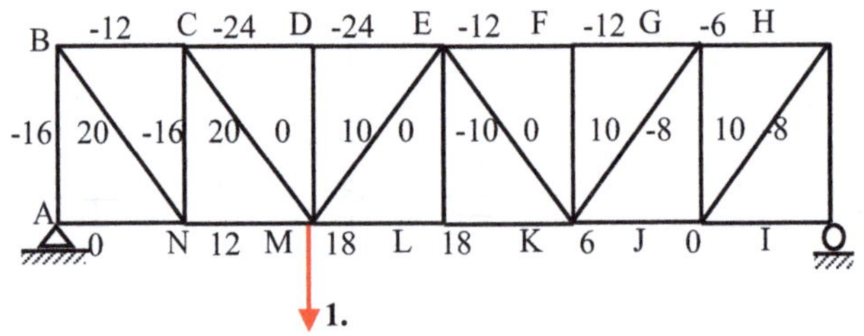

Figure 6.42. *Force on the bars due to a unit force applied at joint M*(1/24)*

The unit force is applied successively to points L, K and J. The forces obtained in the bars for each case are represented in Figures 6.43–6.44.

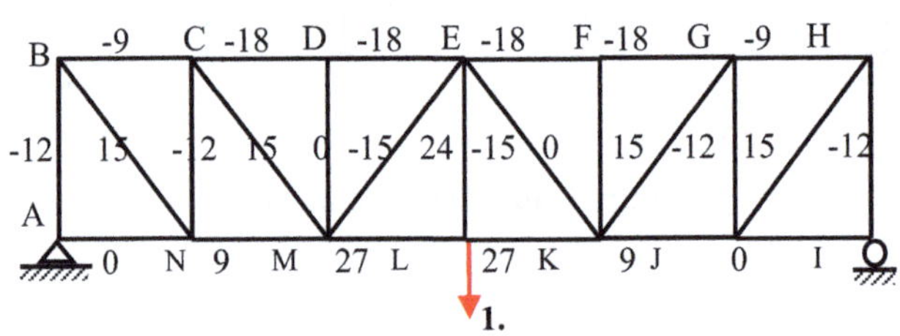

Figure 6.43. *Force on the bars due to a unit force applied at joint L (1/24)*

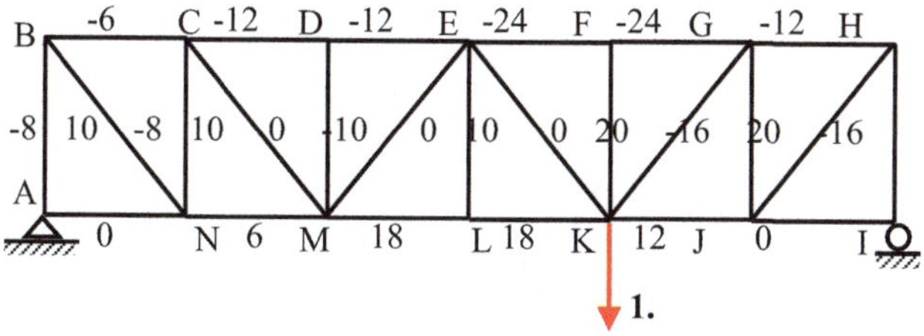

Figure 6.44. *Force on the bars due to a unit force applied at joint K (1/24)*

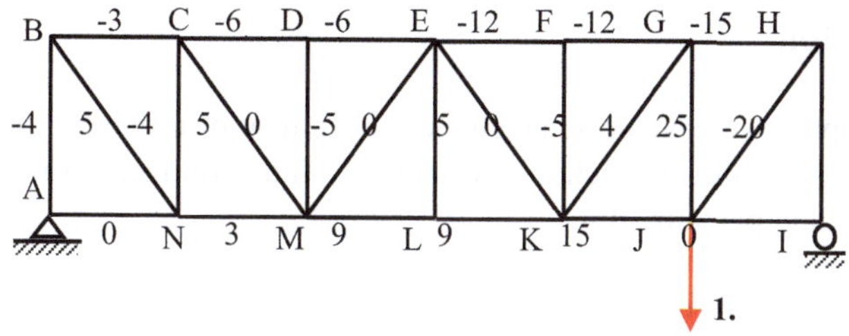

Figure 6.45. *Force on the bars due to a unit force applied at joint J*(1/24)*

The influence lines of support reactions R_L and R_I are given in Figure 7.46.

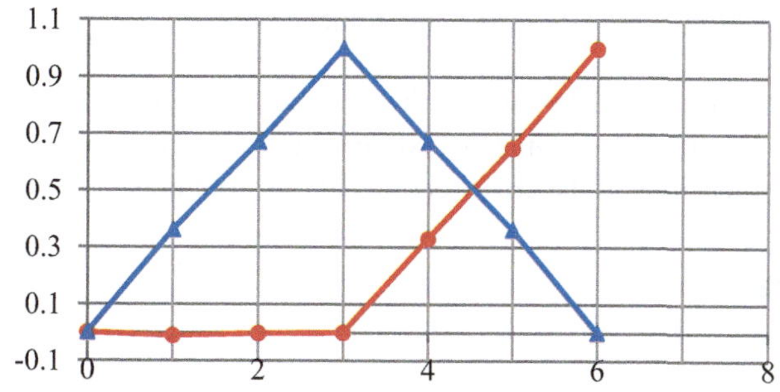

Figure 6.46. *Influence lines of support reactions R_L and R_I*

In the same way, it is possible to draw the influence line of the internal force on bar (DE) by taking the value of the internal force for the different application cases of the unit force going from joint N to joint J (Figure 6.47).

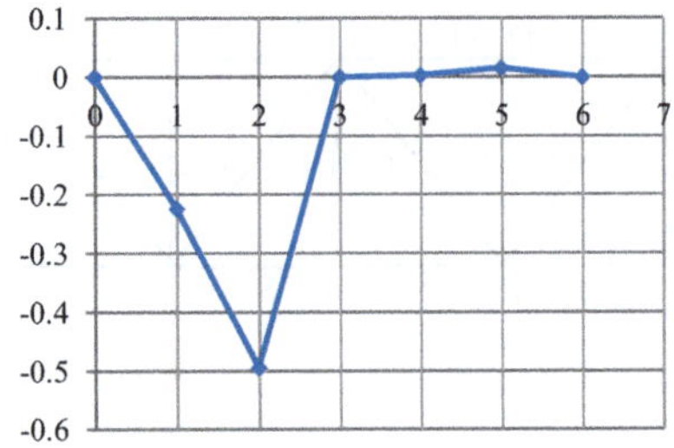

Figure 6.47. *Influence line and internal force on bar (DE)*

6.5. Conclusion

In this chapter, we have presented the establishment of influence lines of support reactions and/or of internal actions of statically indeterminate structures, such as trusses, beams and plane frames.

The influence lines of support reactions and/or internal actions can be deduced by using one of the statically indeterminate structural analysis methods: the method of three moments, the slope-deflection method, the method of forces or the moment-distribution method.

The influence lines of support reactions and/or internal actions can be established by applying a moving unit action. The response obtained gives the support reaction and/or internal action, which describes the influence line.

The study of influence lines allows the designer to highlight the position of the force that can be applied, leading to extreme stresses. This makes it possible to minimize the dimensions of the cross sections of the structures.

6.6. Problems

Exercise 1

Determine the influence lines of support reaction R_A of the following beams. EI is constant for all structures.

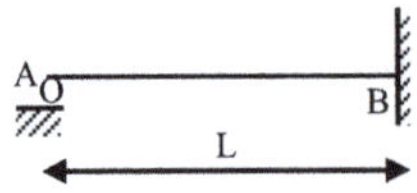

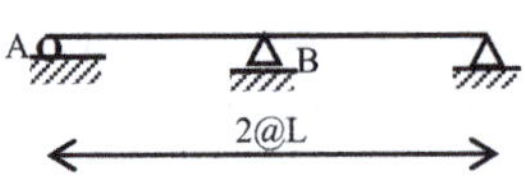

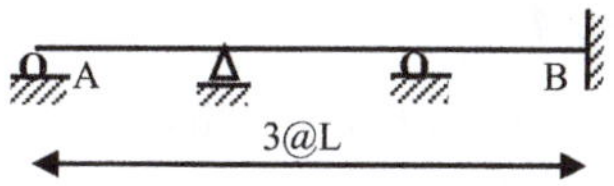

Exercise 2

Determine the influence lines and shear force relating to A and the bending moment M_B for the structures in Exercise 1.

Exercise 3

Determine the influence lines of the support reaction R_A and internal force N_1 of bar 1 for the following trusses. We assume that the unit load is moving on the joints of the upper membrane and E Ω is constant.

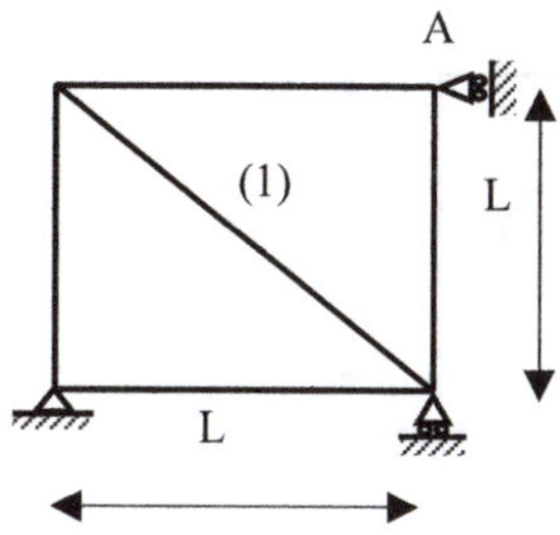

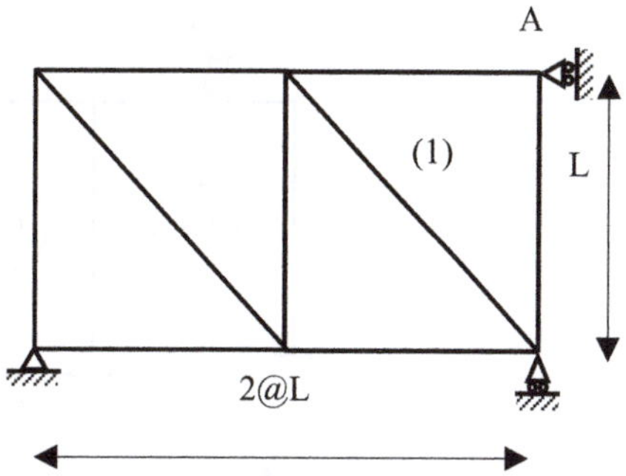

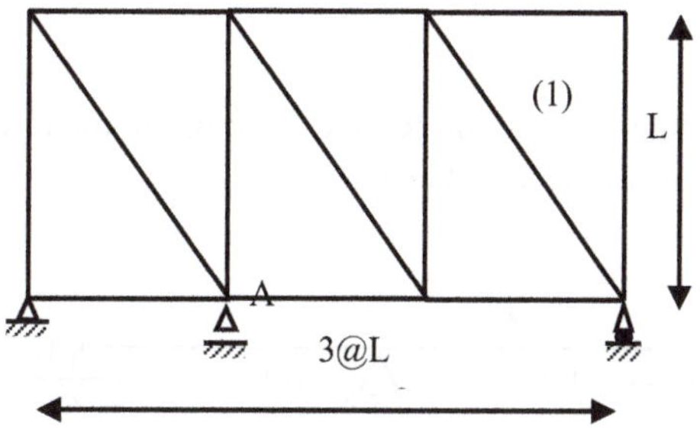

Exercise 4

Draw the influence lines of reaction R_A and bending moment M_B of the following frames:

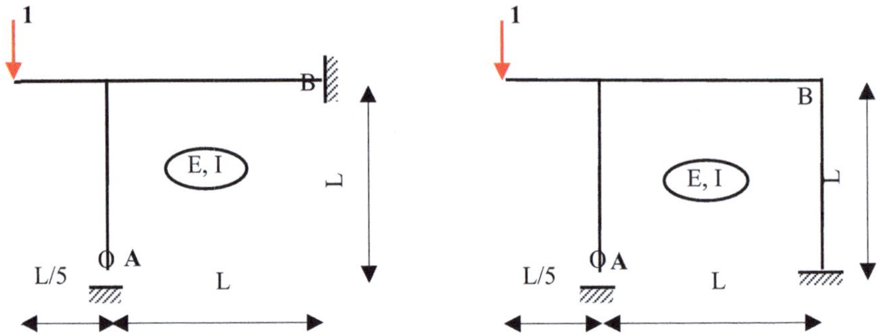

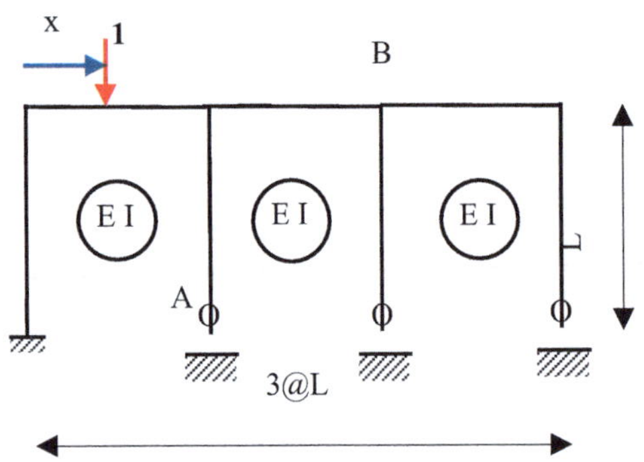

Exercise 5

Draw the influence lines of reaction R_A and bending moment M_B and shear force T_B of the following beam (EI is a constant):

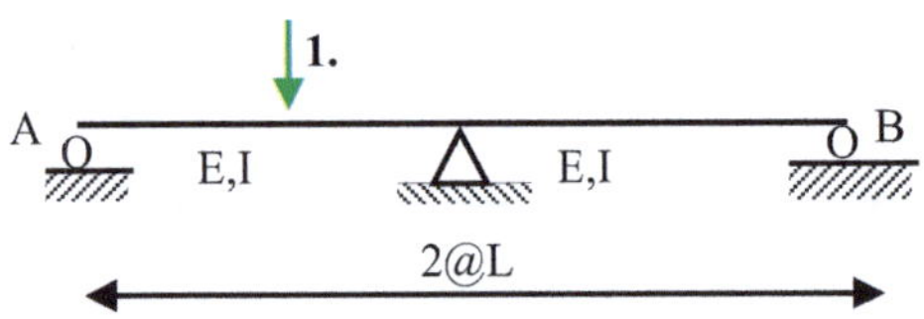

Use the slope-deflection method, the method of forces and the moment-distribution method.

Redo this exercise for the following beam:

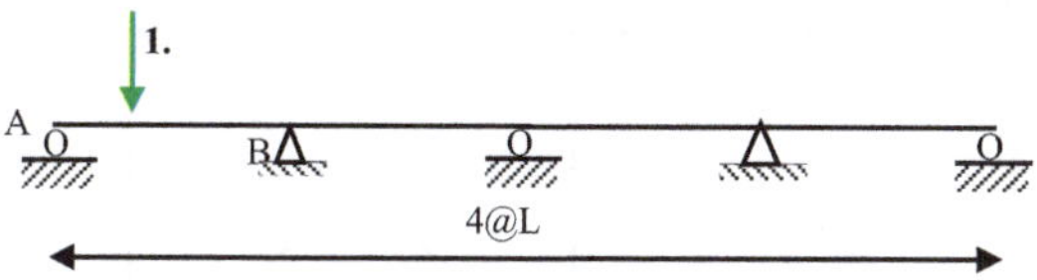

Exercise 6

Draw the influence lines of reaction R_A, bending moment M_B and shear force T_B of the following frames: use the slope-deflection method, the method of forces and the moment-distribution method. EI is a constant.

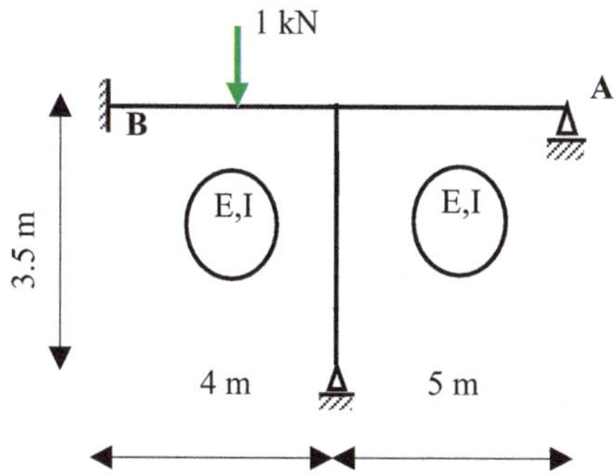

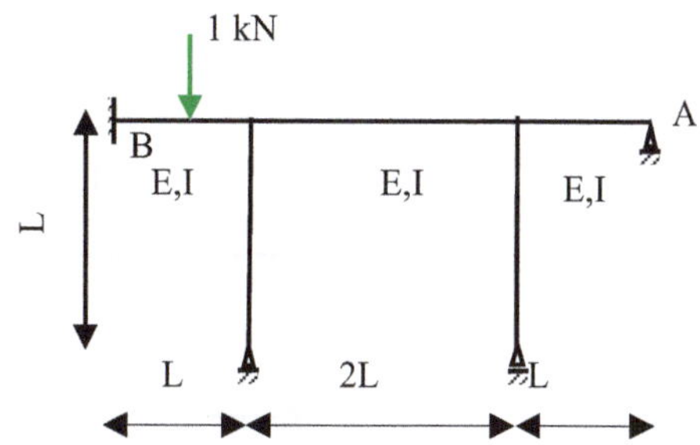

Exercise 7

Draw the influence lines of support reactions R_A and R_B and normal forces N_1 and N_2 of the following truss structures:

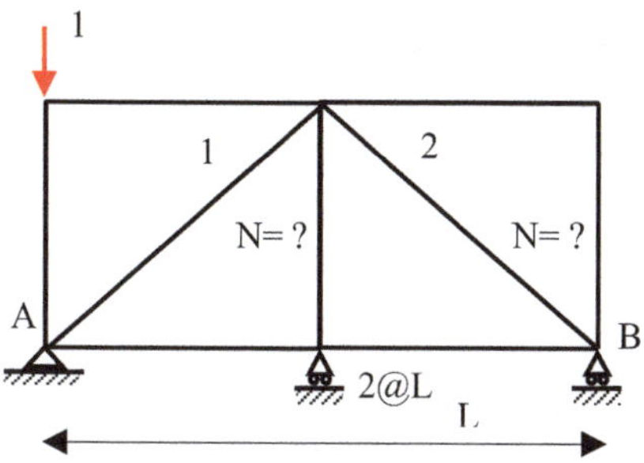

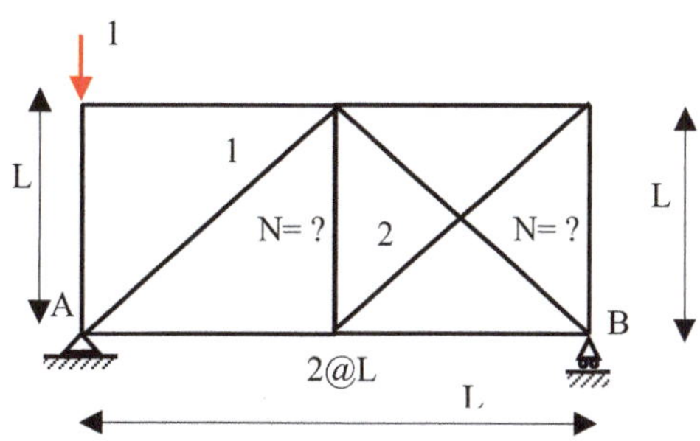

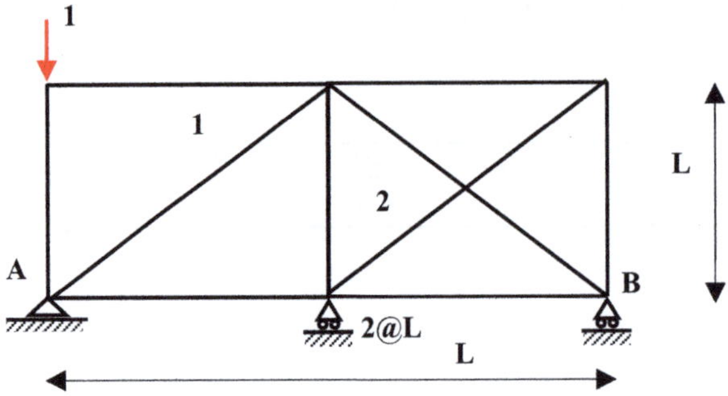

Exercise 8

Draw the influence lines of reaction R_A and bending moment M_A according to the stiffness of spring K_s of the following beam:

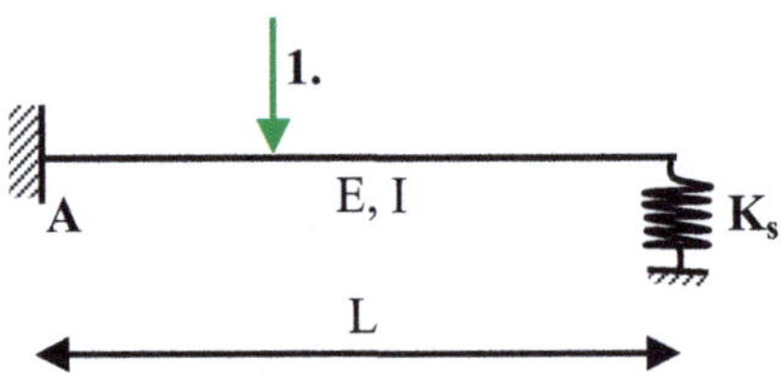

Redo the same study for the following continuous beam.

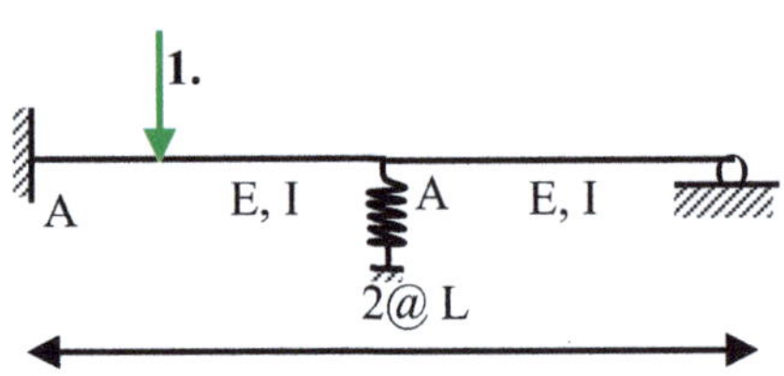

7

Statically Indeterminate Arch Analysis

After studying this chapter, the reader will be able to:

– distinguish between types of arch;

– comprehend the advantages of arch constructions;

– analyze statically indeterminate arches under different loads;

– analyze statically indeterminate tied arches;

– distinguish and analyze arches with any funicular geometry;

– construct influence lines of arches.

In this chapter, we present the fundamental concepts of statically indeterminate arch analysis. Based on geometry, three categories of arches have been studied: (1) semicircular arches, (2) parabolic arches and (3) tied arches. In another way, proceeding from the connection between arches and the external environment, two families of arches are the subject of this analysis: bi-hinged arches and fixed arches. Finally, to show the variation of internal and external actions under a moving load, the influence lines of internal and external actions have been widely established.

7.1. Introduction

The use of arches in construction is very ancient, dating back to the Roman period. They have been used in the construction of bridges as elements resistant to the loads of vaults and slabs; their use is justified by their structural importance and by their high mechanical efficiency.

From a structural point of view, arches can bear very significant compression forces and considerably reduce the bending effect. Initially, they were built using stones or voussoirs. Then, toward the 18th Century, they were transformed into carved stones. Nowadays, arches are constructed of steel, reinforced concrete or prestressed concrete and can withstand strong compression and traction stresses. They are intended to be the main elements of very long bridges.

7.2. Classification of arches

Statically indeterminate arches are classified according to their connection with the external environment. In general, they are grouped into three categories: (1) three-hinged arches, (2) bi-hinged arches and (3) fixed arches. Statically determinate arches are in the first category, while statically indeterminate arches are in the second and third categories. The support reaction can be determined by one of the analysis methods of statically indeterminate structures presented in Chapters 2–5 (Figure 7.1).

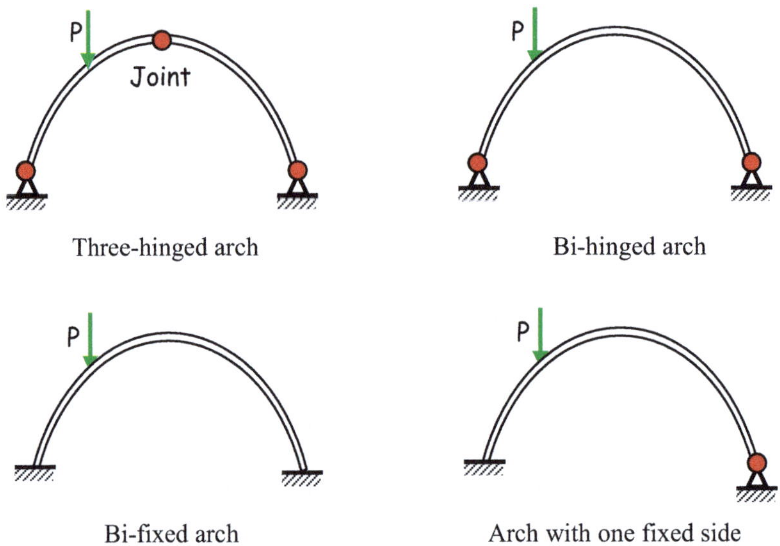

Figure 7.1. *Statically indeterminate arches*[1]

In the same way, it is possible to classify the arches according to their rise or funicular shape. In this case, we distinguish overbowed arches, arches with an average rise and arches with a significant rise (Figure 7.2).

1 All of the figures in this chapter are available to view in full color at www.iste.co.uk/khalfallah/analysis2.zip.

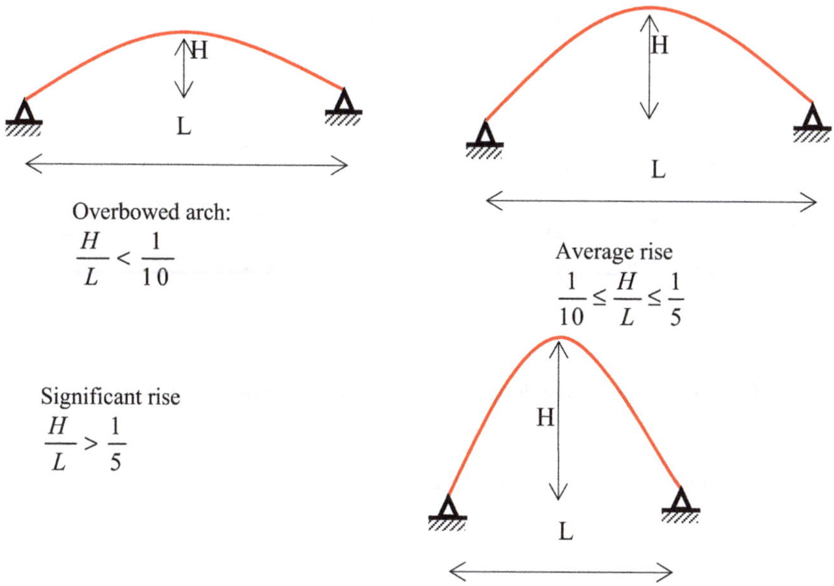

Figure 7.2. *Classes of arch*

7.3. Semicircular arch under concentrated load

We analyze the bi-hinged semicircular arch, which is subjected to a concentrated force of intensity P applied at distance a from support A (Figure 7.3). We assume that the flexural rigidity (EI) is constant.

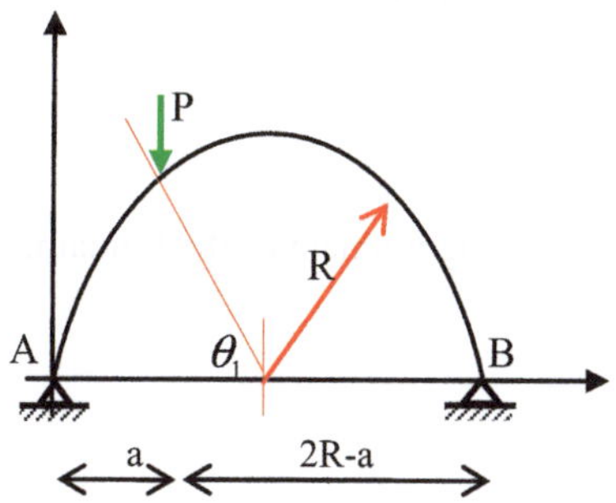

Figure 7.3. *Semicircular arch under concentrated load*

The arch (Figure 7.3) is once statically indeterminate. To analyze it, we use the method of forces (Chapter 5).

Equivalent and fundamental systems (Figure 7.4)

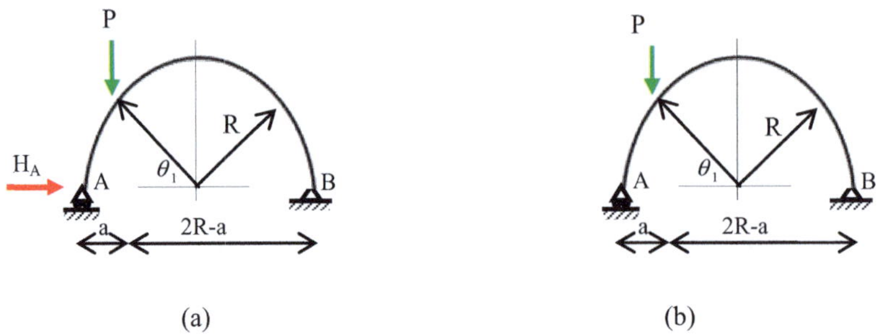

Figure 7.4. *(a) Equivalent system and (b) fundamental system*

The displacement at point A of the given system is null.

$$\delta_A = 0 \qquad [7.1]$$

Using the superposition principle of effects allows us to write

$$\delta_A^0 + H_A \delta_A^1 = 0 \qquad [7.2]$$

δ_A^0 and δ_A^1 are the horizontal displacements at point A, respectively, of the fundamental (Figure 7.4(b)) and unit systems (Figure 7.5).

The support reaction at point A of the fundamental system can be deduced by

$$V_A = P(1 - \frac{a}{2R}) \qquad [7.3]$$

The expressions of the bending moment of the fundamental system are given as

$0 \le x \le a$

$$\mu(x) = P(1 - \frac{a}{2R})x \qquad [7.4a]$$

$a \le x \le 2R$

$$\mu(x) = Pa(1 - \frac{x}{2R}) \qquad [7.4b]$$

The expression of the bending moment of the unit system (Figure 7.5) is given as

$0 \leq x \leq 2R$

$$m(x) = -y \qquad [7.5]$$

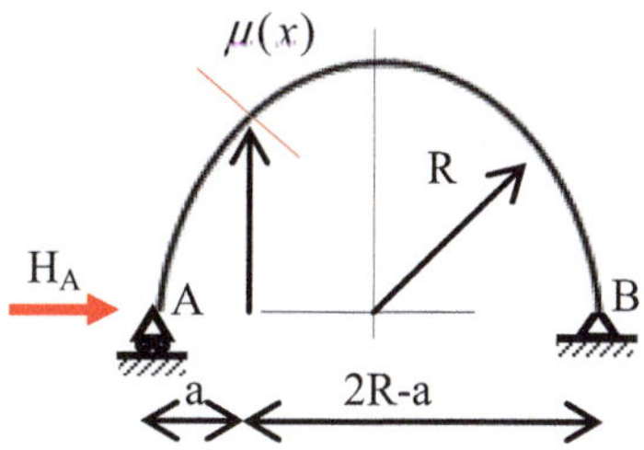

Figure 7.5. *Arch under a unit action*

This time, we use the principle of virtual work to calculate the displacements.

$$\delta_A^0 = \int_0^s \frac{\mu(x)m(x)}{EI} ds \qquad [7.6]$$

$$\delta_A^1 = \int_0^s \frac{m^2(x)}{EI} ds \qquad [7.7]$$

S is the curvilinear length of the arch.

Substituting equations [7.4] and [7.5] with relationships [7.6] and [7.7], we obtain

$$\delta_A^0 = \frac{1}{EI}\left[P(1-\frac{a}{2R})\int_0^a x(-y)ds + Pa\int_a^{2R}(1-\frac{x}{2R})(-y)ds \right] \qquad [7.8]$$

$$\delta_A^1 = \frac{1}{EI}\int_0^s y^2 ds \qquad [7.9]$$

Knowing that $x = R(1-\cos\theta)$, $y = R\sin\theta$ and $ds = Rd\theta$, equations [7.8] and [7.9] are written as

$$\delta_A^0 = \frac{1}{EI}\left[P(1-\frac{a}{2R})\int_0^{\theta_1} R(1-\cos\theta)(-R\sin\theta)Rd\theta \right]$$

$$+\frac{1}{EI}\left[\int_{\theta_1}^{\pi} Pa(1-\frac{x}{2R})(-R\sin\theta)Rd\theta \right] \quad [7.10]$$

Alternatively, it is written as

$$\delta_A^0 = \left[-\frac{PR^3}{2EI}(1-\frac{a}{2R})(\cos^2\theta_1 - 2\cos\theta_1 + 1) - \frac{PaR^2}{2EI}(\frac{1}{2}\cos^2\theta_1 + \cos\theta_1 + \frac{1}{2}) \right]$$

or

$$\delta_A^0 = -\frac{PR^3}{2EI}\left[(\cos^2\theta_1 - 2\cos\theta_1 + 1) + 2\frac{a}{R}\cos\theta_1 \right] \quad [7.11]$$

$$\delta_A^1 = \frac{1}{EI}\int_0^{\pi} R^2 \sin^2\theta . Rd\theta = \frac{\pi R^3}{2EI} \quad [7.12]$$

Substituting equations [7.11] and [7.12] in equation [7.2], we obtain the expression that gives the horizontal reaction at joint A.

$$H_A = \frac{P}{\pi}\left[(\cos^2\theta_1 - 2\cos\theta_1 + 1) + 2\frac{a}{R}\cos\theta_1 \right] \quad [7.13]$$

The reaction at support B can be deduced by

$$H_B = -H_A = -\frac{P}{\pi}\left[(\cos^2\theta_1 - 2\cos\theta_1 + 1) + 2\frac{a}{R}\cos\theta_1 \right] \quad [7.14]$$

The variation of the bending moment is given as

$0 \le x \le a$

$$M(x) = V_A x - H_A y = PR(1-\frac{a}{2R})(1-\cos\theta)$$

$$-\frac{PR}{\pi}\left[(\cos^2\theta_1 - 2\cos\theta_1 + 1) + 2\frac{a}{R}\cos\theta_1 \right]\sin\theta \quad [7.15a]$$

$a \leq x \leq 2R$

$$M(x) = V_A x - H_A y - P(x-a) = \frac{Pa}{2}(1+\cos\theta)$$
$$-\frac{P}{\pi}\left[(\cos^2\theta_1 - 2\cos\theta_1 + 1) + 2\frac{a}{R}\cos\theta_1\right]\sin\theta \qquad [7.15b]$$

In the same way, the variation of the shear force is

$\theta \leq \theta_1$ (Figure 7.6)

$$T(\theta) = V_A \sin\theta - H_A \cos\theta \qquad [7.16]$$

The shear force can also be written another way.

$$T(\theta) = P(1 - \frac{a}{2R})\sin\theta$$
$$-\frac{P}{\pi}\left[(\cos^2\theta_1 - 2\cos\theta_1 + 1) + 2\frac{a}{R}\cos\theta_1\right]\cos\theta \qquad [7.17a]$$

$$T(0) = -\frac{P}{\pi}\left[(\cos^2\theta_1 - 2\cos\theta_1 + 1) + 2\frac{a}{R}\cos\theta_1\right]$$

$$T(\theta_1) = P(1 - \frac{a}{2R})\sin\theta_1 - \frac{P}{\pi}\left[(\cos^2\theta_1 - 2\cos\theta_1 + 1) + 2\frac{a}{R}\cos\theta_1\right]\cos\theta_1$$

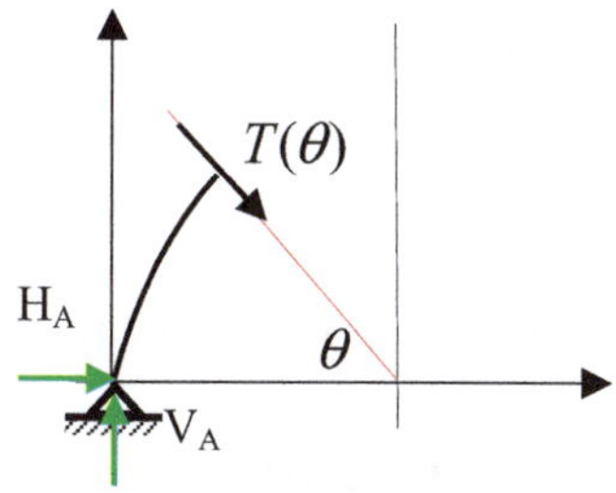

Figure 7.6. *Section 0 ≤ x ≤ a*

$\theta_1 \leq \theta \leq \pi$ (Figure 7.7)

The shear force expression in this section is given as

$$T(\theta) = V_A \sin\theta - H_A \cos\theta - P\sin\theta$$

or

$$T(\theta)=-P(\frac{a}{2R})\sin\theta-\frac{P}{\pi}\left[(\cos^2\theta_1-2\cos\theta_1+1)+2\frac{a}{R}\cos\theta_1\right]\cos\theta \quad [7.17b]$$

$$T(\theta_1)=-P(\frac{a}{2R})\sin\theta_1-\frac{P}{\pi}\left[(\cos^2\theta_1-2\cos\theta_1+1)+2\frac{a}{R}\cos\theta_1\right]\cos\theta_1$$

$$T(\pi)=\frac{P}{\pi}\left[(\cos^2\theta_1-2\cos\theta_1+1)+2\frac{a}{R}\cos\theta_1\right]$$

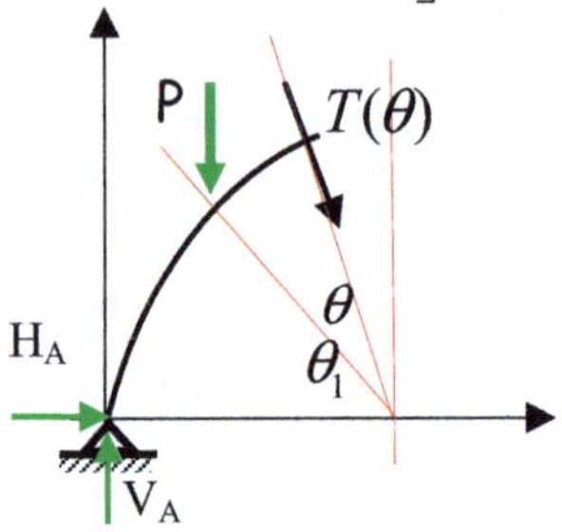

Figure 7.7. *Section* $\theta_1 \leq \theta \leq \pi$

The normal force at section $0 \leq \theta \leq \theta_1$ (Figure 7.8) is given as

$$N(\theta)=-V_A\cos\theta-H_A\sin\theta \quad [7.18]$$

or

$$N(\theta)=-P(1-\frac{a}{2R})\cos\theta-\frac{P}{\pi}\left[(\cos^2\theta_1-2\cos\theta_1+1)+2\frac{a}{R}\cos\theta_1\right]\sin\theta \quad [7.19]$$

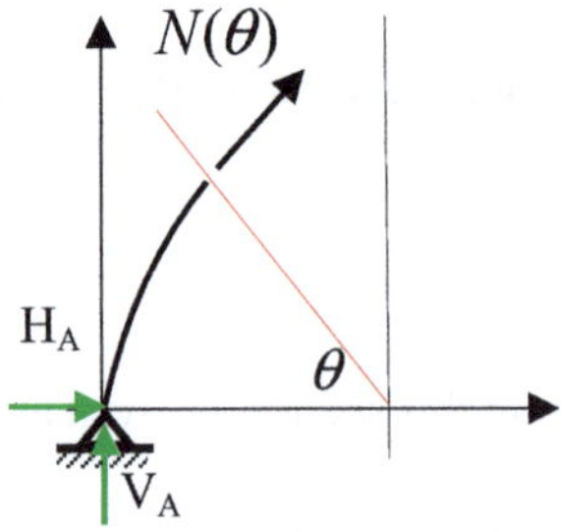

Figure 7.8. *Section* $0 \leq x \leq a$

$$N(0) = -P(1-\frac{a}{2R})$$

$$N(\theta_1) = -P(1-\frac{a}{2R})\cos\theta_1 - \frac{P}{\pi}\left[(\cos^2\theta_1 - 2\cos\theta_1 + 1) + 2\frac{a}{R}\cos\theta_1\right]\sin\theta_1$$

$\theta_1 \leq \theta \leq \pi$ (Figure 7.9)

$$N(\theta) = -V_A\cos\theta - H_A\sin\theta + P\cos\theta$$

$$N(\theta) = P(\frac{a}{2R})\cos\theta - \frac{P}{\pi}\left[(\cos^2\theta_1 - 2\cos\theta_1 + 1) + 2\frac{a}{R}\cos\theta_1\right]\sin\theta$$

$$N(\theta_1) = P(\frac{a}{2R})\cos\theta_1 - \frac{P}{\pi}\left[(\cos^2\theta_1 - 2\cos\theta_1 + 1) + 2\frac{a}{R}\cos\theta_1\right]\sin\theta_1$$

$$N(\pi) = -P(\frac{a}{2R})$$

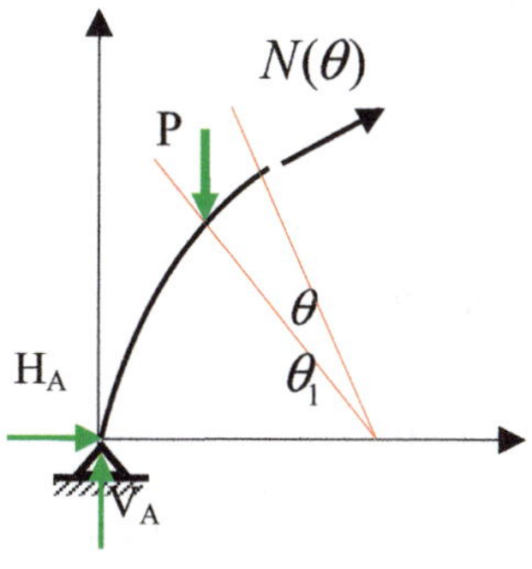

Figure 7.9. *Section* $\theta_1 \leq \theta \leq \pi$

Numerical application

$$\frac{a}{R} = \frac{1}{2},\ \theta_1 = \frac{\pi}{3},\ \cos\theta_1 = \frac{1}{2} \text{ and } \sin\theta_1 = \frac{\sqrt{3}}{2}$$

$$H_A = P$$

$$H_B = -P$$

$$V_A = 0.75P$$

$$V_B = 0.25P$$

The bending moment expressions are given as:

$0 \le x \le a$

$$M(x) = PR(0.75 - 0.75\cos\theta - \sin\theta)$$

$$\theta = 0: \quad M(0) = 0.$$

$$\theta = \frac{\pi}{4}: \quad M(0) = -0.487PR$$

$$\theta = \theta_1: \quad M(\theta_1) = -0.491PR$$

$a \le x \le 2R$

$$M(x) = PR(0.25 + 0.25\cos\theta - \sin\theta)$$

$$\theta = \theta_1: \quad M(\theta_1) = -0.491PR$$

$$\theta = \frac{\pi}{2}: \quad M(\frac{\pi}{2}) = -0.75PR$$

$$\theta = \frac{3\pi}{4}: \quad M(\frac{3\pi}{4}) = -0.633PR$$

$$\theta = \pi: \quad M(\pi) = 0.$$

The variation of shear force for $0 \le \theta \le \theta_1$ is given as

$$T(\theta) = V_A \sin\theta - H_A \cos\theta$$

Or, $T(\theta) = 0.75P\sin\theta - P\cos\theta$

$$T(0) = -P$$

$$T(\theta) = -0.176P$$

$$T(\theta_1) = 0.149P$$

$\theta_1 \leq \theta \leq \pi$

$$T(\theta) = V_A \sin\theta - H_A \cos\theta - P\sin\theta$$
$$T(\theta) = -0.25P\sin\theta - P\cos\theta$$

Hence

$$T(\theta_1) = -0.716P$$

$$T(\frac{\pi}{2}) = -0.25P$$

$$T(\frac{3\pi}{4}) = 0.53P$$

$$T(\pi) = P$$

In the same way, the variation of the normal force is given as

$$N(\theta) = -V_A \cos\theta - H_A \sin\theta = -0.75P\cos\theta - P\sin\theta$$

$$\theta = 0: \ N(0) = -0.75P$$

$$\theta = \frac{\pi}{4}: \ T(\frac{\pi}{4}) = -1.237P$$

$$\theta = \theta_1: \ N(\theta_1) = -1.082P$$

$\theta_1 \leq \theta \leq \pi$

$$N(\theta) = -V_A \cos\theta - H_A \sin\theta + P\cos\theta = 0.25P\cos\theta - P\sin\theta$$
$$\theta = \theta_1: \ N(\theta_1) = -0.582P$$
$$\theta = \frac{\pi}{2}: \ N(\frac{\pi}{2}) = -P$$
$$\theta = \frac{3\pi}{4}: \ N(\frac{3\pi}{4}) = -0.883P$$
$$\theta = \pi: \ N(\pi) = -0.25P$$

The diagrams of the bending moment, the shear force and the normal force are shown, respectively, in Figures 7.10–7.12.

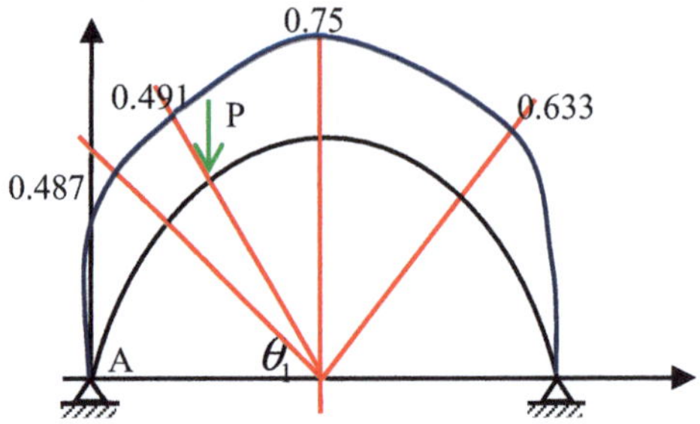

Figure 7.10. *Bending moment diagram (*PR)*

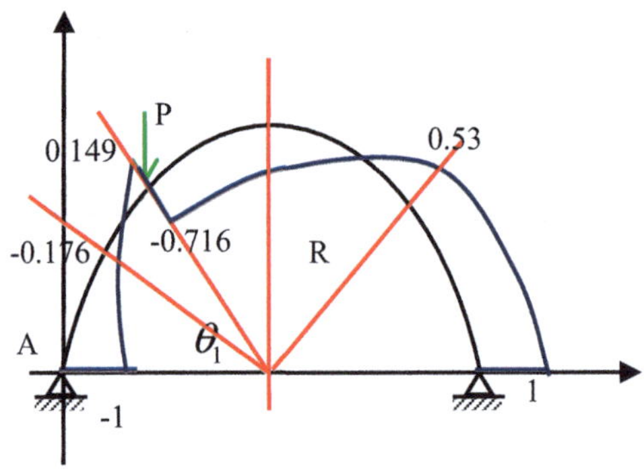

Figure 7.11. *Shear force diagram *(P)*

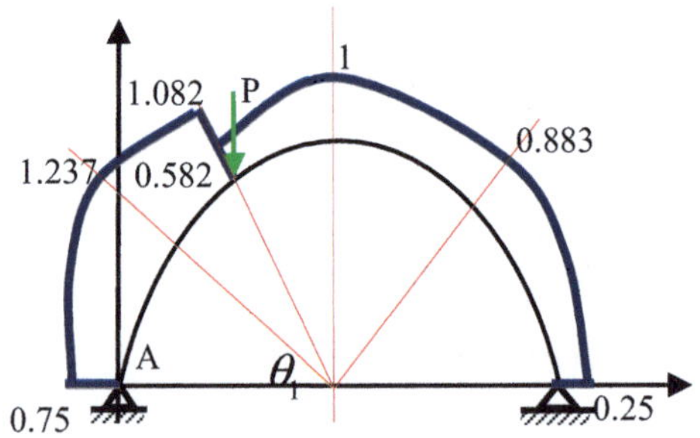

Figure 7.12. *Normal force diagram *(-P)*

7.4. Parabolic arch under concentrated load

We analyze the parabolic arch in Figure 7.13. We assume that the flexural rigidity (EI) is constant and the Cartesian equation is $y = 2\frac{h}{L}x - \frac{h}{L^2}x^2$.

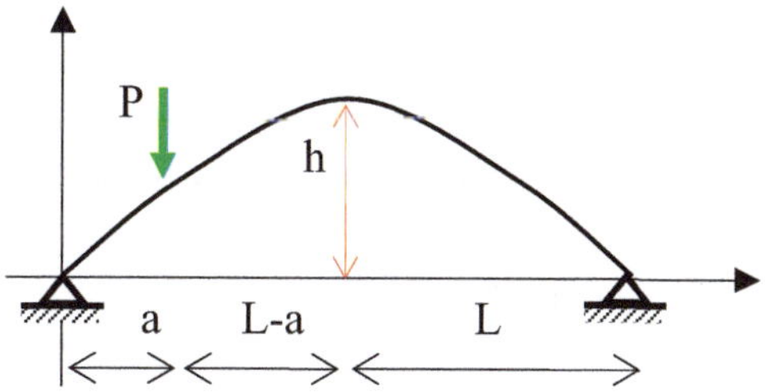

Figure 7.13. *Parabolic arch*

Analysis of the arch (Figure 7.13) requires the use of the method of forces. The fundamental system leads to determining the displacement δ_A^0 (Figure 7.14(a)) and the unit system allows us to calculate the displacement δ_A^1 (Figure 7.14(b)).

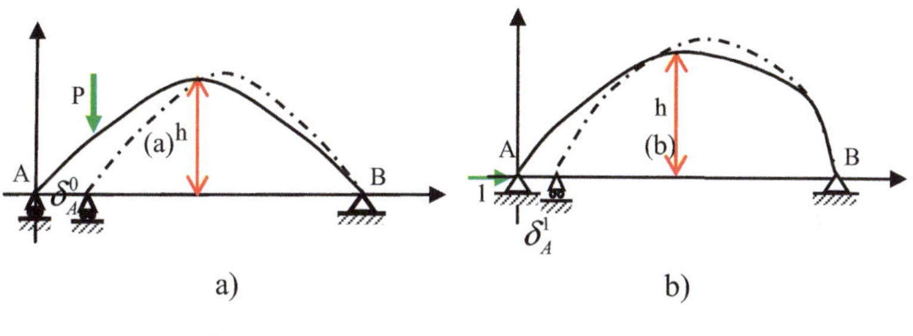

Figure 7.14. *Fundamental and unit arches*

The horizontal displacement at point A of the initial system is 0.

$$\delta_A = 0$$

This relationship is written in the following form:

$$\delta_A^0 + H_A \delta_A^1 = 0$$

δ_A^0 and δ_A^1 are the displacements of the fundamental system (Figure 7.14(a)) and the unit system (Figure 7.14(b)).

The support reaction at point A of the fundamental system is given as

$$V_A = P(1-\frac{a}{2L})$$

The expressions of the bending moment of the fundamental system are given by

$$0 \le x \le a \qquad \mu(x) = P(1-\frac{a}{2L})x$$

$$a \le x \le 2L \quad \mu(x) = P(1-\frac{a}{2L})x - P(x-a) = Pa(1-\frac{x}{2L})$$

The expression of the moment of the unit system is given by

$$0 \le x \le 2L \qquad m(x) = -y = -(2\frac{h}{L}x - \frac{h}{L^2}x^2)$$

The displacement calculation is carried out using the method of virtual work.

$$\delta_A^0 = \int_0^s \frac{\mu(x)m(x)}{EI}ds$$

$$\delta_A^1 = \int_0^s \frac{m^2(x)}{EI}ds$$

By introducing the expressions of the bending moments, we obtain

$$\delta_A^0 = \frac{1}{EI}\left[P(1-\frac{a}{2L})\int_0^a x(-y)dx + Pa\int_a^{2L}(1-\frac{x}{2L})(-y)dx\right]$$

$$\delta_A^0 = \frac{1}{EI}\left[-P(1-\frac{a}{2L})\int_0^a x(2\frac{h}{L}x - \frac{h}{L^2}x^2)dx\right] - \frac{1}{EI}\left[Pa\int_a^{2L}(1-\frac{x}{2L})(2\frac{h}{L}x - \frac{h}{L^2}x^2)dx\right]$$

This can also be written as

$$\delta_A^0 = -\frac{PahL}{EI}\left(\frac{2}{3} - \frac{a^2}{3L^2} + \frac{1}{12}\frac{a^3}{L^3}\right)$$

$$\delta_A^1 = \frac{1}{EI}\int_0^s y^2 ds = \frac{1}{EI}\int_0^{2L}\left(2\frac{h}{L}x - \frac{h}{L^2}x^2\right)^2 dx = \frac{16h^2L}{15EI}$$

We obtain the horizontal reaction using the compatibility equation of displacements at point A.

$$H_A = \frac{5}{8}\frac{Pa}{h}\left(1 - \frac{1}{2}\frac{a^2}{L^2} + \frac{1}{8}\frac{a^3}{L^3}\right)$$

The expression of the bending moment is given by

$0 \le x \le a$

$$M(x) = V_A x - H_A y = P\left(1 - \frac{a}{2L}\right)x - \frac{15}{16}\frac{Pa}{h}\left(\frac{2}{3} - \frac{1}{3}\frac{a^2}{L^2} + \frac{1}{12}\frac{a^3}{L^3}\right)\left(2\frac{h}{L}x - \frac{h}{L^2}x^2\right)$$

$$M(0) = 0.$$

$$M(a) = V_A a - H_A y(x=a) = Pa\left(1 - \frac{14}{8}\frac{a}{L} + \frac{5}{8}\frac{a^2}{L^2} + \frac{5}{8}\frac{a^3}{L^3} - \frac{5}{32}\frac{a^4}{L_4} + \frac{5}{64}\frac{a^5}{L^5}\right)$$

$a \le x \le 2L$

$$M(x) = V_A x - H_A y - P(x-a) = P\left(1 - \frac{a}{2L}\right)x - \frac{15}{16}\frac{Pa}{h}\left(\frac{2}{3} - \frac{1}{3}\frac{a^2}{L^2} + \frac{1}{12}\frac{a^3}{L^3}\right)\left(2\frac{h}{L}x - \frac{h}{L^2}x^2\right) + Pa - Px$$

$$M(x=a) = Pa\left(1 - \frac{14}{8}\frac{a}{L} + \frac{5}{8}\frac{a^2}{L^2} + \frac{5}{8}\frac{a^3}{L^3} - \frac{5}{32}\frac{a^4}{L_4} + \frac{5}{64}\frac{a^5}{L^5}\right)$$

$$M(x=2L) = 0.$$

Numerical application

To interpret the obtained results, we consider $a = L$.

$$V_A = \frac{P}{2}$$

$$H_A = \frac{25}{64}\frac{PL}{h}$$

$0 \le x \le L$

$$M(x) = V_A x - H_A y = -\frac{9}{32}Px + \frac{25}{64}P\frac{x^2}{L}$$

$$M(0) = 0.$$

$$M(L) = \frac{7}{64}PL$$

$L \le x \le 2L$

$$M(x) = V_A x - H_A y - P(x-a) = -\frac{41}{32}Px + \frac{25}{64}\frac{Px^2}{L} + PL$$

$$M(x = L) = \frac{7}{64}PL$$

$$M(x = 2L) = 0.$$

The bending moment diagram of the arch is shown in Figure 7.15.

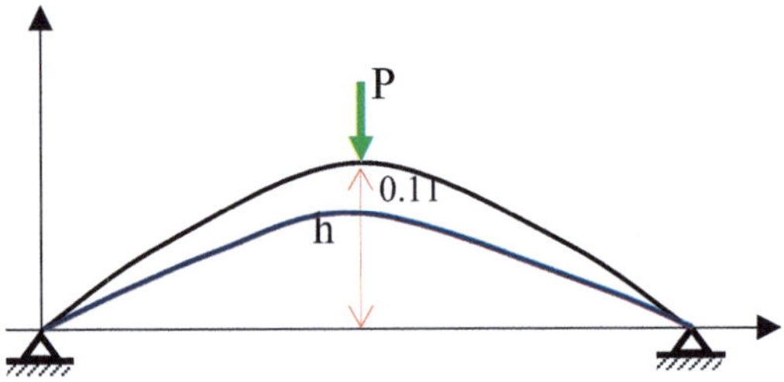

Figure 7.15. *Bending moment diagram *(PL)*

$0 \leq x \leq L$

$$T(x) = \frac{dM(x)}{dx} = -\frac{9}{32}P + \frac{25}{32}P\frac{x}{L}$$

$$T(0) = -\frac{9}{32}P$$

$$T(L) = \frac{1}{2}P$$

$L \leq x \leq 2L$

$$T(x) = -\frac{41}{32}P + \frac{25}{32}P\frac{x}{L}$$

$$T(x = L) = -\frac{1}{2}P$$

$$T(x = 2L) = \frac{9}{32}P$$

The shear force diagram is shown in Figure 7.16.

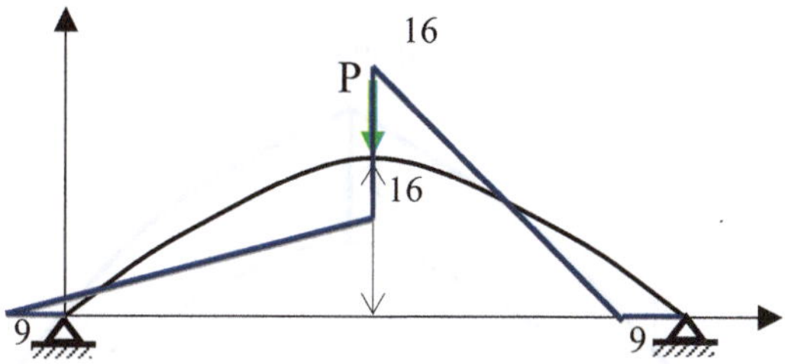

Figure 7.16. *Shear force diagram *(P)*

We cut the arch at an angle θ (Figure 7.17).

The variation in normal force is

$$N(\theta) = -V_A \cos\theta - H_A \sin\theta = -\frac{P}{2}\cos\theta - \frac{25}{64}\frac{PL}{h}\sin\theta$$

$$N(0) = -\frac{P}{2}$$

$$N(\frac{\pi}{2}) = -\frac{25}{64}\frac{PL}{h}$$

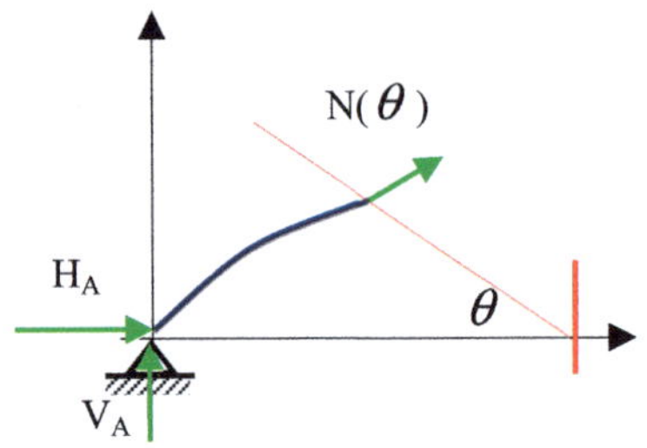

Figure 7.17. *Section $0 \le x \le a$*

The symmetry of the beam makes it possible to deduce the shear force diagram (Figure 7.18).

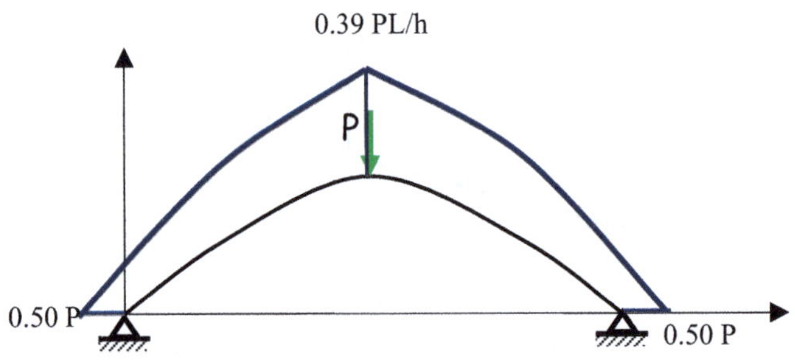

Figure 7.18. *Shear force diagram*

7.5. Semicircular arch under distributed load

Calculate the support reactions and draw diagrams of the bending moment, shear force and normal force of a semicircular statically indeterminate arch of radius R and constant flexural rigidity (EI) along the arch. An arch is stressed by a uniformly distributed load of intensity q (Figure 7.19).

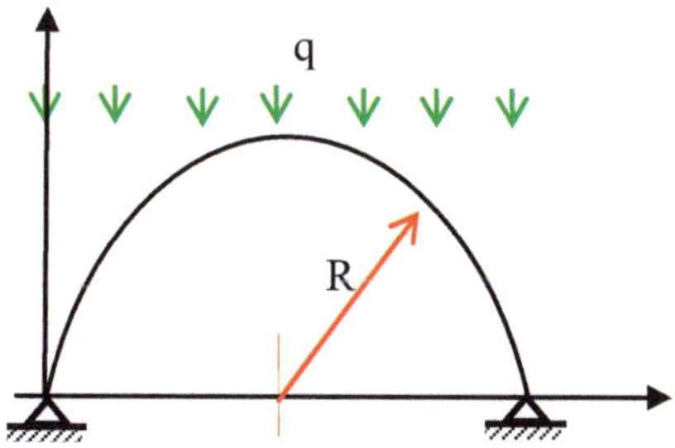

Figure 7.19. *Semicircular arch under uniform load*

The fundamental and unit arch structures are shown in Figures 7.20 and 7.21. Point A is displaced from δ_A^0 in the fundamental system and from δ_A^1 in the unit system.

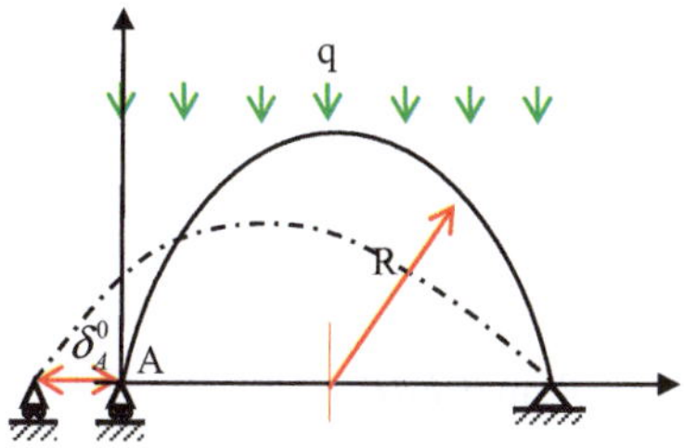

Figure 7.20. *Deflected fundamental system*

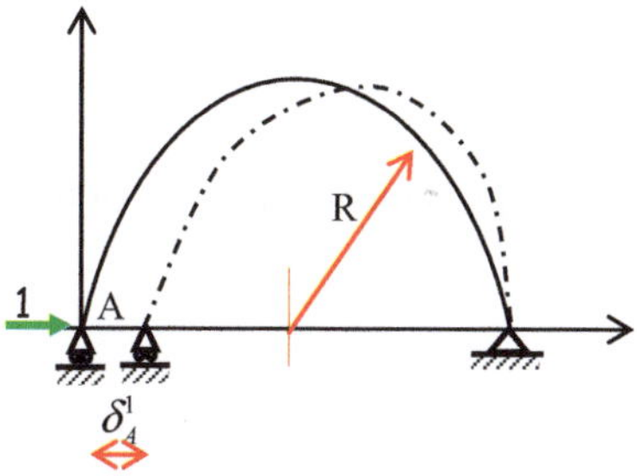

Figure 7.21. *Deflected unit system*

The horizontal displacement at point A of the given structure is given as

$$\delta_A^0 + H_A \delta_A^1 = 0 \qquad [7.20]$$

Support reaction A of the fundamental system is deduced by

$$V_A = \frac{qL}{4} \qquad [7.21]$$

The expression of the bending moment of the fundamental system is given by

$$0 \le x \le 2R \quad \mu(x) = qRx - \frac{q}{2}x^2 \qquad [7.22]$$

The expression of the bending moment of the unit system is given by

$$0 \le x \le 2R \quad m(x) = -R\sin\theta \qquad [7.23]$$

By substituting relationships [7.22] and [7.23] with [7.20], we obtain the horizontal displacement of point A of the fundamental system.

$$\delta_A^0 = \int_0^s \frac{\mu(x)m(x)}{EI} ds \qquad [7.24]$$

or

$$\delta_A^0 = \frac{1}{EI}\int_0^\pi (qRx - \frac{q}{2}x^2).(R\sin\theta)Rd\theta \qquad [7.25]$$

with $x = R(1-\cos\theta)$

$$\delta_A^0 = \frac{2qR^4}{3EI} \qquad [7.26]$$

Similarly, the horizontal displacement at point A of the unit structure is given as

$$\delta_A^1 = \int_0^s \frac{m^2(x)}{EI} ds \qquad [7.27]$$

$$\delta_A^1 = \int_0^R \frac{R^3 \sin^2\theta}{EI} d\theta = \frac{\pi R^3}{2EI} \qquad [7.28]$$

By substituting [7.26] and [7.8] with relationship [7.20], we obtain the horizontal displacement of joint A by

$$H_A = -\frac{\delta_A^0}{\delta_A^1} = \frac{4}{3}\frac{qR}{\pi} \tag{7.29}$$

And the horizontal reaction at joint B is deduced by

$$H_B = -H_A = -\frac{4}{3}\frac{qR}{\pi} \tag{7.30}$$

The expression of the bending moment is

$0 \leq x \leq 2R$

$$M(x) = V_A x - \frac{1}{2}qx^2 - H_A y = qRx - \frac{1}{2}qx^2 - \frac{4}{3}\frac{qR}{\pi}y \tag{7.31}$$

$0 \leq \theta \leq \pi$

$$M(\theta) = qR^2(1-\cos\theta) - \frac{1}{2}qR^2(1-\cos\theta)^2 - \frac{4}{3}\frac{qR}{\pi}R\sin\theta = qR^2(\frac{\sin^2\theta}{2} - \frac{4}{3\pi}\sin\theta) \tag{7.32}$$

$$M(\theta) = qR^2(\frac{\sin^2\theta}{2} - \frac{4}{3\pi}\sin\theta) \tag{7.33}$$

The bending moment diagram of the arch is shown in Figure 7.22.

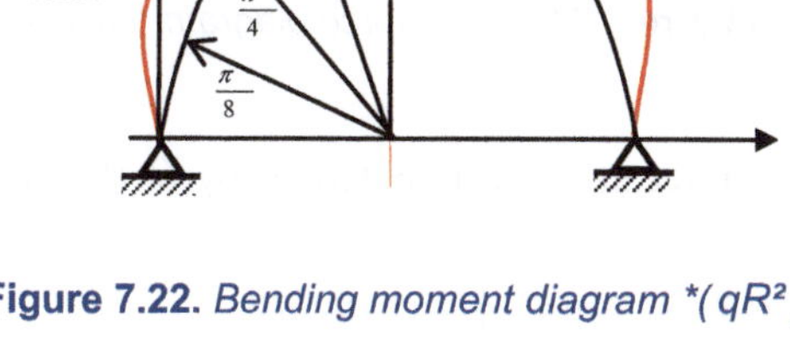

Figure 7.22. *Bending moment diagram *(qR²)*

The variation of the shear force can be deduced by carrying out the radial equilibrium for each section defined by the angle θ (Figure 7.23).

$$T(\theta) = V_A \sin\theta - H_A \cos\theta - \int_0^\theta q \sin\theta . R d\theta \quad [7.34]$$

Alternatively, the shear force is written as

$$T(\theta) = qR(\sin\theta - \frac{4}{3\pi}\cos\theta + \cos\theta - 1) \quad [7.35]$$

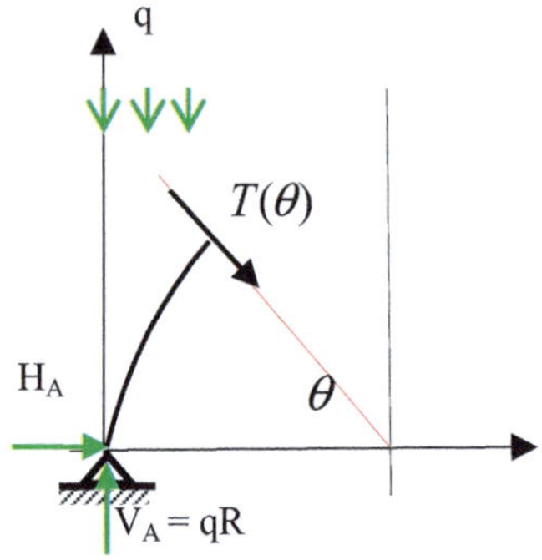

Figure 7.23. *Shear force*

The shear force diagram can be obtained by taking into account the symmetry of the arch (Figure 7.24).

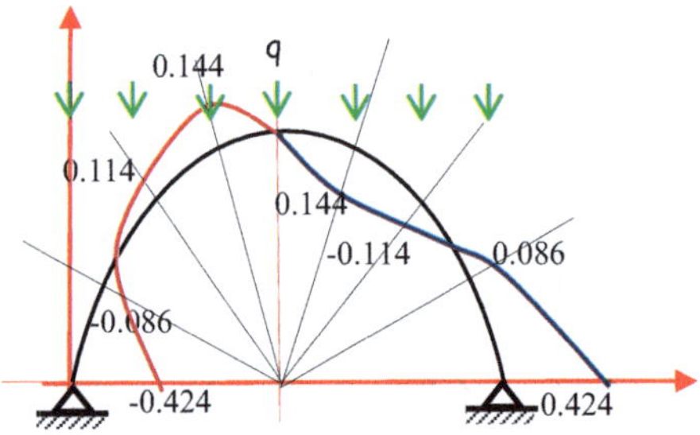

Figure 7.24. *Shear force diagram *(qR²)*

Finally, the normal force at the section defined by θ (Figure 7.5) is given by

$$N(\theta) = -V_A \cos\theta - H_A \sin\theta + \int_0^\theta q \cos\theta . R d\theta \quad [7.36]$$

$$N(\theta) = qR(-\cos\theta - \frac{4}{3\pi}\sin\theta + \sin\theta) \quad [7.37]$$

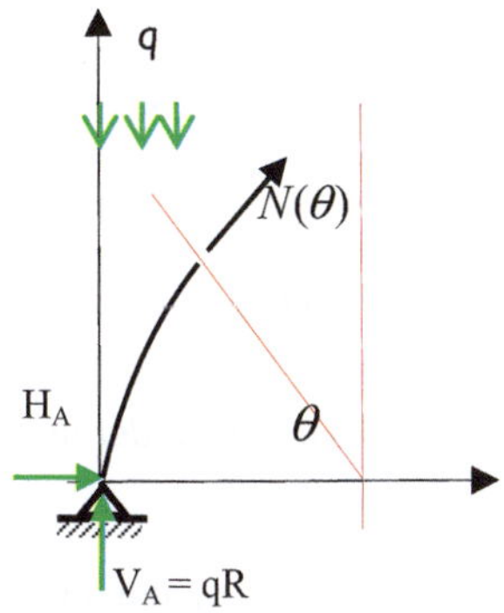

Figure 7.25. *Normal force*

The shear force diagram is shown in Figure 7.26.

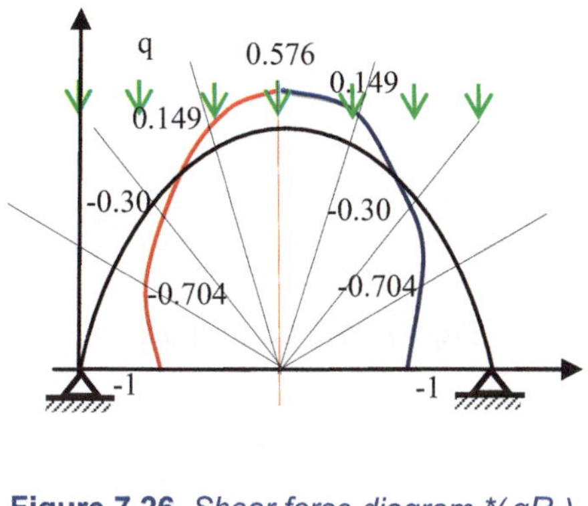

Figure 7.26. *Shear force diagram *(qR)*

7.6. Parabolic arch under distributed load

An arch of constant flexural rigidity (EI) is defined by the equation $y = 2\frac{h}{L}x - \frac{h}{L^2}x^2$. It is stressed by a uniform load of intensity q distributed along length L (Figure 7.27).

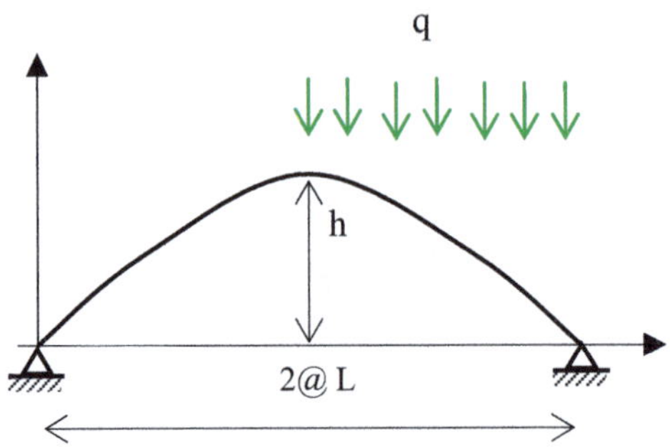

Figure 7.27. *Parabolic arch under uniform load*

We construct the fundamental (Figure 7.28) and unit (Figure 7.29) systems.

The fundamental system leads to determining the displacement δ_A^0 and the unit system allows us to deduce the displacement δ_A^1.

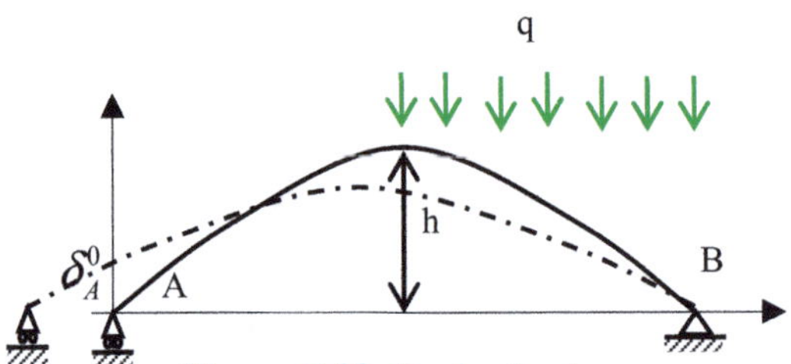

Figure 7.28. *Basic structure*

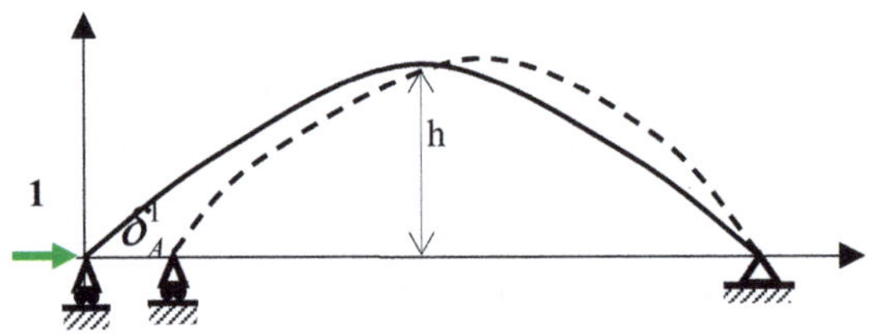

Figure 7.29. *Unit structure*

The horizontal displacement at point A of the given system is given as

$$\delta_A^0 + H_A \delta_A^1 = 0$$

The support reaction at point A of the fundamental system is given by

$$V_A = \frac{qL}{4}$$

The expressions of the bending moment of the fundamental system are

$$0 \le x \le L \quad \mu(x) = \frac{qL}{4}x$$

$$L \le x \le 2L \quad \mu(x) = \frac{qL}{4}x - \frac{q}{2}(x-L)^2 = \frac{5qL}{4}x - \frac{qx^2}{2} - \frac{qL^2}{2}$$

The expression of the bending moment of the unit system is given by

$$0 \le x \le 2L \qquad m(x) = -y = -(\frac{2h}{L}x - \frac{h}{L^2}x^2)$$

The displacement calculations are carried out using the method of virtual work.

$$\delta_A^0 = \int_0^s \frac{\mu(x)m(x)}{EI}ds$$

$$\delta_A^1 = \int_0^s \frac{m^2(x)}{EI}ds$$

The introduction of the equations above makes it possible to obtain

$$\delta_A^0 = \frac{1}{EI}\left[\int_0^L -\frac{qL}{4}x(\frac{2h}{L}x - \frac{h}{L^2}x^2)dx - \int_L^{2L}(\frac{5qL}{4}x - \frac{qx^2}{2} - \frac{qL^2}{2})(\frac{2h}{L}x - \frac{h}{L^2}x^2)dx\right]$$

$$\delta_A^1 = \frac{1}{EI}\int_0^s y^2 ds = \frac{1}{EI}\int_0^{2L} y^2 dx$$

$$\delta_A^0 = -\frac{9}{40}\frac{qhL^3}{EI}$$

$$\delta_A^1 = \frac{1}{EI}\int_0^{2L}(2\frac{h}{L}x - \frac{h}{L^2}x^2)^2 = \frac{16h^2L}{15EI}$$

The compatibility equation of the displacements at point A makes it possible to calculate

$$H_A = -\frac{\delta_A^0}{\delta_A^1} = \frac{27}{128}\frac{qL^2}{h}$$

The expression of the bending moment is

$0 \le x \le L$

$$M(x) = V_A x - H_A y = \frac{qL}{4}x - \frac{27}{128}\frac{qL^2}{h}(2\frac{h}{L}x - \frac{h}{L^2}x^2)$$

$$M(x) = \frac{27}{128}qx^2 - \frac{11}{64}qLx$$

$L \le x \le 2L$

$$M(x) = V_A x - H_A y - \frac{q(x-L)^2}{2} = -\frac{37}{128}qx^2 + \frac{53}{64}qLx - \frac{1}{2}qL^2$$

The bending moment diagram of the arch is shown in Figure 7.30.

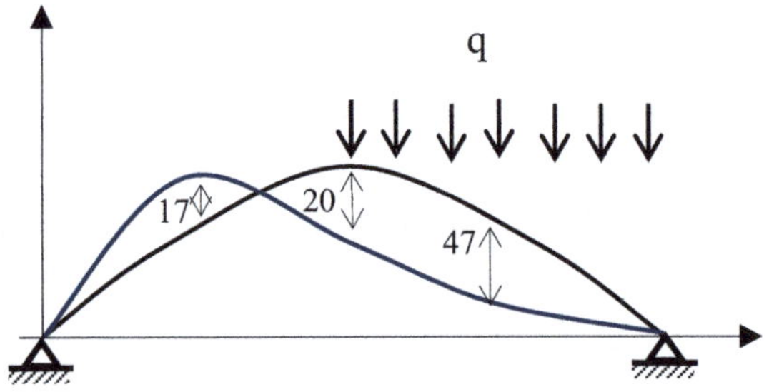

Figure 7.30. *Bending moment diagram* $*(\frac{qL^2}{512})$

$0 \le x \le L$

$$T(x) = \frac{dM(x)}{dx} = \frac{27}{64}qx - \frac{11}{64}qL$$

$$T(0) = -\frac{11}{64}qL$$

$$T(L) = \frac{1}{4}qL$$

$L \le x \le 2L$

$$T(x) = -\frac{37}{64}qx + \frac{53}{64}qL$$

$$T(x = L) = \frac{1}{4}qL$$

$$T(x = 2L) = \frac{-21}{64}qL$$

The shear force diagram is shown in Figure 7.31.

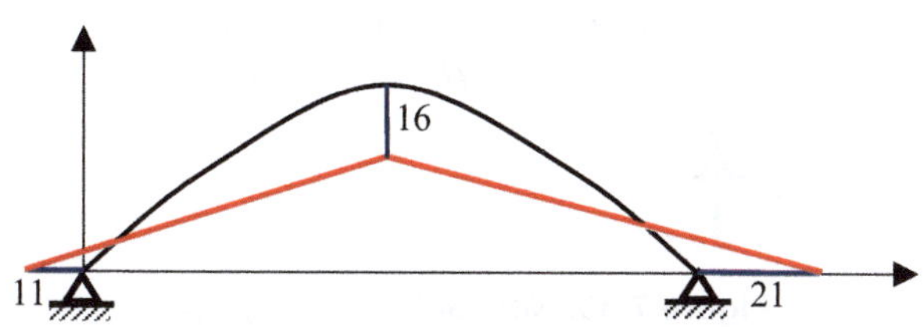

Figure 7.31. *Shear force diagram* $*(\frac{qL}{64})$

The normal force at any section (Figure 7.32) is given by

$$N(\theta) = -V_A \cos\theta - H_A \sin\theta = -\frac{qL}{4}\cos\theta - \frac{27}{128}\frac{qL^2}{h}\sin\theta$$

$$\theta = 0 \qquad N(0) = -\frac{qL}{4}$$

$$\theta = \frac{\pi}{2} \qquad N(\frac{\pi}{2}) = -\frac{27}{128}qL$$

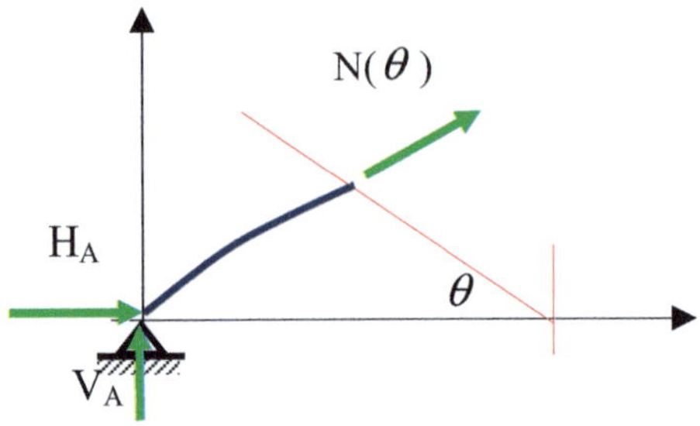

Figure 7.32. *Normal force* $0 \leq \theta \leq \frac{\pi}{2}$

$\frac{\pi}{2} \leq \theta \leq \pi$ (Figure 7.33).

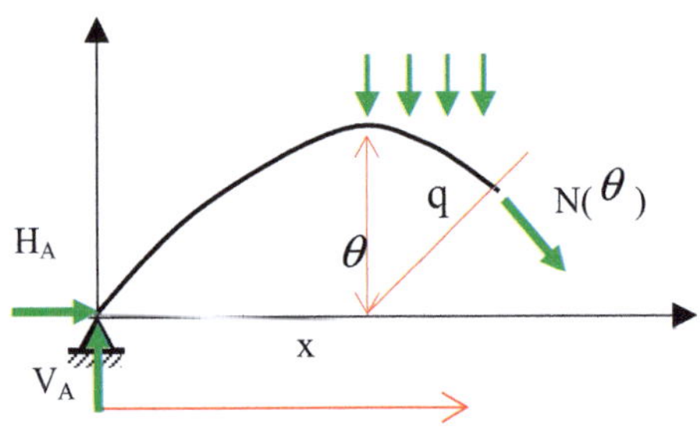

Figure 7.33. *Normal force* $\frac{\pi}{2} \leq \theta \leq \pi$

Knowing that the curve of the arch $\rho(\theta) = \frac{1}{y''(\mathrm{x})}$ and $y''(\mathrm{x})$ is the second derivative of the deflection:

$$N(\theta) = -V_A \cos\theta - H_A \sin\theta - \int_{\pi/2}^{\theta} q \cos\theta.\rho(\theta).d\theta$$

$$N(\theta) = -\frac{qL}{4}\cos\theta - \frac{27}{128}\frac{qL^2}{h}\sin\theta + \frac{qL^2}{2h}(\sin\theta - 1)$$

$$N(\frac{\pi}{2}) = -\frac{27}{128}\frac{qL^2}{h}$$

$$N(\pi) = \frac{qL}{4} - \frac{qL^2}{h} = \frac{qL^2}{4h}(\frac{h}{L} - 2)$$

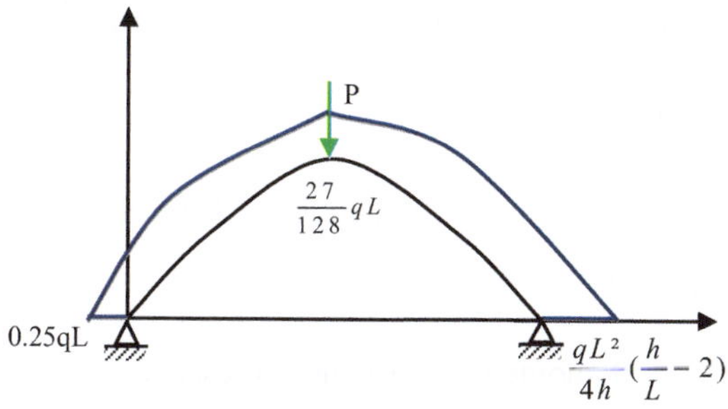

Figure 7.34. *Normal force diagram*

7.7. Semicircular arch fixed under concentrated load

A semicircular arch fixed at one end and simply supported at the other is stressed by a concentrated load P applied at a distance R from the left support (Figure 7.35). Knowing that the flexural rigidity (EI) is constant along the arch, we draw the diagram of the bending moment using the different statically indeterminate structural analysis methods.

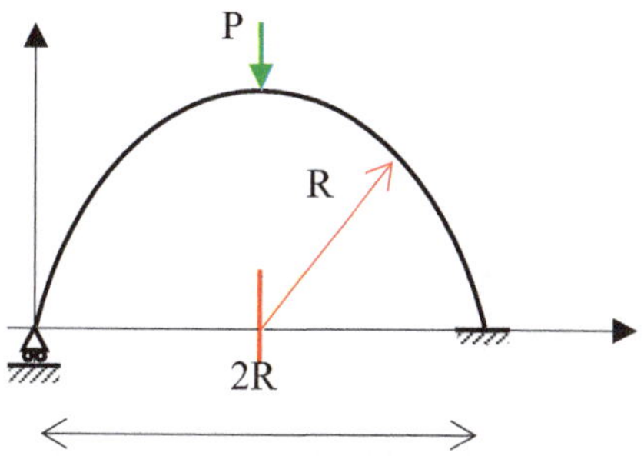

Figure 7.35. *Arch fixed at an end*

To enrich the structural analysis, this time we use three distinct methods: (1) the method of three moments, (2) the method of forces and (3) the slope-deflection method.

– *Method of three moments*

The initial slope at joint B (Figure 7.32) is 0.

$$\omega_B = 0 \qquad [7.38]$$

The equation of three moments makes it possible to write

$$\omega_B = \omega_B^{"} - aM_A + cM_B = 0 \tag{7.39}$$

where a and c in the relationship [7.39] are the mechanical constants of the arch and $\omega_B^{"}$ is the slope at point B of the fundamental system.

Knowing that $M_A = 0$, moment M_B has the expression

$$M_B = -\frac{\omega_B^{"}}{c} \tag{7.40}$$

To calculate the displacement $\omega_B^{"}$, we use the method of virtual work.

$$\omega_B^{"} = \int_0^s \frac{\mu(x)m(x)}{EI} ds \tag{7.41}$$

with

$$\mu(x) = \frac{Px}{2} \quad 0 \le \theta \le \frac{\pi}{2} \tag{7.42a}$$

$$\mu(x) = PR - \frac{Px}{2} \quad \frac{\pi}{2} \le \theta \le \pi \tag{7.42b}$$

The expression of the bending moment of the unit system (Figure 7.36) is given by

$$0 \le \theta \le \pi \qquad m(x) = \frac{x}{2R} \tag{7.43}$$

Integrating the relationships [7.42] and [7.43], the slope $\omega_B^{"}$ is given by

$$\omega_B^{"} = \frac{1}{EI}\left[\int_0^R \frac{P}{2} x . \frac{x}{2R} dx + \int_R^{2R} (PR - \frac{P}{2} x) . \frac{x}{2R} dx\right] \tag{7.44a}$$

$$\omega_B^{''} = \frac{1}{EI}\left[\int_0^{\pi/2} \frac{PR^2}{2}(1-\cos\theta)\frac{1-\cos\theta}{2}d\theta + \int_{\pi/2}^{\pi} \frac{PR^2}{2}(1+\cos\theta)\frac{1-\cos\theta}{2}d\theta\right] \qquad [7.44b]$$

$$\omega_B^{''} = \frac{PR^2}{4EI}(\pi - 2) \qquad [7.44c]$$

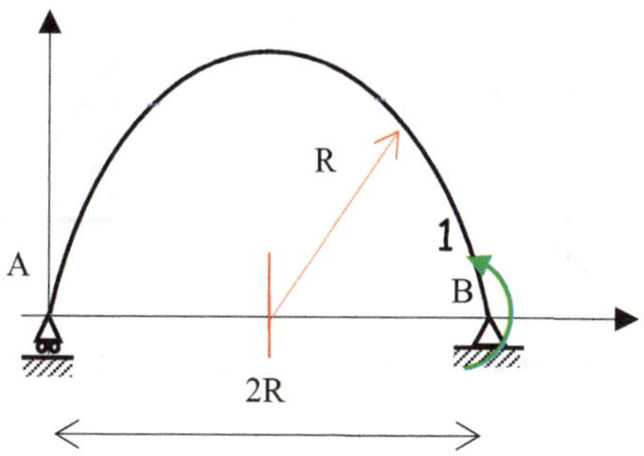

Figure 7.36. *Unit system*

The mechanical constant c is given in polar coordinates by

$$c = \frac{1}{EI}\int_0^L (\frac{x}{L})^2 dx = \frac{1}{EI}\int_0^{\pi} (\frac{R^2(1-\cos\theta)^2}{4R^2})R.d\theta = \frac{3\pi R}{8EI} \qquad [7.45]$$

$$M_B = -\frac{\omega_B^{''}}{C} = -\frac{PR(2\pi - 4)}{3\pi} \qquad [7.46]$$

The variation of the bending moment at each interval is given by

$0 \leq x \leq R$

$$M(\theta) = \frac{PR}{2}(1-\cos\theta) + \frac{PR(4-2\pi)}{3\pi}\frac{1-\cos\theta}{2} \qquad [7.47a]$$

$R \leq x \leq 2R$

$$M(\theta) = \frac{PR}{2}(1+\cos\theta) + \frac{PR(4-2\pi)}{3\pi}\frac{1-\cos\theta}{2} \qquad [7.47b]$$

The bending moment diagram is shown in Figure 7.37.

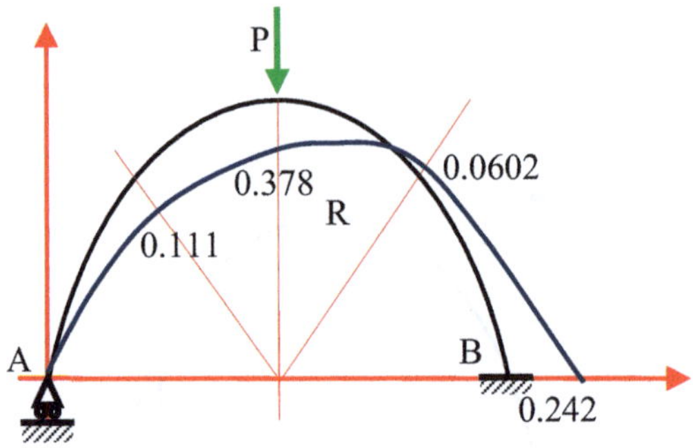

Figure 7.37. *Bending moment diagram (*PR)*

– *Method of forces*

The fundamental structure is shown in Figure 7.38.

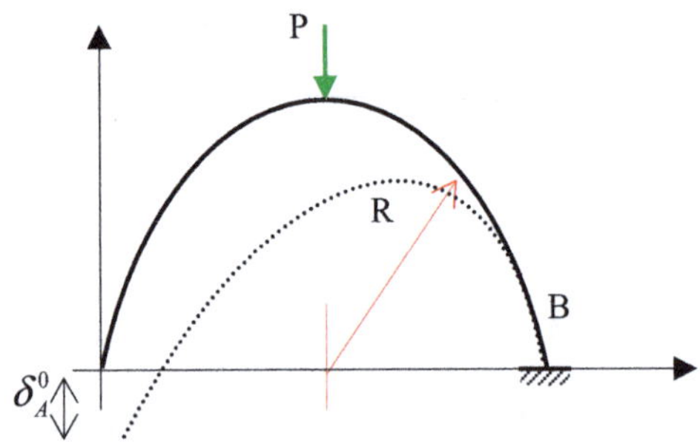

Figure 7.38. *Fundamental arch*

$\mu(x) = 0.$ $\qquad 0 \le x \le R$

$\mu(x) = -P(x-R) = PR\cos\theta$ $\qquad R \le x \le 2R$ $\qquad R \le x \le 2R$

The unit system is shown in Figure 7.39.

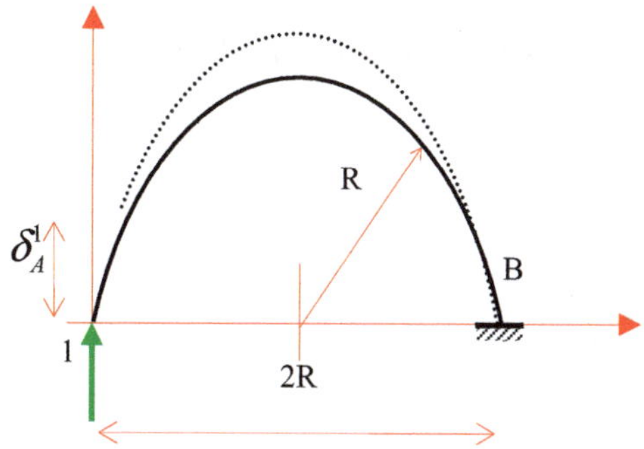

Figure 7.39. *Unit system*

$$0 \le x \le 2R \qquad m(x) = 1.x = R(1-\cos\theta)$$

The initial structure (Figure 7.35) is defined by

$$\delta_A = 0$$

Using the superposition of effects principle allows us to write

$$\delta_A = \delta_A^0 + V_A \delta_A^1 = 0$$

The displacement calculation is carried out using the principle of virtual work.

$$\delta_A^0 = \int_0^s \frac{\mu(x)m(x)}{EI} ds$$

$$\delta_A^0 = \frac{1}{EI} \int_{\pi/2}^{\pi} PR\cos\theta.R(1-\cos\theta).R.d\theta$$

$$\delta_A^0 = \frac{-PR^3}{EI}(1+\frac{\pi}{4})$$

$$\delta_A^1 = \int_0^s \frac{m^2(x)}{EI} ds = \frac{1}{EI}\int_0^{\pi} R^2(1-\cos\theta)^2 Rd\theta = \frac{3\pi R^3}{2EI}$$

The reaction at support A is written as

$$V_A = -\frac{\delta_A^0}{\delta_A^1} = \frac{4+\pi}{6\pi} P$$

The rotational equilibrium leads to the calculation of M_B.

$$M_B = V_A.2R - PR$$

or

$$M_B = \frac{4-2\pi}{3\pi} PR$$

The expression of the bending moment is written as

$$M(x) = \mu(x) + V_A m(x)$$

The variation of the bending moment at each interval is given by

$0 \le x \le R$

$$M(\theta) = \frac{4+\pi}{6\pi} PR(1-\cos\theta)$$

$R \le x \le 2R$

$$M(\theta) = PR(\cos\theta + \frac{4+\pi}{6\pi}(1-\cos\theta))$$

The bending moment diagram is established by varying the angle θ of 0 to π (Figure 7.40).

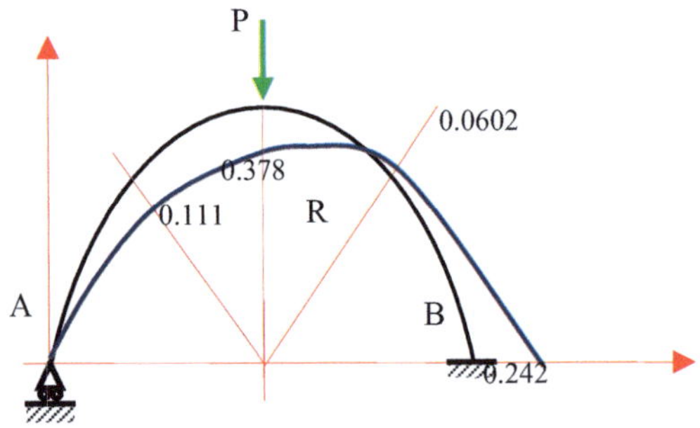

Figure 7.40. *Bending moment diagram (*PR)*

– *Slope-deflection method*

The slope-deflection method provides the fixed-end moments at the ends of a bar by

$$M_{AB} = \gamma_{AB} + K_{AB}\omega_A + K_{AB}\lambda_{AB}\omega_B - K_{AB}(1+\lambda_{AB})\beta_{AB} \qquad [7.48]$$

$$M_{BA} = \gamma_{BA} + K_{BA}\omega_B + K_{BA}\lambda_{BA}\omega_A - K_{BA}(1+\lambda_{BA})\beta_{BA} \qquad [7.49]$$

In this case $\omega_A = 0$ and $\omega_B \neq 0$, the kinematic conditions to consider are

$M_{AB} = 0.$, $\lambda_{BA} = 0.$, $\omega_B = 0.$ <u>and</u> $\beta_{AB} = \beta_{BA} = 0.$

Hence,

$$M_{BA} = \gamma_{BA}$$

Using the method of three moments makes it possible to write

$$\omega_B = \omega_B'' - a\gamma_{AB} + c\gamma_{BA} = 0 \qquad [7.50]$$

The fixed-end moment γ_{BA} is deduced by

$$\gamma_{BA} = -\frac{\omega_B''}{c} \qquad [7.51]$$

Equation [7.51] is similar to the result obtained using the method of three moments [7.46].

Variations of the shear force and the normal force can be deduced by carrying out the radial and orthoradial equilibrium for each section defined by an angle θ.

$0 \leq \theta \leq \frac{\pi}{2}$ (Figure 7.41)

$$T(\theta) = V_A \sin\theta = \frac{4+\pi}{6\pi} P \sin\theta$$

$$N(\theta) = -V_A \cos\theta = -\frac{4+\pi}{6\pi}\cos\theta$$

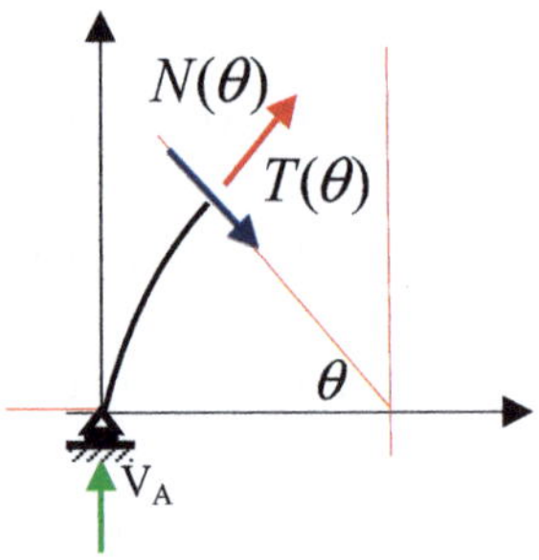

Figure 7.41. *Section* $0 \leq \theta \leq \dfrac{\pi}{2}$

$\dfrac{\pi}{2} \leq \theta \leq \pi$ (Figure 7.42)

$$T(\theta) = V_A \sin\theta - P\sin\theta = \frac{4-5\pi}{6\pi}P\sin\theta$$

$$N(\theta) = -V_A \cos\theta + P\cos\theta = \frac{5\pi-4}{6\pi}P\cos\theta$$

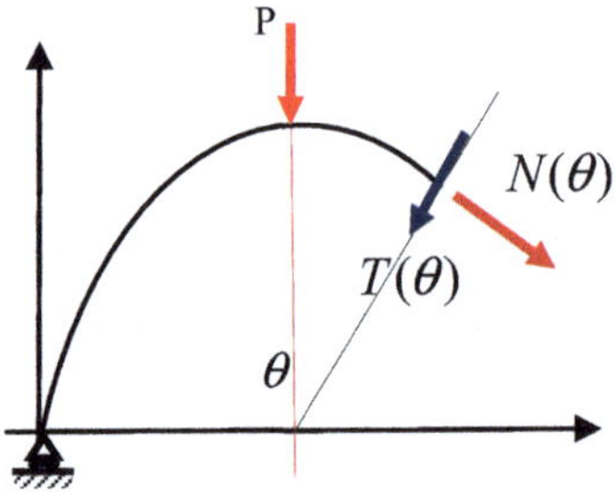

Figure 7.42. *Section* $\dfrac{\pi}{2} \leq \theta \leq \pi$

The variation of the shear force and the normal force are shown in Figures 7.43 and 7.44.

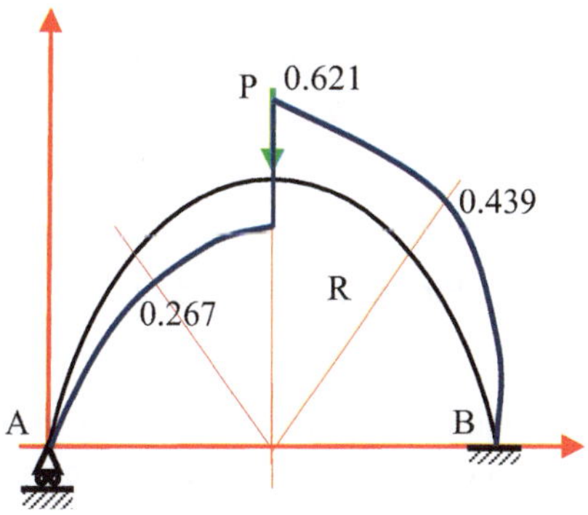

Figure 7.43. *Shear force diagram *(P)*

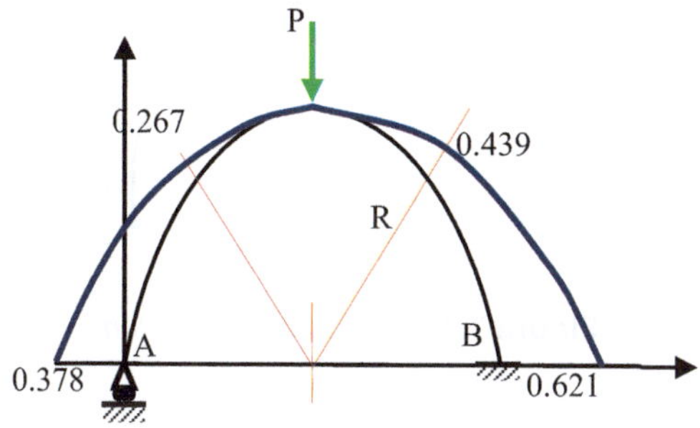

Figure 7.44. *Normal force diagram *(P)*

7.8. Statically indeterminate tied arch

A tie (DB) is associated with a semicircular arch (ABC). The arch is stressed by a force applied at point C (Figure 7.45). We assume that the flexural rigidity (EI) and membrane rigidity (E Ω) are constant. We must:

– calculate the tension in the cable (DB);

– draw diagrams of the bending moment and the shear force;

– deduce the vertical displacement at point C.

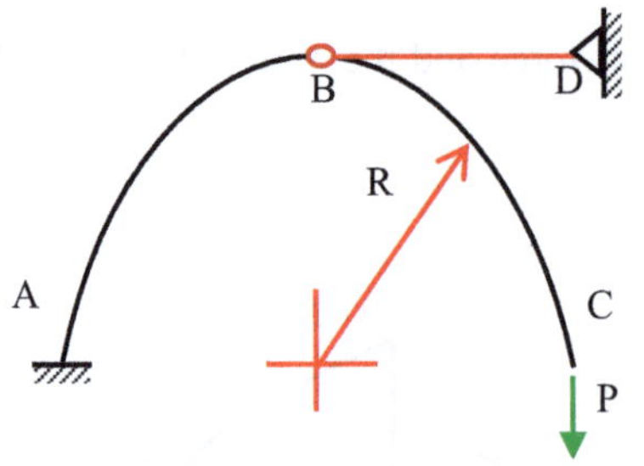

Figure 7.45. *Tied arch*

This type of structure requires the method of forces to be used for its analysis. The fundamental system and the unit system are presented, respectively, by Figure (7.46) and Figure (7.47).

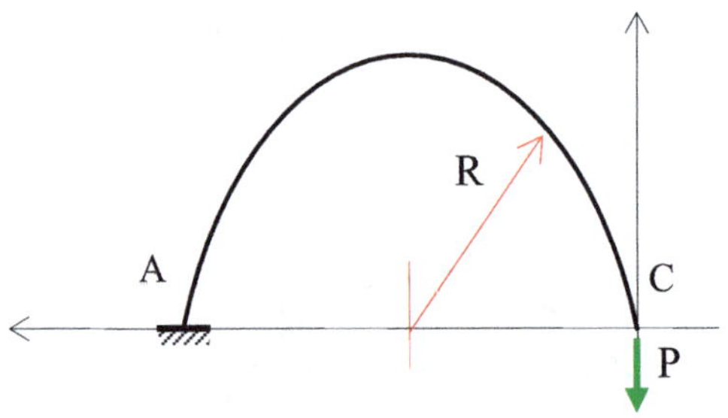

Figure 7.46. *Fundamental system*

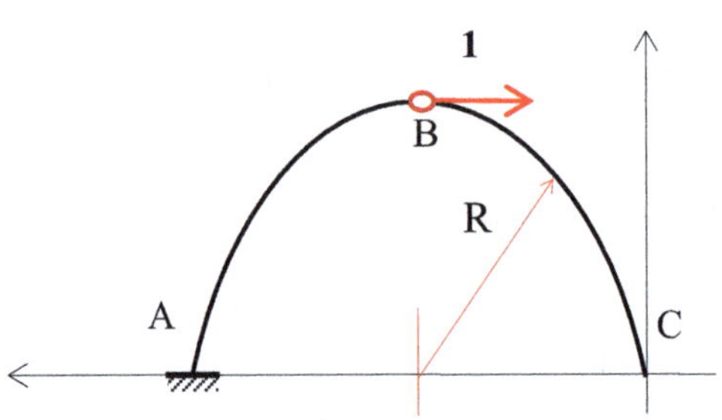

Figure 7.47. *Unit system*

The horizontal displacement of hinge B is null.

$$\delta_B^H = 0$$

Using the superposition principle, the horizontal displacement at point B can be broken down into the following

$$\delta_B^0 + T_{BD}(\delta_B^1 + \delta_B^{11}) = 0$$

The expressions of the bending moment of the fundamental system (Figure 7.46) are given by

$$0 \le x \le 2R \qquad \mu(x) = -PR(1-\cos\theta)$$

Similarly, the expression of the bending moment of the unit system (Figure 7.47) is given by

$$R \le x \le 2R \qquad m(x) = -R(1-\cos\theta)$$

The horizontal displacement at point B of the fundamental system is given as

$$\delta_B^0 = \int_0^s \frac{\mu(x)m(x)}{EI} ds$$

$$\delta_B^0 = \frac{1}{EI}\int_{\frac{\pi}{2}}^{\pi} PR^3(1-\cos\theta)^2 d\theta = \frac{PR^3}{2EI}(\pi - 1)$$

In the same way, the horizontal displacement at joint B of the unit structure is given as

$$\delta_B^1 = \int_0^s \frac{m^2(x)}{EI} ds = \int_0^{\pi/2} \frac{R^3(1-\cos\theta)^2}{EI} d\theta = \frac{(-7+3\pi)R^3}{4EI}$$

$$\delta_B^{11} = \frac{R}{E\Omega}$$

The tension in cable (BD) can be deduced by

$$T_{BD} = -\frac{\delta_B^0}{\delta_B^1 + \delta_B^{11}} = -\frac{\frac{\pi - 1}{2EI} PR^3}{\frac{R}{E\Omega} + \frac{R^3}{4EI}(3\pi - 7)}$$

or

$$T_{BD} = -\frac{\pi - 1}{2(R^2\frac{I}{\Omega} + \frac{3\pi - 7}{4})}P$$

The reaction at support D can be deduced by

$$H_D = -T_{BD} = \frac{\pi - 1}{2(R^2\frac{I}{\Omega} + \frac{3\pi - 7}{4})}P$$

The expression of the bending moment is given by

$$\theta \leq \frac{\pi}{2}$$

$$M(\theta) = \mu(\theta) + T_{BD}m(\theta) = -PR(1 - \cos\theta)$$

$$M(0) = \mu(0) = 0.$$

$$M(\frac{\pi}{2}) = \mu(\frac{\pi}{2}) = -PR$$

$$\frac{\pi}{2} \leq \theta \leq \pi$$

$$M(\theta) = \mu(\theta) + T_{BD}m(\theta) = -PR(1 - \cos\theta)$$

$$M(\frac{\pi}{2}) = \mu(\frac{\pi}{2}) = -PR$$

$$M(\pi) = \mu(\pi) - T_{BD}m(\pi) = -2PR - \frac{\pi - 1}{2(R^2\frac{I}{\Omega} + \frac{3\pi - 7}{4})}PR$$

The bending moment diagram is shown in Figure 7.48.

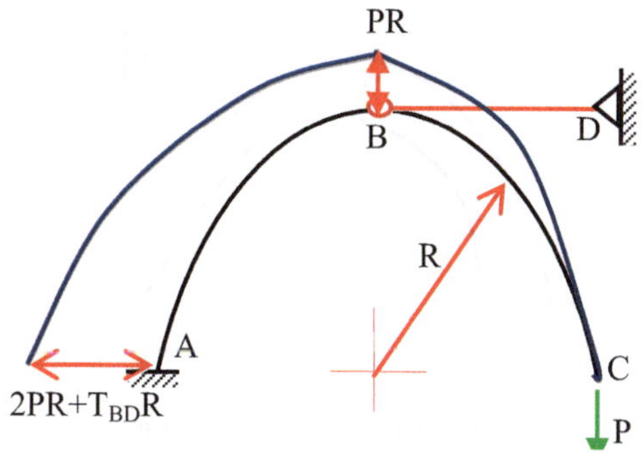

Figure 7.48. *Bending moment diagram*

In the same way, the expression of the shear force can be deduced by

$$\theta \leq \frac{\pi}{2}$$

$$T(\theta) = P\sin\theta$$

$$T(\theta) = 0.$$

$$T(\theta) = P$$

$$\frac{\pi}{2} \leq \theta \leq \pi$$

$$T(\theta) = P\sin\theta - T_{BD}\cos\theta$$

$$T(\theta) = P$$

$$T(\pi) = T_{BD} = \frac{\pi - 1}{2(R^2\frac{I}{\Omega} + \frac{3\pi - 7}{4})}P$$

The shear force diagram is shown in Figure 7.49.

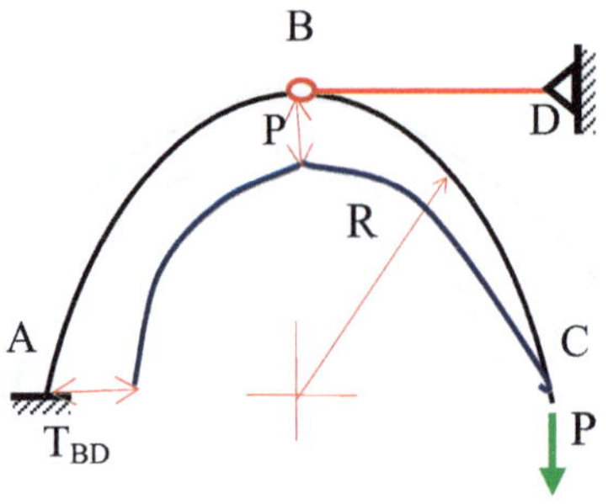

Figure 7.49. *Shear force diagram*

7.9. Arch with many degrees of freedom

A bi-hinged semicircular arch is stressed by a uniformly distributed load of intensity q. It is assumed that the flexural rigidity (EI) is constant and analysis of this arch is required (Figure 7.50).

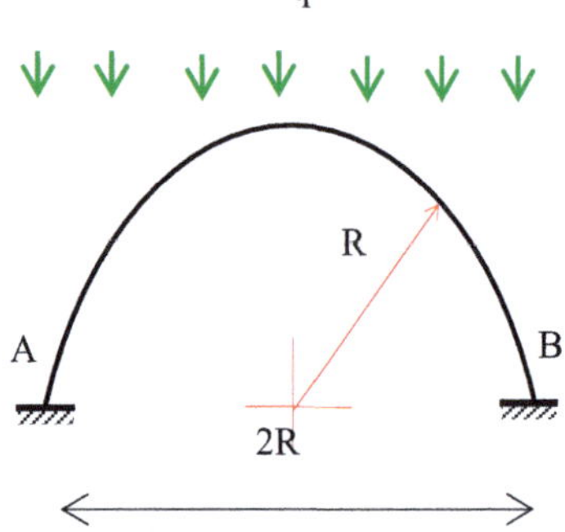

Figure 7.50. *Bi-fixed arch*

The fundamental and the unit systems are plotted, respectively, on Figure 7.51 and Figure 7.52.

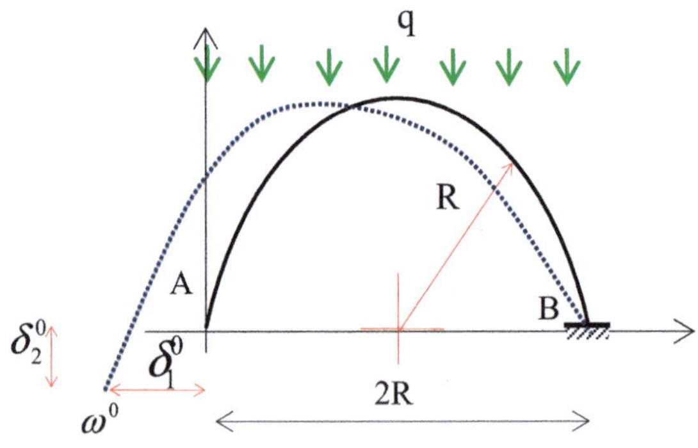

Figure 7.51. *Deflected fundamental arch*

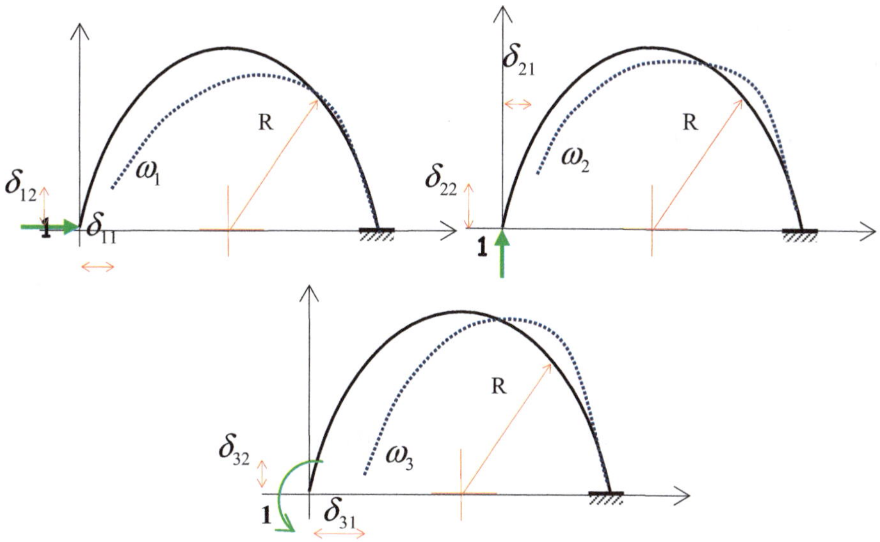

Figure 7.52. *Arch under unit actions*

The displacements at joint A for the given arch (Figure 7.50) are

$$\delta_A^x = 0 \tag{7.52a}$$

$$\delta_A^y = 0 \tag{7.52b}$$

$$\omega_B = 0 \tag{7.52c}$$

The displacements at joint A can be written as

$$\delta_A^x = \delta_1^0 + H_A\delta_{11} + V_A\delta_{21} + M_A\omega_1 = 0 \tag{7.53a}$$

$$\delta_A^y = \delta_2^0 + H_A\delta_{21} + V_A\delta_{22} + M_A\omega_2 = 0 \tag{7.53b}$$

$$\omega_A = \omega^0 + H_A\delta_{31} + V_A\delta_{32} + M_A\omega_3 = 0 \tag{7.53c}$$

Resolving the system of equations [7.53] leads to removing the degrees of static indeterminacy of the structure.

The expression of the bending moment of the fundamental arch (Figure 7.51) is given by

$$\mu(x) = -\frac{1}{2}qx^2 = -\frac{1}{2}qR^2(1-\cos\theta)^2 \quad [7.54]$$

$$m_1(x) = -1.y = -R\sin\theta \quad [7.55a]$$

$$m_2(x) = 1.x = R(1-\cos\theta) \quad [7.55b]$$

$$m_3(x) = -1. \quad [7.55c]$$

The displacements calculation is done by the method of virtual work.

$$\delta_{ij} = \int_0^s \frac{\mu_i(x)m_j(x)}{EI}ds \quad [7.56]$$

The displacements at point A of the statically determinate structure are given as

$$\delta_1^0 = \int_0^s \frac{\mu(x)m_1(x)}{EI}ds = \frac{1}{EI}\int_0^\pi -\frac{1}{2}qR^2(1-\cos\theta)^2(-R\sin\theta)Rd\theta = \frac{4qR^4}{3EI}$$

$$\delta_2^0 = \int_0^s \frac{\mu(x)m_2(x)}{EI}ds = \frac{1}{EI}\int_0^\pi -\frac{1}{2}qR^2(1-\cos\theta)^2R(1-\cos\theta)Rd\theta = -\frac{5\pi}{4}\frac{qR^4}{EI}$$

$$\omega^0 = \int_0^s \frac{\mu(x)m_3(x)}{EI}ds = \frac{1}{EI}\int_0^\pi -\frac{1}{2}qR^2(1-\cos\theta)^2(-1)Rd\theta = \frac{3\pi qR^3}{4EI}$$

The displacements of the unit systems can be evaluated by

$$\delta_{11} = \int_0^s \frac{m_1^2(x)}{EI}ds = \frac{1}{EI}\int_0^\pi R^2\sin^2\theta.Rd\theta = \frac{R^3}{EI}\int_0^\pi \sin^2\theta.d\theta = \frac{\pi R^3}{2EI}$$

$$\delta_{22} = \int_0^s \frac{m_2^2(x)}{EI}ds = \frac{1}{EI}\int_0^\pi R^2(1-\cos\theta)^2.Rd\theta = \frac{R^3}{EI}\int_0^\pi (1-\cos\theta)^2d\theta = \frac{3\pi R^3}{2EI}$$

$$\omega^1 = \int_0^s \frac{m_3^2(x)}{EI}ds = \frac{1}{EI}\int_0^\pi (-1)^2.Rd\theta = \frac{\pi R}{EI}$$

$$\delta_{12} = \int_0^s \frac{m_1(x)m_2(x)}{EI} ds = \frac{1}{EI}\int_0^{\pi} -R\sin\theta.R(1-\cos\theta)Rd\theta = -\frac{2R^3}{EI}$$

$$\delta_{13} = \int_0^s \frac{m_1(x)m_3(x)}{EI} ds = \frac{1}{EI}\int_0^{\pi} -R\sin\theta.(-1)Rd\theta = \frac{2R^2}{EI}$$

$$\delta_{23} = \int_0^s \frac{m_2(x)m_3(x)}{EI} ds = \frac{1}{EI}\int_0^{\pi} R(1-\cos\theta)(-1)Rd\theta = -\frac{\pi R^2}{EI}$$

We construct the system of equations by

$$\frac{4qR^4}{3EI} + \frac{\pi R^3}{2EI}H_A - \frac{2R^3}{EI}V_A + \frac{2R^2}{EI}M_A = 0 \qquad [7.57a]$$

$$-\frac{5\pi qR^4}{4EI} - \frac{2R^3}{EI}H_A + \frac{3\pi R^3}{2EI}V_A - \frac{\pi R^2}{EI}M_A = 0 \qquad [7.57b]$$

$$\frac{3\pi qR^3}{4EI} + \frac{2R^2}{EI}H_A - \frac{\pi R^2}{EI}V_A + \frac{\pi R}{EI}M_A = 0 \qquad [7.57c]$$

Resolving the system of equations [7.57] leads to evaluating the redundant reactions.

$$H_A = 0.56qR$$

$$V_A = qR$$

$$M_A = -0.106qR^2$$

Now, it is possible to analyze the statically indeterminate arch by the variation of the bending moment, the shear force and the normal force (Figure 7.53).

$0 \le \theta \le \pi$

$$M(x) = V_A x - H_A y + M_A - \frac{1}{2}qx^2$$

With $x = R(1-\cos\theta)$ and $y = R\sin\theta$

$$M(x) = \frac{qR^2}{2}(1-\cos^2\theta) - 0.56qR^2\sin\theta + 0.106.qR^2$$

The variation of the shear force can be deduced by carrying out the radial equilibrium for a section defined by the angle θ (Figure 7.53).

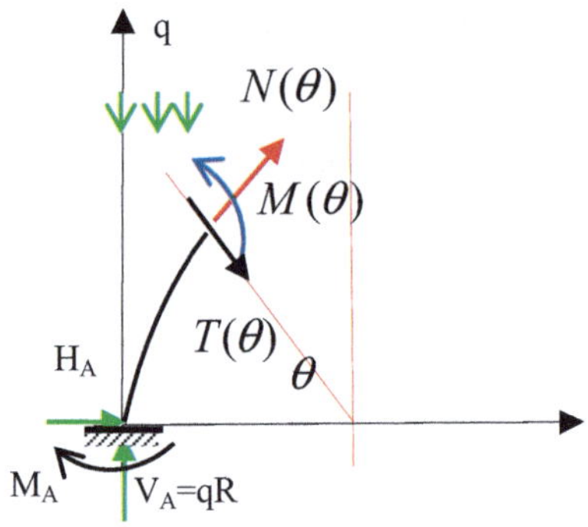

Figure 7.53. *Section $0 \leq \theta \leq \pi$*

$0 \leq \theta \leq \dfrac{\pi}{2}$

$$T(\theta) = V_A \sin\theta - H_A \cos\theta - \int_0^\theta q\sin\theta.Rd\theta$$

Alternatively, the shear force is written as

$$T(\theta) = qR(\sin\theta + 0.44\cos\theta - 1)$$

In the same way, the normal force of a section defined by a slope θ is (Figure 7.53)

$$N(\theta) = -V_A \cos\theta - H_A \sin\theta + \int_0^\theta q\cos\theta.Rd\theta = qR(-\cos\theta + 0.44\sin\theta)$$

The internal actions diagrams are shown in Figures 7.54–7.56.

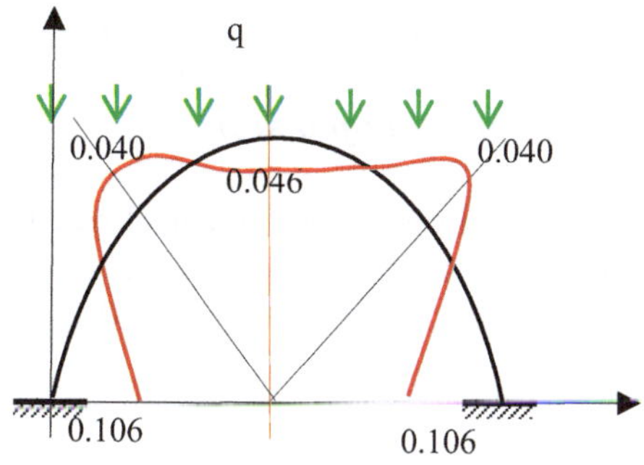

Figure 7.54. *Bending moment diagram *(qR^2)*

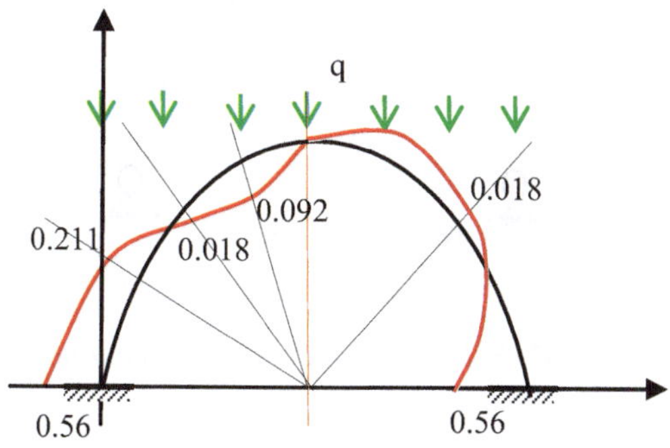

Figure 7.55. *Shear force diagram *(qR)*

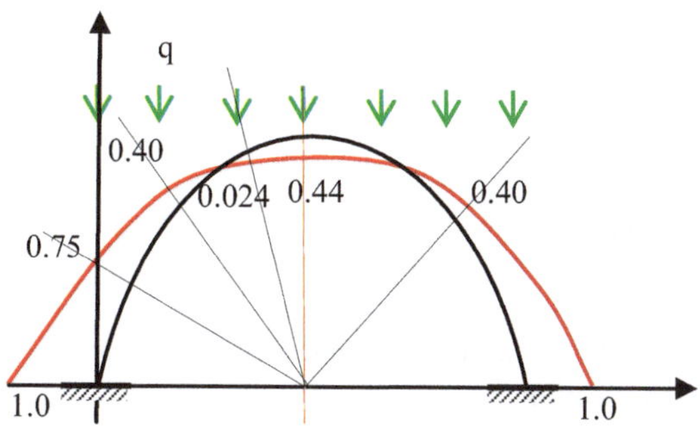

Figure 7.56. *Normal force diagram *(qR)*

7.10. Influence lines of statically indeterminate arch

7.10.1. *Influence lines of bi-hinged arch*

Determine the influence lines of the horizontal reaction at joint A, the vertical reaction of support B and the internal actions of the arch (Figure 7.57) in relation to any section. We assume that the flexural rigidity (EI) is constant. The unit force moves along the arch.

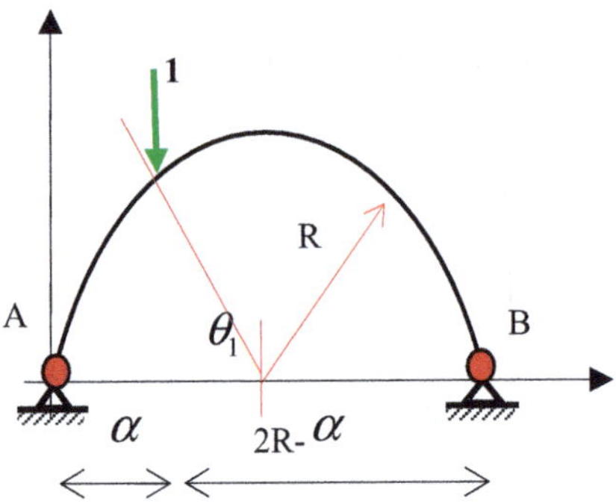

Figure 7.57. *Bi-hinged arch under concentrated load*

The horizontal reaction at support A is given by relationship [7.13]. Especially for P = 1, the expression of the reaction becomes

$$H_A = \frac{1}{\pi}(1 - \cos^2\theta)$$

The rotational equilibrium makes it possible to deduce vertical reaction V_B.

$$V_B = \frac{1}{2}(1 - \cos\theta)$$

The expressions of the bending moment depend on the locus of the moving force and the position of the section.

$\alpha \leq x$ (Figure 7.58)

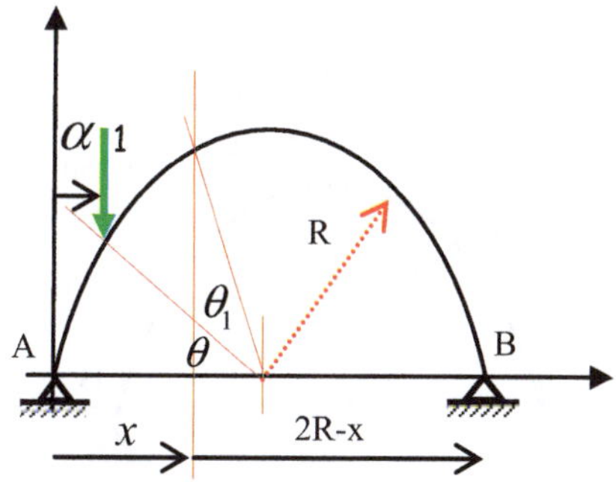

Figure 7.58. *Concentrated force $\alpha \leq x$*

$$M(x) = V_A x - H_A y - 1.(x-\alpha) =$$
$$\frac{R}{2}(1-\cos\theta)(1-\cos\theta_1) - \frac{R}{\pi}(1-\cos^2\theta)\sin\theta_1 - R(\cos\theta - \cos\theta_1)$$

$\alpha \geq x$ (Figure 7.59)

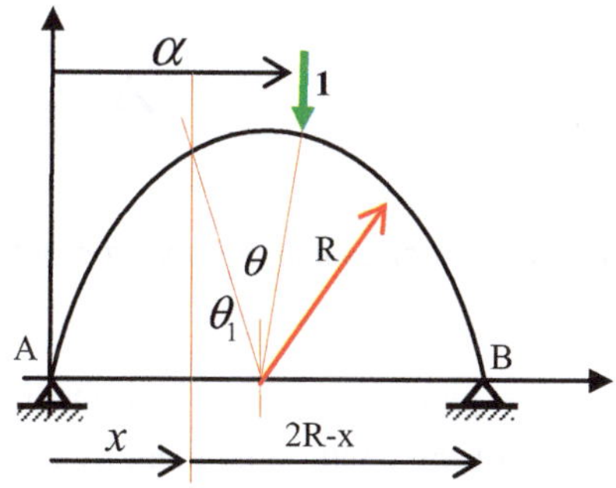

Figure 7.59. *Concentrated force $\alpha \leq x$*

$$M(x) = V_A x - H_A y = \frac{R}{2}(1-\cos\theta)(1-\cos\theta_1) - \frac{R}{\pi}(1-\cos^2\theta)\sin\theta_1$$

The influence lines of support reactions and internal actions are shown in Figures 7.60 and 7.61, respectively.

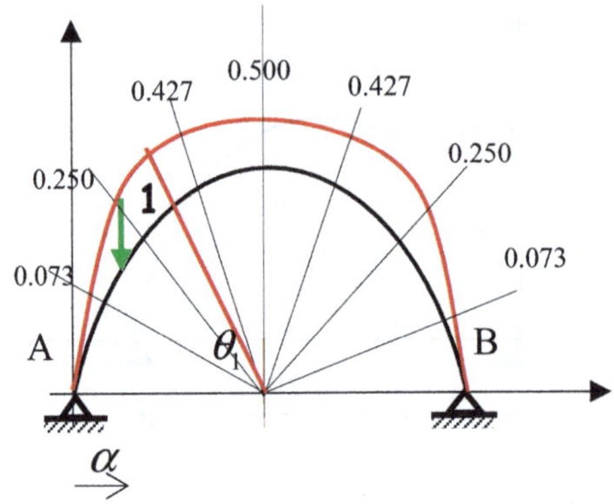

Figure 7.60. *Influence line of reaction H_A*

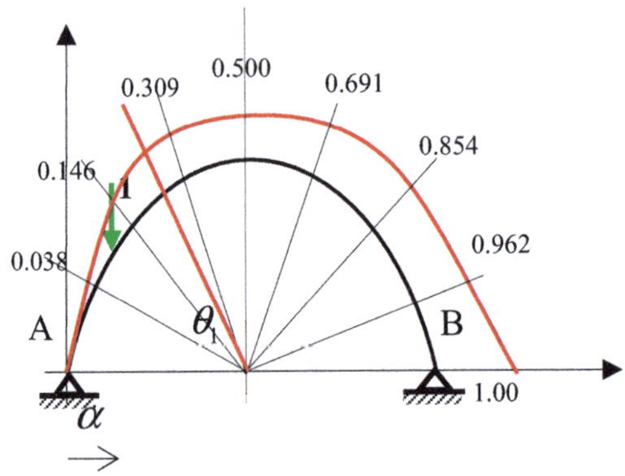

Figure 7.61. *Influence line of support reaction V_B*

The previous expressions give the variation of the bending moment, which represents the influence lines of the bending moment by any section θ_1 . To plot the influence lines of the bending moment, it is necessary to position this section. For example $\theta_1 = \dfrac{\pi}{2}$.

In this case, the bending moment expressions are:

$\alpha \leq x \ (\theta \leq \theta_1)$

$$M(\theta) = R(0.1817 - 0.5\cos\theta + 0.318\cos^2\theta)$$

$\alpha \geq x \ (\theta_1 \leq \theta)$

$$M(\theta) = R(0.1817 - 0.5\cos\theta + 0.318\cos^2\theta)$$

The influence line of the bending moment in relation to the defined section $\theta_1 = \dfrac{\pi}{2}$ is shown in Figure 7.62.

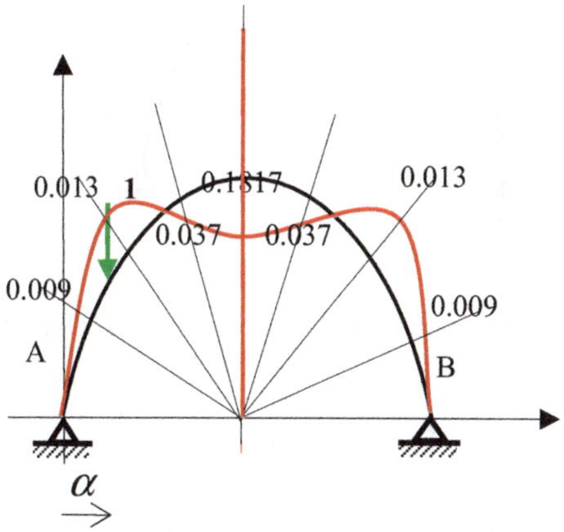

Figure 7.62. *Influence line of a bending moment *(R) in relation to* $\theta_1 = \dfrac{\pi}{2}$

The influence lines of internal actions are shown in Figures 7.63 and 7.64.

$\alpha \leq x \ (\theta \leq \theta_1)$

$$T(\theta) = \frac{1}{2}(1+\cos\theta)\sin\theta_1 - \frac{1}{\pi}(1-\cos^2\theta)\cos\theta_1 - 1.\sin\theta_1$$

$$N(\theta) = -\frac{1}{2}(1+\cos\theta)\cos\theta_1 - \frac{1}{\pi}(1-\cos^2\theta)\sin\theta_1 + 1.\cos\theta_1$$

$\alpha \geq x \ (\theta_1 \leq \theta)$

$$T(\theta) = \frac{1}{2}(1+\cos\theta)\sin\theta_1 - \frac{1}{\pi}(1-\cos^2\theta)\cos\theta_1$$

$$N(\theta) = -\frac{1}{2}(1+\cos\theta)\cos\theta_1 - \frac{1}{\pi}(1-\cos^2\theta)\sin\theta_1$$

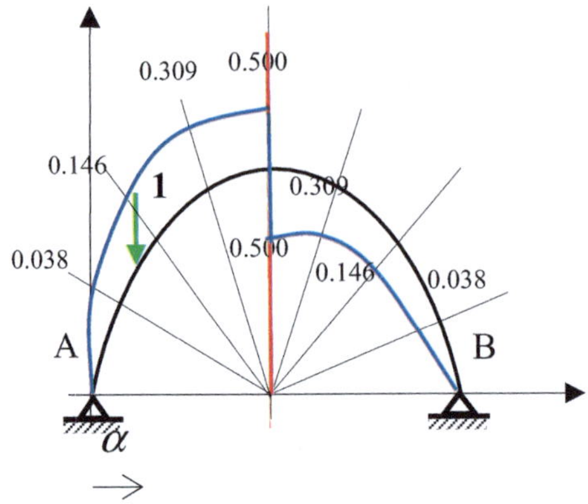

Figure 7.63. *Influence line of a shear force in relation to $\theta_1 = \frac{\pi}{2}$*

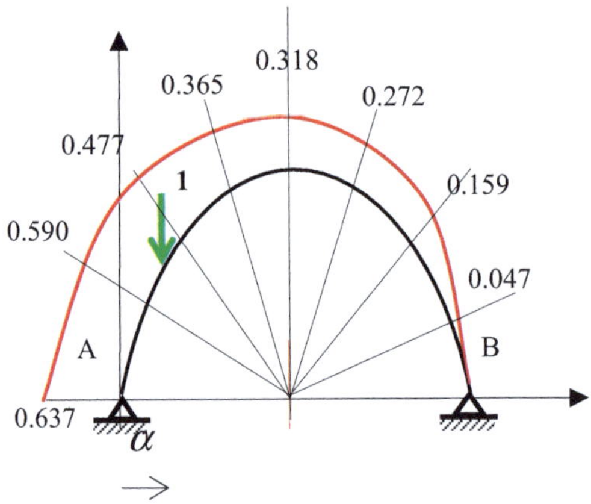

Figure 7.64. *Influence line of a normal force in relation to $\theta_1 = \frac{\pi}{2}$*

7.10.2. *Influence line of fixed-end arch*

We determine the influence lines of the vertical reaction of support V_A, bending moment M_B and the bending moment in relation to point C of the arch (Figure 7.65). We assume that the flexural rigidity (EI) is constant. The applied unit force moves along the arch.

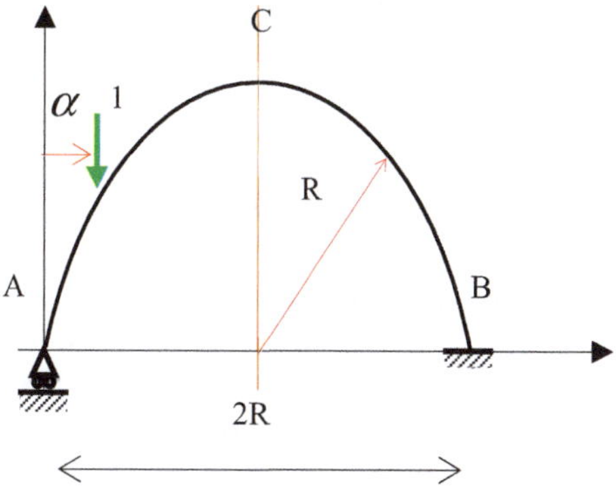

Figure 7.65. *Arch fixed at an end*

We apply the method of forces to calculate reaction V_A.

The horizontal displacement at point A of the initial system is 0.

$$\delta_A^H = 0$$

Using the superposition of effects principle leads to

$$\delta_A^0 + V_A \delta_A^1 = 0$$

δ_A^0 and δ_A^1 are the vertical displacements at point A, respectively, of the fundamental and unit systems. We construct the fundamental system ignoring the effect of reaction VA (Figure 7.66).

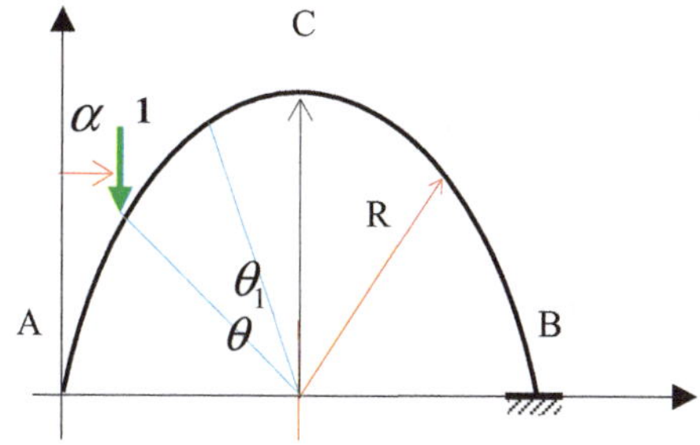

Figure 7.66. *Fundamental system*

The expressions of the bending moment of the fundamental system are given by

$$0 \le x \le \alpha \qquad \mu(x) = 0.$$

$$\alpha \le x \le 2R \qquad \mu(x) = -R(\cos\theta - \cos\theta_1)$$

Similarly, the expression of the bending moment of the unit system is shown in Figure 7.67.

$$0 \le x \le 2R \qquad m(x) = R(1 - \cos\theta_1)$$

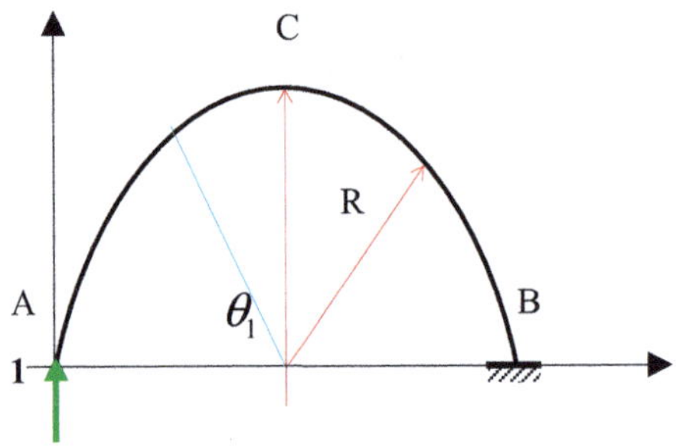

Figure 7.67. *Unit system*

The displacements are calculated using the method of virtual work.

$$\delta_A^0 = \int_0^s \frac{\mu(x)m(x)}{EI} ds$$

$$\delta_A^1 = \int_0^s \frac{m^2(x)}{EI} ds$$

By introducing the bending moment expressions of the fundamental and unit systems, we obtain

$$\delta_A^0 = \frac{1}{EI}\left[\int_\theta^\pi -R(\cos\theta - \cos\theta_1)R(1 - \cos\theta_1)Rd\theta_1\right]$$

$$\delta_A^0 = \frac{-R^3}{EI}\left[\frac{1}{2}(\pi - \theta)(1 + 2\cos\theta) + s\text{in}\,\theta + \frac{1}{2}\sin 2\theta\right]$$

$$\delta_A^1 = \frac{1}{EI}\int_0^\pi R^2(1 - \cos\theta_1)^2.Rd\theta_1 = \frac{3\pi R^3}{2EI}$$

Hence, the vertical reaction V_A becomes

$$V_A = \frac{1}{3\pi}[(\pi-\theta)(1+2\cos\theta)+2s\text{in}\,\theta+\sin 2\theta]$$

The expression of the fixed-end moment is given by

$$M_B = V_A(2R)-(2R-\alpha)$$

or

$$M_B = \frac{2R}{3\pi}\left[-\frac{\pi}{2}-\frac{\pi}{2}\cos\theta-\theta(1+\cos\theta)+2s\text{in}\,\theta+\sin 2\theta\right]$$

The bending moment at point C is expressed as

$$0 \le \theta \le \frac{\pi}{2}$$

$$M(\theta) = V_A R - 1.(R-\alpha) = (V_A - \cos\theta).R$$

$$\frac{\pi}{2} \le \theta \le \pi$$

$$M(\theta) = V_A R$$

The influence lines of reaction V_A and fixed-end moment M_B are shown in Figures 7.68 and 7.69.

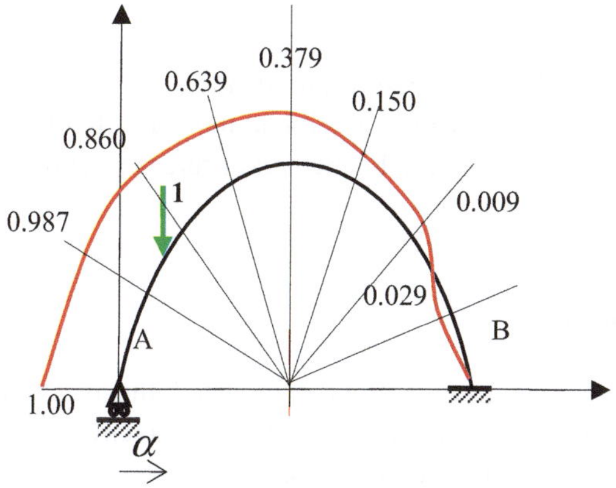

Figure 7.68. *Influence line of reaction V_A*

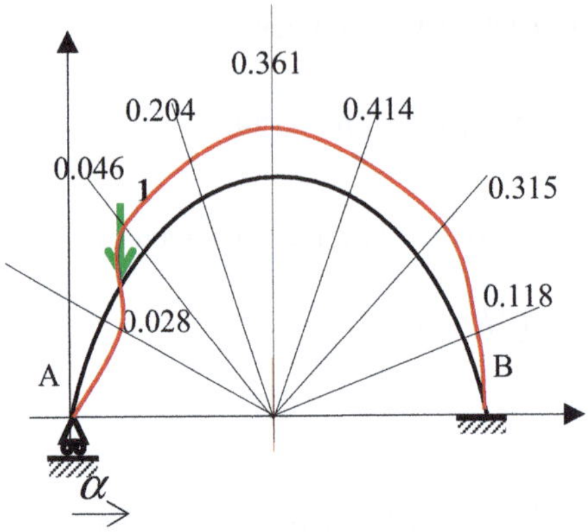

Figure 7.69. *Influence line of fixed-end moment M_B*(R)*

Finally, the influence line of the bending moment at point C is shown in Figure 7.70.

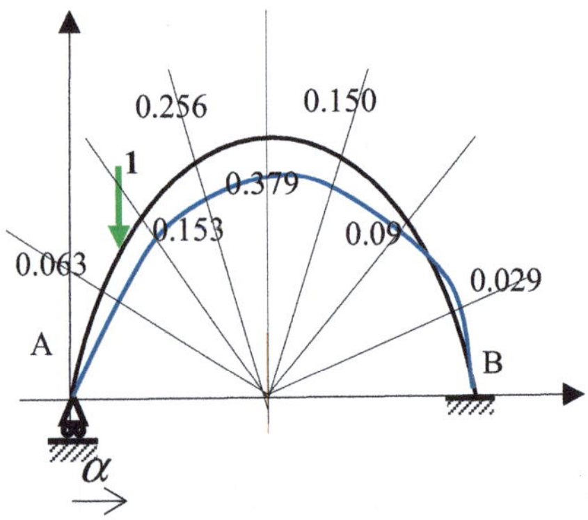

Figure 7.70. *Influence line of a bending moment in relation to $\theta_1 = \frac{\pi}{2}$ *(R)*

Finally, we note that these results are identical to those obtained in the previous sections. It is possible to calculate fixed-end moment M_B of relationship [7.46] and vertical reaction V_A by applying the method of forces (section 7.7) replacing force P with a unit force.

7.11. Conclusion

In this chapter, we have presented in detail the methodology for analyzing statically indeterminate arches. In this context, semicircular arches, parabolic arcs and tied arches are the objects of analysis.

In general, the arches studied are statically indeterminate and the method of forces is widely used. In the same concept, we also used the method of three moments, the slope-deflection method and the moment-distribution method.

This study was used to master the analysis methodology of semicircular arches, parabolic arches and tied arches. The internal actions and their diagrams are developed according to the angle of the curve. The diagrams of internal actions make it possible to derive the extreme values and consequently the dimensioning of the cross-sections of the arches.

In this chapter, only arches with constant inertia have been used; in practice, arches with variable inertia are widely seen and they are studied in the same way.

The last part of this chapter details the influence lines of internal and external actions of arches subjected to a unit force moving along the arch.

7.12. Problems

Exercise 1

Determine the support reactions of the following semicircular arches. We assume that EI is constant.

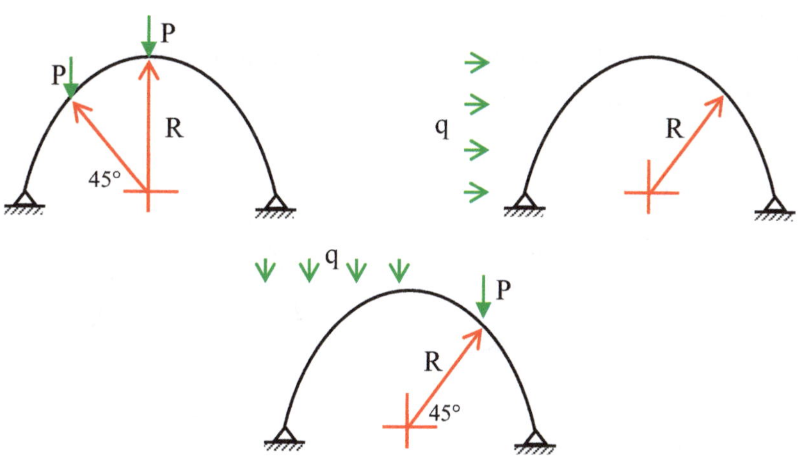

Exercise 2

Determine the support reactions of parabolic arches of equation $y(x)=\frac{4h}{L^2}x(L-x)$, where L and h are, respectively, the length and height of the arch. EI is constant.

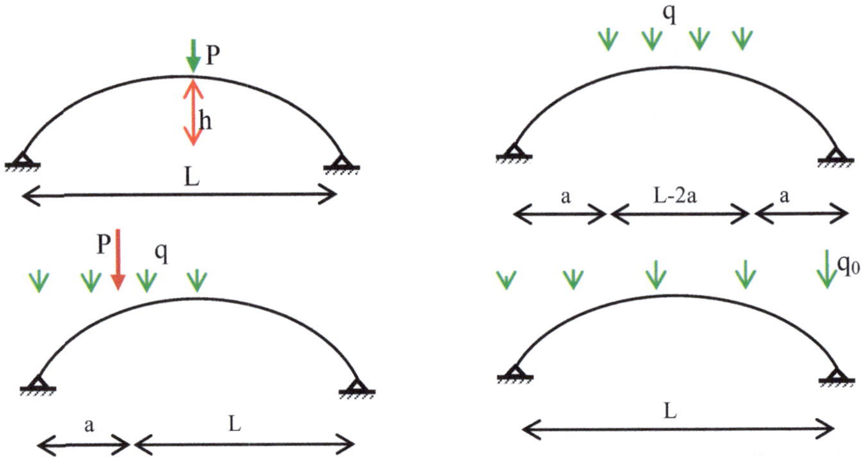

Exercise 3

Calculate the fixed-end moment of the following arches. EI is a constant and $y(x)=\frac{4h}{L^2}x(L-x)$ for parabolic arches.

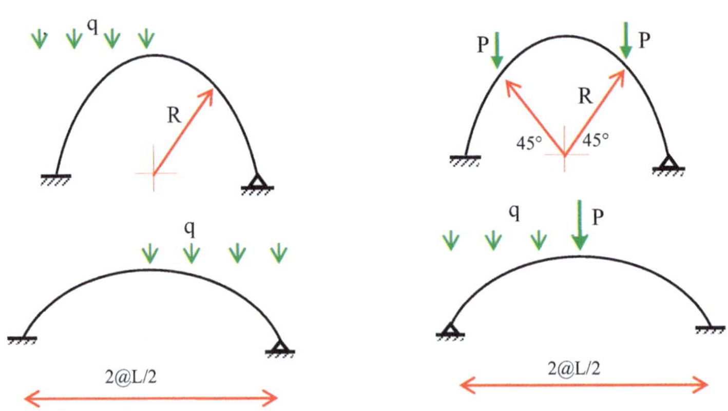

Exercise 4

Determine the variations in the internal actions of the following arches. EI is a constant.

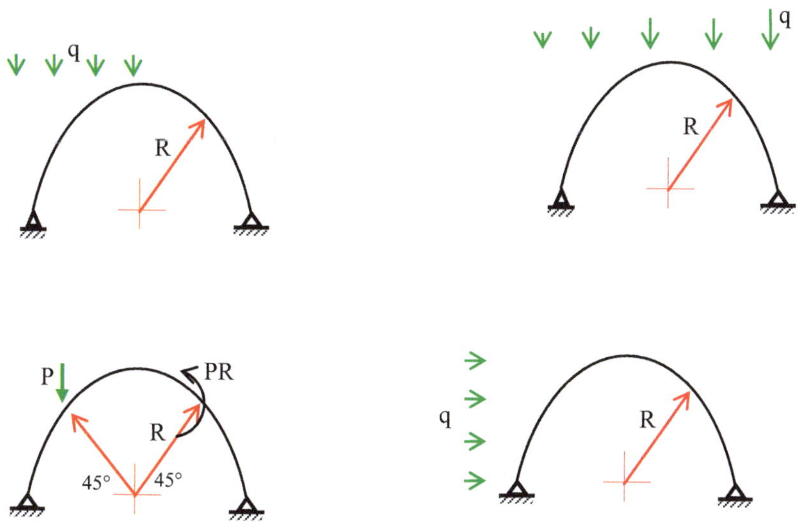

Exercise 5

Determine the variations in the internal actions. EI is a constant and $y(x)=\frac{4h}{L^2}x(L-x)$.

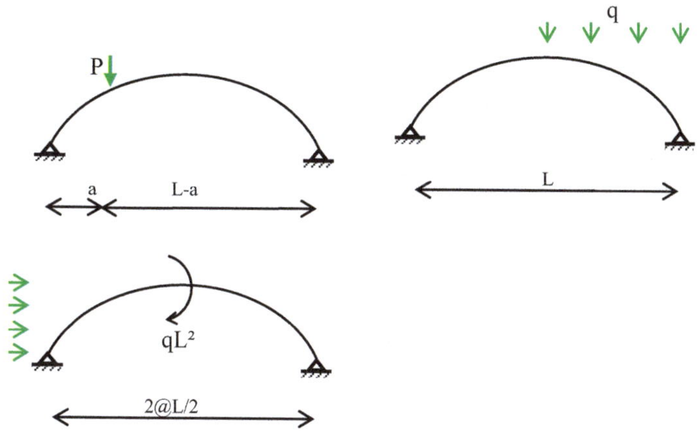

Exercise 6

Determine the variations in the internal actions and draw their diagrams. EI is a constant.

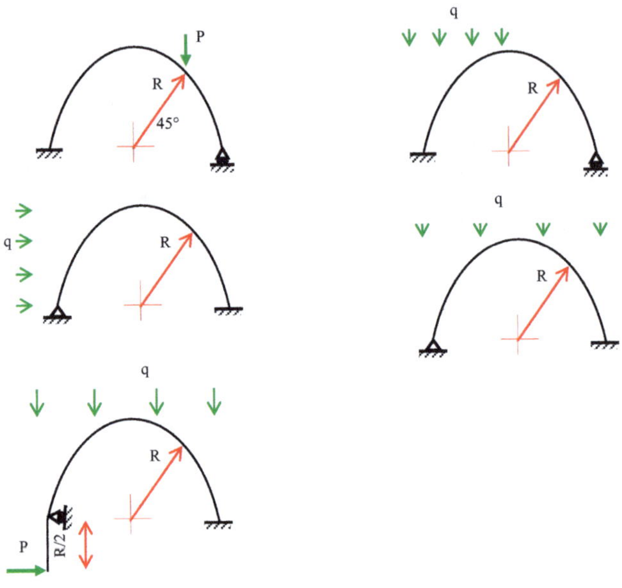

Exercise 7

Determine the variations in the internal actions and draw their diagrams. EI is a constant and $y(x)=\frac{4h}{L^2}x(L-x)$.

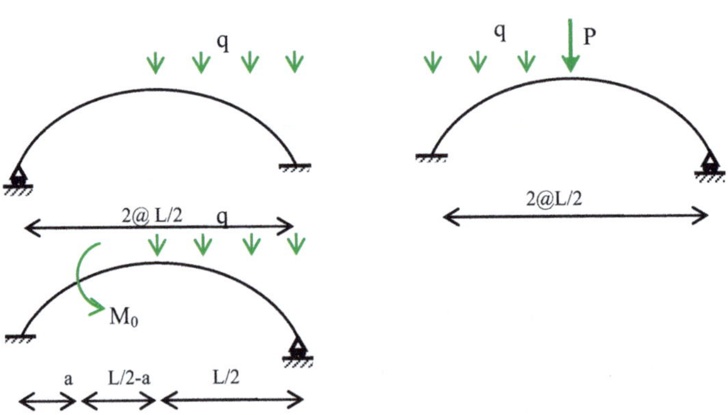

Exercise 8

Determine the variations in the internal actions and draw their diagrams. EI is a constant.

Calculate the tension in the spring and deduce the displacement at its end.

Calculate the tension in the cable of the following structures.

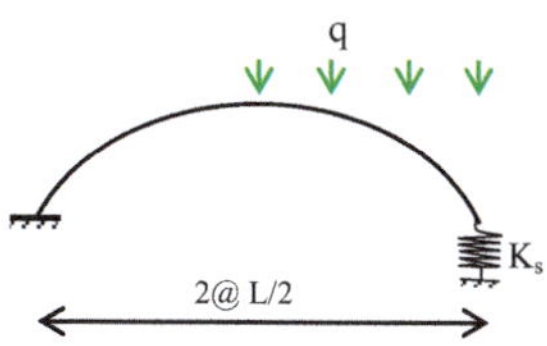

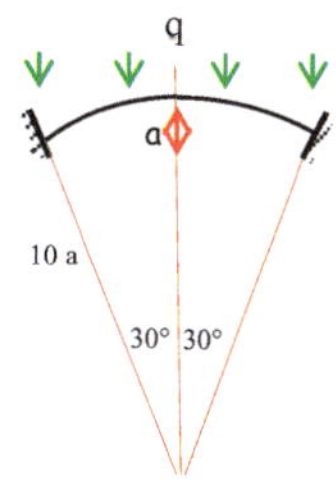

Exercise 9

Draw the influence lines of external actions mentioned for each arch stressed by a moving unit load.

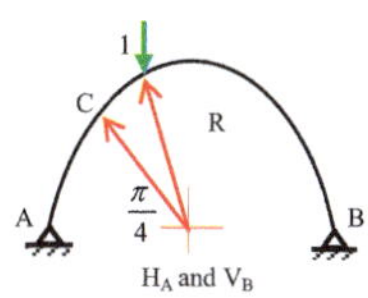

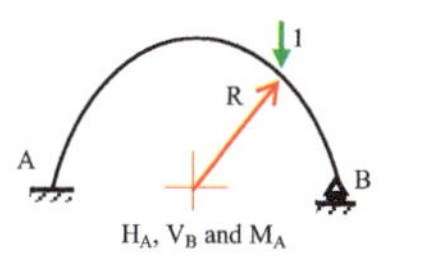

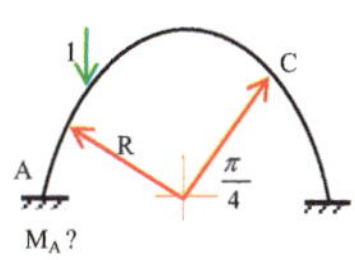

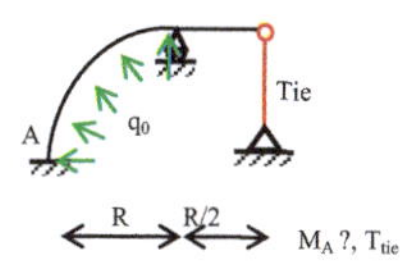

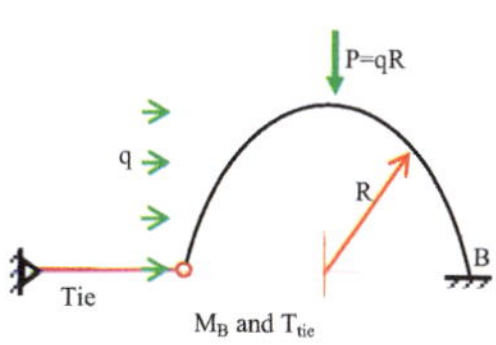

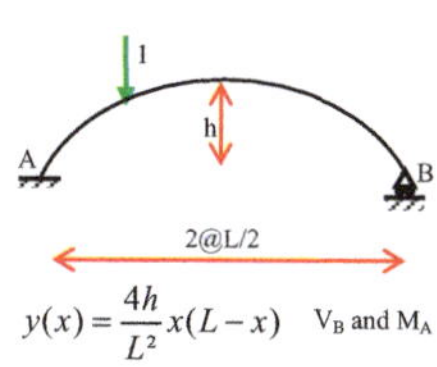

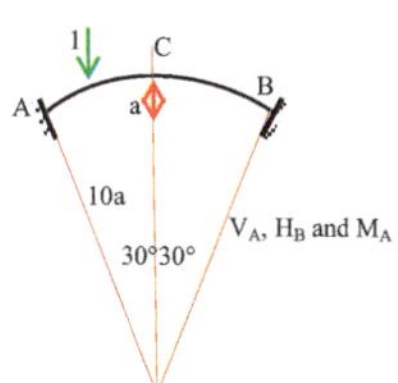

Draw the influence lines of internal actions relating to point C for each example.

Appendix

A.1. Standard structural deflections

The beams have a constant flexural rigidity EI and a length L.

Structure	Deformation
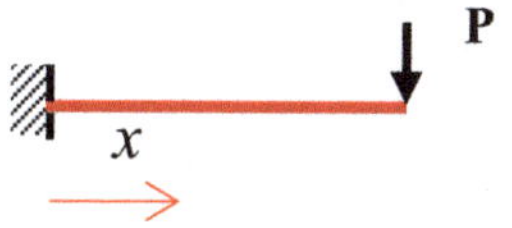	$v(x) = -\frac{PL^3}{6EI}(x^3 - 3Lx)$
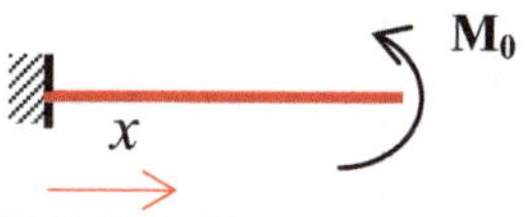	$v(x) = \frac{M_0}{2EI}x^2$
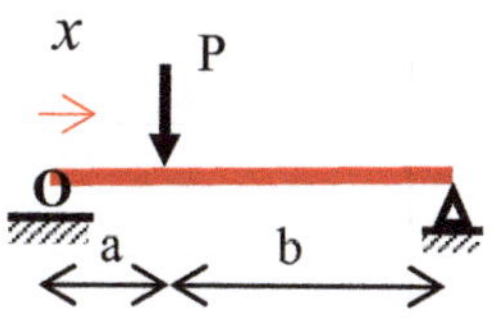	$v(x) = -\frac{Pbx}{6EI}(L^2 - b^2 - x^3)$ $0 \leq x \leq a$
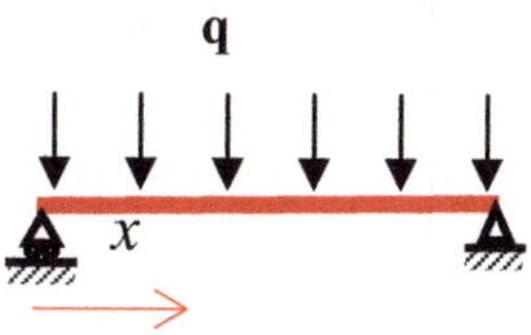	$v(x) = -\frac{qx}{24EI}(x^3 - 2Lx^2 + L^3)$

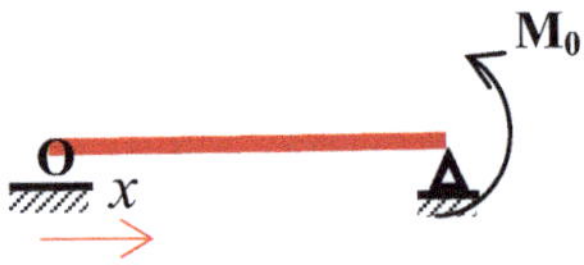	$v(x) = -\frac{M_0 x}{6EIL}(L^2 - x^2)$
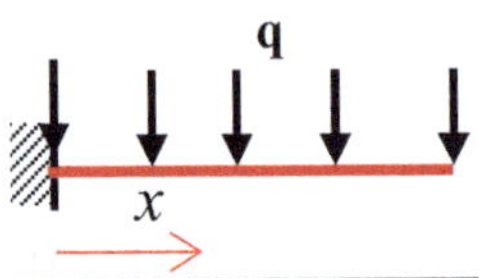	$v(x) = -\frac{q}{24EI}(x^4 - 4Lx^3 + 6L^2x^2)$

A.2. Fixed-end moments

The beam		γ_{AB}
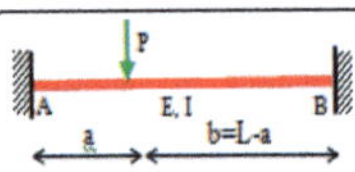	$P\frac{ab^2}{L^2}$	$-P\frac{a^2b}{L^2}$
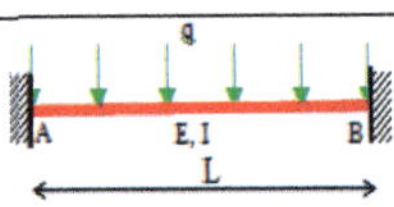	$\frac{1}{12}qL^2$	$-\frac{1}{12}qL^2$
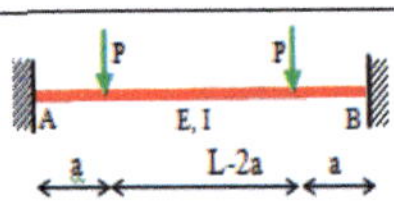	$p\frac{a}{L}(L-a)$	$-p\frac{a}{L}(L-a)$
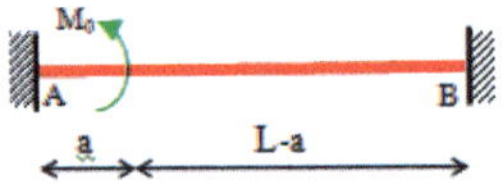	$M_0\frac{(L-a)(L-3a)}{L^2}$	$-M_0\frac{a(2L-3a)}{L^2}$
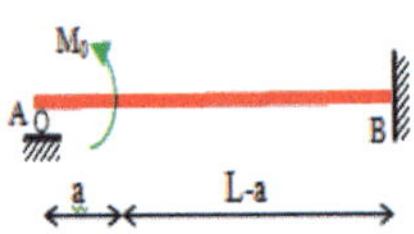	0	$-M_0\frac{(L-a)(L-3a)}{L^2}$
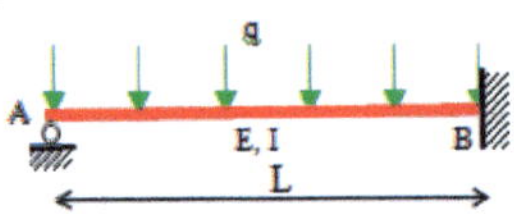	0	$-\frac{1}{8}qL^2$
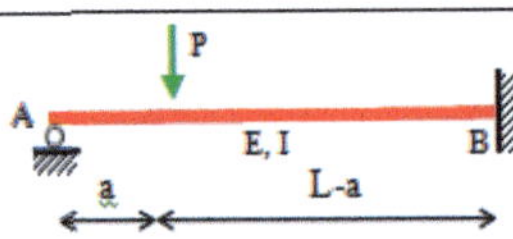	0	$-\frac{3}{16}PL$

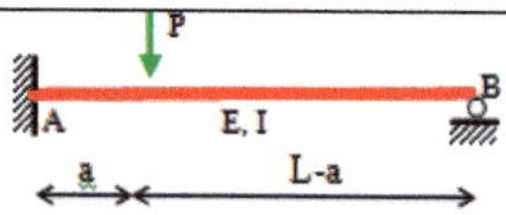	$\frac{3}{16}PL$	0
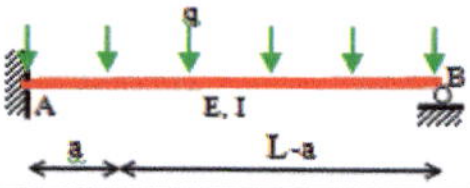	$\frac{1}{8}qL^2$	0

Bibliography

[BHA 13] BHAVIKATTI S.S., *Structural Analysis-II*, Vikas Publishing House, New Delhi, 2013.

[BOU 17] BOUVET C., *Mechanics of Aeronautical Solids Material and Structures*, ISTE Ltd, London and John Wiley & Sons, New York, 2017.

[BRO 94] BROUGHTON P., NDUMBARO P., *The Analysis of Cable and Centenary Structures*, Thomas Telford, London, 1994.

[CRO 30] CROSS H., "Analysis of continuous frames by distributing fixed-end moments", *Proceedings of the American Society of Civil Engineers,* vol. 56, pp. 919–928, 1930.

[DEL 12] DEL PEDRO M., GMÜR T., BOTSIS J., *Introduction à la mécanique des solides et des structures*, Presses Polytechniques et Universitaires Romandes Lausanne, 2012.

[DEL 15] DELAPLACE A., GATUINGT F., RAGUENEAU F., *Aide-mémoire Mécanique des structures*, Dunod, Paris, 2015.

[EIS 11] EISLEY J.G. WAAS A.M., *Analysis of Structures: An Introduction Including Numerical Methods*, John Wiley & Sons, New York, 2011.

[ERI 47] ERICKSEN B., *Influence Lines for Thrust and Bending Moments in Fixed Arcs*, Concrete Publication Limited, London, 1947.

[GHA 09] GHALI A., NEVILLE M., BROWN T. G., *Structural Analysis: A Unified Classical and Matrix Approach*, Spon Press, Abingdon, 2009.

[GHA 15] GHANEM S., NAJM H., "Structural performance of two-hinged Islamic arches", *Journal of Engineering and Architecture*, vol. 3, no. 1, pp. 38–52, 2015.

[GOU 11] GOURINAT Y., *Exercices et problèmes de mécanique des solides et des structures,* Dunod, Paris, 2011.

[HIB 17] HIBBELER R.C., *Structural Analysis*, Pearson, London, 2017.

[IGO 12] IGOR A.K., *Theory of Arched Structures: Strength, Stability, Vibration*, Springer, New York, 2012.

[IRV 92] IRVINE H.M., *Cables Structures*, MIT Press, Cambridge, 1992.

[KAR 09] KARL-EUGEN K., *The History of the Theory of Structures: From Arch Analysis to Computational Mechanics*, Ernst & Sohn Verlag, Berlin, 2009.

[KEN 17] KENNETH M., LEET E., CHIA-MING U., *Fundamentals of Structural Analysis*, McGraw-Hill Education, London, 2017.

[KHA 15] KHALFALLAH S., *Méthodes d'analyse des structures hyperstatiques*, Edilivre, Paris, 2015.

[KHA 16] KHALFALLAH S., *Analyse des structures hyperstatiques: recueil d'exercices résolus*, Edilivre, Paris, 2016.

[KAS 11] KASSIMALI A., *Structural Analysis*, Cengage Learning, Boston, 2011.

[LEE 05] LEET K.M., UANG C.M., *Fundamentals of Structural Analysis*, McGraw Hill, London, 2005.

[LES 14] LESTUZZI P., PLUG L., *Analyse des structures et milieu continu: Structures en barres et poutres*, Presse Polytechnique et Universitaire Romande, Lausanne, 2014.

[MEG 05] MEGSON T.H.G., *Structural and Stress Analysis*, Elsevier, Oxford, 2005.

[NOR 76] NORRIS C.H., WILBUR J.B., UTKU S., *Elementary Structural Analysis*, McGraw-Hill, New York, 1976.

[NAY 04] NAYFEH A.H., PAI F.R., *Linear and Nonlinear Structural Mechanics*, Wiley Interscience, Hoboken, 2004.

[PAN 07] PANDIT G.S., GUPTA S.P., *Structural Analysis: A Matrix Approach*, McGraw-Hill, New Delhi, 2007.

[PEV 79] PEVROT A.H., GOULOIS A.M., "Analysis of cable structures", *Computers & Structures*, vol. 10, no. 5, pp. 805–813, 1979.

[PIC 92] PICARD A., *Analyse des structures*, Elsevier-Masson, Paris, 1992.

[POP 68] POPOV E.P., *Introduction to Mechanics of Solids*, Prentice Hall, Upper Saddle River, 1968.

[PRZ 12] PRZEMIENIECKI J.S., *Theory of Matrix Structural Analysis*, Dover Publication, Mineola, 2012.

[RED 94] REDDY C.S., *Basic Structural Analysis*, McGraw-Hill, London, 1994.

[SAT 98] SATHYAMOORTHEY M., *Nonlinear Analysis of Structures*, CRC Press, Boca Raton, 1998.

[SMI 88] SMITH J.C., *Structural Analysis*, Harper & Row, New York, 1988.

[SON 14] SONAVANE T., Analysis of Arches, Master's thesis, Virginia University, 2014.

[TOD 81] TODD J.D., *Structural Theory and Analysis*, Palgrave, London, 1981.

[THE 14] THEORIN A., FORET G., *Calcul des structures*, Ecole Polytechnique Universitaire, Paris, 2014.

[VAI 04] VAIDYANATHAN R., PERUMAL P., *Structural Analysis, Volume II*, Laxmi Publications, Bangalore, 2004.

[VOL 08] VOLDOIRE F., BAMBERGER Y., *Mécanique des structures Initiation – Approfondissements – Applications*, Presses des Ponts, Paris, 2008.

[WIL 09] WILLIAMS A., *Structural Analysis: In Theory and Practice*, Elsevier, Oxford, 2009.

[WES 89] WEST H.H., *Analysis of Structures: An Integration of Classical and Modern Methods,* Wiley, New York, 1989.

[YAN 94] YANG Y.B., KUO S.R., *Nonlinear Framed Structures*, Prentice Hall, Upper Saddle River, 1994.

Index

F

I, K

M

R, S

T, U

Other titles from

ISTE

in

Civil Engineering and Geomechanics

2018

KHALFALLAH Salah
Structural Analysis 1: Statically Determinate Structures

VERBRUGGE Jean-Claude, SCHROEDER Christian
Geotechnical Correlations for Soils and Rocks

2017

LAUZIN Xavier
Civil Engineering Structures According to the Eurocodes

PUECH Alain, GARNIER Jacques
Design of Piles Under Cyclic Loading: SOLCYP Recommendations

SELLIER Alain, GRIMAL Étienne, MULTON Stéphane, BOURDAROT Éric
Swelling Concrete in Dams and Hydraulic Structures: DSC 2017

2016

BARRE Francis, BISCH Philippe, CHAUVEL Danièle *et al.*
Control of Cracking in Reinforced Concrete Structures: Research Project CEOS.fr

PIJAUDIER-CABOT Gilles, LA BORDERIE Christian, REESS Thierry, CHEN Wen, MAUREL Olivier, REY-BETHBEDER Franck, DE FERRON Antoine
Electrohydraulic Fracturing of Rocks

TORRENTI Jean-Michel, LA TORRE Francesca
Materials and Infrastructures 1
Materials and Infrastructures 2

2015

AÏT-MOKHTAR Abdelkarim, MILLET Olivier
Structure Design and Degradation Mechanisms in Coastal Environments

MONNET Jacques
In Situ Tests in Geotechnical Engineering

2014

DAÏAN Jean-François
Equilibrium and Transfer in Porous Media – 3-volume series
Equilibrium States – Volume 1
Transfer Laws – Volume 2
Applications, Isothermal Transport, Coupled Transfers – Volume 3

2013

AMZIANE Sofiane, ARNAUD Laurent
Bio-aggregate-based Building Materials: Applications to Hemp Concretes

BONELLI Stéphane
Erosion in Geomechanics Applied to Dams and Levees

CASANDJIAN Charles, CHALLAMEL Noël, LANOS Christophe, HELLESLAND Jostein
Reinforced Concrete Beams, Columns and Frames: Mechanics and Design

GUÉGUEN Philippe
Seismic Vulnerability of Structures

HELLESLAND Jostein, CHALLAMEL Noël, CASANDJIAN Charles, LANOS Christophe

Reinforced Concrete Beams, Columns and Frames: Section and Slender Member Analysis

LALOUI Lyesse, DI DONNA Alice
Energy Geostructures: Innovation in Underground Engineering

LEGCHENKO Anatoly
Magnetic Resonance Imaging for Groundwater

2012

BONELLI Stéphane
Erosion of Geomaterials

JACOB Bernard *et al.*
ICWIM6 – Proceedings of the International Conference on Weigh-In-Motion

OLLIVIER Jean-Pierre, TORRENTI Jean-Marc, CARCASSES Myriam
Physical Properties of Concrete and Concrete Constituents

PIJAUDIER-CABOT Gilles, PEREIRA Jean-Michel
Geomechanics in CO_2 Storage Facilities

2011

BAROTH Julien, BREYSSE Denys, SCHOEFS Franck
Construction Reliability: Safety, Variability and Sustainability

CREMONA Christian
Structural Performance: Probability-based Assessment

HICHER Pierre-Yves
Multiscales Geomechanics: From Soil to Engineering Projects

IONESCU Ioan R. *et al.*
Plasticity of Crystalline Materials: from Dislocations to Continuum

LOUKILI Ahmed
Self Compacting Concrete

MOUTON Yves
Organic Materials for Sustainable Construction

NICOT François, LAMBERT Stéphane
Rockfall Engineering

PENSÉ-LHÉRITIER Anne-Marie
Formulation

PIJAUDIER-CABOT Gilles, DUFOUR Frédéric
Damage Mechanics of Cementitious Materials and Structures

RADJAI Farhang, DUBOIS Frédéric
Discrete-element Modeling of Granular Materials

RESPLENDINO Jacques, TOUTLEMONDE François
Designing and Building with UHPFRC

2010

ALSHIBLI A. Khalid
Advances in Computed Tomography for Geomechanics

BUZAUD Eric, IONESCU Ioan R., VOYIADJIS Georges
Materials under Extreme Loadings / Application to Penetration and Impact

LALOUI Lyesse
Mechanics of Unsaturated Geomechanics

NOVA Roberto
Soil Mechanics

SCHREFLER Bernard, DELAGE Pierre
Environmental Geomechanics

TORRENTI Jean-Michel, REYNOUARD Jean-Marie, PIJAUDIER-CABOT Gilles
Mechanical Behavior of Concrete

2009

AURIAULT Jean-Louis, BOUTIN Claude, GEINDREAU Christian
Homogenization of Coupled Phenomena in Heterogenous Media

CAMBOU Bernard, JEAN Michel, RADJAI Fahrang
Micromechanics of Granular Materials

MAZARS Jacky, MILLARD Alain
Dynamic Behavior of Concrete and Seismic Engineering

NICOT François, WAN Richard
Micromechanics of Failure in Granular Geomechanics

2008

BETBEDER-MATIBET Jacques
Seismic Engineering

CAZACU Oana
Multiscale Modeling of Heterogenous Materials

HICHER Pierre-Yves, SHAO Jian-Fu
Soil and Rock Elastoplasticity

JACOB Bernard *et al.*
HVTT 10

JACOB Bernard *et al.*
ICWIM 5

SHAO Jian-Fu, BURLION Nicolas
GeoProc2008

2006

BALAGEAS Daniel, FRITZEN Claus-Peter, GÜEMES Alfredo
Structural Health Monitoring

DESRUES Jacques *et al.*
Advances in X-ray Tomography for Geomaterials

FSTT
Microtunneling and Horizontal Drilling

MOUTON Yves
Organic Materials in Civil Engineering

2005

PIJAUDIER-CABOT Gilles, GÉRARD Bruno, ACKER Paul
Creep Shrinkage and Durability of Concrete and Concrete Structures CONCREEP – 7

Printed and bound by CPI Group (UK) Ltd, Croydon, CR0 4YY
15/09/2026
14962346-0002